CLIMATE CHANGE
AND THE BUILT ENVIRONMENT

CLIMATE CHANGE
AND THE BUILT ENVIRONMENT

EDITED BY PATRICIA B. GARY AND LISA L. CHURCHILL

ACEC
American Council of Engineering Companies

Printed in the United States of America

ISBN 978-1-7376473-0-0

The American Council of Engineering Companies ("ACEC") is a Section 501(c)(6) tax exempt organization dedicated to the business and advocacy interests of engineering companies.

ACEC
American Council of Engineering Companies

American Council of Engineering Companies
1015 15th Street, NW, 8th Floor
Washington, DC 20005-2605
(202) 347-7474
(202) 898-0068
www.acec.org

Design by Michael Brady Design

Front cover, Matthias Ibeler (top) and ; p. iii, Matthias Ibeler; p. xvii, WDNet Agency; p. 6, Maria Jose Vinas (NASA); p. 7, Australian Institute of Marine Science; p. 11, NASA; p. 13, Charles Knight; p. 14, BabelStone [user name]; p. 15, National Park Service; p. 16, No. 57 [Wikimedia user name]; p. 19, Royal Institution of Great Britain; p. 21, Operation Excellence; p. 29, Mal Langsdon, Reuters; p. 32, European Union; p. 46, Roberta Price; p. 47, iStock/Franck Reporter; United Nations; p. 49, Oliver Niblett; p. 58, Halkin/Mason Photography; p. 67, Greg Cooper/Shutterstock; p. 69, Boston Planning and Development Agency; p. 73, NOAA; p. 99, AIR Worldwide; p. 107, Patrick Perkins; p. 131, Stephanie Kruel; p. 146, City of Boston; p. 148, Stephanie Kruel; p. 199, Diego Delso; p. 232, Stantec; p. 233, Michael Elkan Photography; pp. 246–261, Stantec; p. 283, NOAA/GOES; pp. 285–290, Silman; p. 294, Muller; pp. 298–303, Muller; p. 305, Weston & Sampson; p. 306, Boston Public Library; p. 306, Boston Globe; pp. 308–313, Weston & Sampson; p. 314, Bret Clancy.

Contents

Preface

Climate change is one of the most complex issues of our time. In 1992, the United Nations Framework Convention on Climate Change (UNFCCC) was established following the Rio Earth Summit, with the aim of "stabilization of greenhouse gas concentrations in the atmosphere at a level that would prevent dangerous anthropogenic interference with the climate system." Almost 30 years have passed, and the world has still not met this challenge. Atmospheric CO_2 levels are continuing to climb unabated, temperatures and sea levels are rising, glaciers are melting, drought and wildfires are increasing, and extreme weather events are intensifying.

There is scientific consensus that greenhouse gas emissions are causing climate change, but the question of what to do about it is still debated. The built environment contributes to carbon emissions on many levels, due to the way infrastructure is constructed, operated, and located. Buildings and construction activities are responsible for 39% of all carbon emissions in the world, with operational emissions accounting for 28% and embodied carbon emissions 11%. The entire design and construction industry has a role to play in moving to a low-carbon pathway, and design professionals are uniquely positioned to lead the way in both decarbonizing and building climate resilience into the built environment.

The year 2020 tied with 2016 as the hottest year on record since record keeping began, and in 2020, wildfires burned more acres in the U.S. than ever before in modern records. The impacts of sea level rise, storm surge, higher temperatures, and extreme weather events are affecting the reliability of essential infrastructure such as transportation networks and water and energy systems. Adaptation measures, including strategic use of green space and green infrastructure solutions, are essential to reducing the harmful effects of climate change.

This book is intended to be a primer for design professionals or anyone seeking to understand the complex problem of climate change and the built environment. A collaborative effort, the selected subjects, topics, and case studies are written by authors from a variety of professional backgrounds and geographical regions of the country. The scope of the problem and the need for action require a diversity of voices and perspectives, and we are grateful to our colleagues for offering their experience and solution sets from which others can draw. The goal of this book is to both inform and inspire, recognizing the urgency to act.

Abbie Goodman, EXECUTIVE DIRECTOR, ACEC OF MASSACHUSETTS

Acknowledgments

We would like to acknowledge ACEC for the opportunity to create this work, especially Abbie Goodman and Marie Ternieden, for their support and encouragement during its development. We are indebted to all of the authors who have been generous in their contributions of time and expertise. The strength of this volume lies within the variety of perspectives and experiences that each has brought and we are so proud to be able to collaborate with such an inspiring group of women, truly looking to make the world a better place.

We extend our sincere thanks to David J. Hatem, P.C. for his legal review of Chapter 7 and to James F. Bramante, PhD for his science review of Chapters 1 and 2. We especially thank Michael Brady for his contribution as the copyeditor in bringing this book to completion.

Finally, we are grateful to our family and friends who have been supportive and patient with our efforts.

Patricia B. Gary, DONOVAN HATEM LLP
Lisa L. Churchill, CLIMATE ADVISORY LLC

About the Authors

Patricia B. Gary, Esq. is Of Counsel at Donovan Hatem LLP in Boston, Massachusetts. She focuses her practice on the defense of professional liability actions against architects and engineers. She provides legal advice to construction project participants in drafting and negotiating their design and construction contracts. An experienced appellate attorney, she also concentrates her practice in the area of insurance coverage litigation, and advice. Patricia earned her B.A. from Middlebury College and her J.D. from Suffolk University Law School. She is actively involved with ACEC in speaking and writing about risk management issues related to climate change.

Lisa Churchill, PG is the founder of Climate Advisory LLC and a climate change expert, working closely at the intersection of the built and natural environments. She has 25 years of experience in the A&E field, starting her career as a transportation planner at MaineDOT before turning her attention to the carbon markets and, more recently, to alternative financing and investment options related to resilience. She leads climate projects for the public and private sectors, and has given presentations to the Pentagon and a Congressional briefing on climate security and investments in resilient infrastructure. Her training as a paleontologist, with a focus on mass extinctions, gives her a unique perspective on resilience.

Rachel Bannon-Godfrey is the Sustainability Discipline Leader for Stantec Buildings, and heads their Carbon Impact Team. Rachel has over 15 years' experience in the design, construction and analysis of high-performance, net zero buildings and sites. With a passion for multidisciplinary engagement, community resilience, climate justice and expanding the boundaries of sustainability, she serves on a range of local and national councils such as the Colorado Resource Council and AIA 2030 Working Group.

Dr. Luce Bassetti has more than 17 years in the management and technical study of regional and international projects related to port and coastal engineering. Many of the projects Luce undertakes integrate natural and nature-based solutions to improve the resilience of coastal communities and facilities. Luce focuses on coastal resilience, climate change, and sustainable development to adapt solutions to global challenges. She believes that it is essential to understand the economic, environmental, and social equity aspects of projects, in order to achieve successful and resilient outcomes tailored to each community.

Nicole Boothman-Shepard is AECOM's Vice President and Senior Director for Resilience + Recovery. Nicole is passionate about partnering with communities on strategic risk reduction and amplifying outcomes for equity, sustainability, resilience, climate adaptation, and fiscal stewardship. Nicole provides decision support to government executives on best-value building and infrastructure investments that systematically reduce long-term risk exposure to stresses and shocks. Post-disaster, she drives rapid response operations to truncate downtime, restore lifelines, and reopen Main Street in parallel with resilient reconstruction and adaptation approaches that leverage durable co-benefits for people and communities, environmental sustainability, and economic stability and growth. Nicole is a highly-sought expert in federal disaster policy, funding, and audit risk mitigation for state and local governments and has facilitated awards totaling $7B in FEMA, FHWA, and HUD funds across the US and CA.

Elizabeth Bradford, MS, ENV SP, is the Vice President, National Resilience Lead for Michael Baker International. In this capacity Elizabeth supports Michael Baker's integrated sustainability and resilience services across all sectors, markets, disciplines, geographies, and project life cycles—both as part of the company's infrastructure, environmental, community, and Federal projects, as well as stand-alone or complementary services. She brings over 27 years of industry experience across a broad range of sectors and services, to inform the development of the implementation frameworks and tools that translate sustainability and resilience strategy into individual ownership and action throughout an organization. She has been recognized with the 2017 *Environmental Business Journal*'s Industry Leadership Award for her work developing the Resiliency and Sustainability Framework, and the 2019 City & State New York Responsible 100 Infrastructure Award.

Julie Eaton Ernst, PE is a project manager and climate resiliency lead at Weston & Sampson. A Professional Engineer and MVP Certified Provider, she is responsible for providing targeted engineering services and advice related to resiliency and adaptation, including multi-hazard risk assessments, resilience planning, and the identification of flexible, engineered adaptation strategies. Julie approaches climate resilience from the dual perspectives of policy/planning and design/engineering to communicate the need for planning and preparedness, evaluate potential impacts to vulnerable infrastructure and natural resources, and develop a range of resilient design strategies that emphasize flexible approaches, manage risk and uncertainty, and create value. She specializes in using a risk-based approach to assess, identify, and prioritize adaptation strategies. In 2018, Julie received the Ascending Leader Award from the Environmental Business Council of New England.

Raine Gardner, PE is a Team Leader at MSA Professional Services Inc. located throughout the Upper Midwest. Raine leads a large team of diverse professionals who specialize in public works and parks and recreation services. Her diverse background in these areas has allowed her to develop more resilient and equitable projects for communities to sustain them well into the future, which is her passion and drive in partnering with them. In 2018, Raine won the ACEC Young Professional of the Year award. Raine is also a member of MSA's Board of Directors.

Abbie Goodman is the Executive Director of the American Council of Engineering Companies of Massachusetts, a business association dedicated to assisting its members in achieving higher professional, ethical, business and economic standards to provide quality consulting engineering services for their clients. Member firms range in size from single employee entrepreneurs to large, locally-based firms and branch offices and headquarters of national firms. Abbie is a tireless advocate for promoting the appointment of women and people of diverse cultures to public boards and commissions. She is also a strong advocate for improving infrastructure and increasing science, technology, engineering, and math (STEM) education and career awareness opportunities for students.

Karin Holland leads Haley & Aldrich's sustainability and resilience practice, and helps organizations across the built environment lifecycle grow and adapt to an ever-changing world. With over 15 years of experience in resilience consulting for large campuses in the United Kingdom, Europe, and the United States, Karin has recognized expertise in strategic planning, multi-stakeholder facilitation and systems thinking. Karin has recently focused on helping address some of the societal challenges to holistic resilience programs.

Amy Hwang is an Associate and Technical Development Director at Silman. She joined Silman in 2000 and in more than two decades of experience, she has worked on many of the firm's most complex and challenging projects, including the new Whitney Museum of American Art in New York City and the Institute for Contemporary Art at VCU in Richmond, VA. Immediately following the devastation caused by

Superstorm Sandy in 2012, Amy led an emergency contract for post-disaster assessments issued by the NYC Department of Buildings. She worked directly with the city and organized volunteer teams of engineers to provide rapid and detailed evaluations of buildings across the city's boroughs. In her current role, Amy leads an internal Technical Advancement Group, which is a rotating group of engineers dedicated to developing knowledge capture and lessons learned about resources and processes.

Stephanie Kruel is an environmental entitlement permitting and coastal regulation/policy specialist at VHB, a Boston-based engineering and planning firm. She provides guidance to real estate developers, institutions, and government agencies on planning and building for future climate conditions. Stephanie has prior experience as a conservation agent, floodplain manager, community planner, and zoning reviewer in the public and non-profit sectors. She currently holds leadership positions with Boston Harbor Now's Climate Round Table and NAIOP's Climate Change Resiliency Committee. Stephanie maintains an Envision™ Sustainability Professional certification and holds a Master of City Planning degree from Georgia Institute of Technology, as well as a Bachelor of Arts degree in Anthropology and Psychology from Emory University.

Amanda Ludlow is a designer of nature-based solutions who has spent her 22-year career focused on the development of innovative sustainable solutions to solve environmental problems including engineered wetlands, phytoremediation, natural media filtration, green infrastructure, green and sustainable remediation, and slope and shoreline stabilization. She has applied these nature-based alternatives to a diverse range of impacted environments, including groundwater and surface water contamination, stormwater runoff, landfill leachate, mine drainage, contaminated soils and sediments, and industrial/municipal wastewaters.

Kelly Maiorana is the Environmental Planning Manager at Muller Engineering Company in Lakewood, Colorado. She has 22 years of experience in NEPA evaluations and associated technical studies, land use planning, regulatory negotiations, and public involvement. In response to the 2013 flood event that devastated Colorado, Kelly provided comprehensive program support to the Colorado Department of Transportation's Incident Command Center, which was established after severe flooding affected hundreds of miles of Colorado's state highways and interstates. This experience provided a valuable perspective to apply resiliency applications to future planning studies.

Elena Mihaly is a Senior Attorney at Conservation Law Foundation (CLF), where she works in CLF's Clean Air and Water Program and Healthy and Resilient Communities Program. Elena's work focuses on developing and implementing law and policy solutions to protect public health, enhance community climate preparedness, and ensure all communities have equal access to a clean and healthy environment. She received her J.D., summa cum laude, and Master's in Environmental Law and Policy from Vermont Law School and is licensed in Vermont, New Hampshire, and Massachusetts. She holds a B.A. in Environmental Science from Colorado College.

Deanna Moran is Director of Environmental Planning at Conservation Law Foundation (CLF) where she works in CLF's Healthy and Resilient Communities Program. Deanna's work focuses on climate change in the built environment through planning, development, and regulation. She oversees CLF's climate adaptation activities and manages the Boston Harbor Climate Preparedness and Public Access Initiative. Deanna received a Master's in City and Regional Planning and a Master's in Public Policy from the Bloustein School at Rutgers University. She holds a B.A. in Environmental Design from the University at Buffalo.

Nancy M. Rigassio is Vice President and Executive Claims Counsel for AXA XL, the P&C and specialty risk division of AXA. She has 20 years of experience in the claims end of the professional liability insurance industry. Her focus is on the defense of high-profile, large-loss claims against design professionals and risk management in negotiating contracts. Before a career in insurance, she practiced trial law in New Jersey defending professional negligence actions. She graduated with a B.A. in economics from Douglass College, and earned a Juris Doctor degree from Rutgers School of Law.

April Schneider is a civil engineer and urban planner driven by a desire to plan and design better futures for cities. She works across sectors to incorporate sustainability and resilience into the built environment. For the Lower Manhattan Coastal Resiliency project at the Battery, she is working with her co-author, Amy Seek, to explore adaptable design of a waterfront wharf that must withstand 80 years of rising seas while maintaining accessibility and existing character in the near term. She is also the Northeast Practice Lead for Stantec's Smart Mobility team, providing holistic mobility solutions that leverage existing and emerging modes to improve urban livability.

Amy Seek is a landscape architect and writer whose work includes large-scale public parks, event spaces, waterfronts, and their intersection with other city systems, from mobility to resilience infrastructure. Her recent work is focused on addressing uncertainty to create adaptable, multi-benefit landscapes in challenging urban environments. She currently serves as Design Director of Stantec's NYC landscape architecture studio.

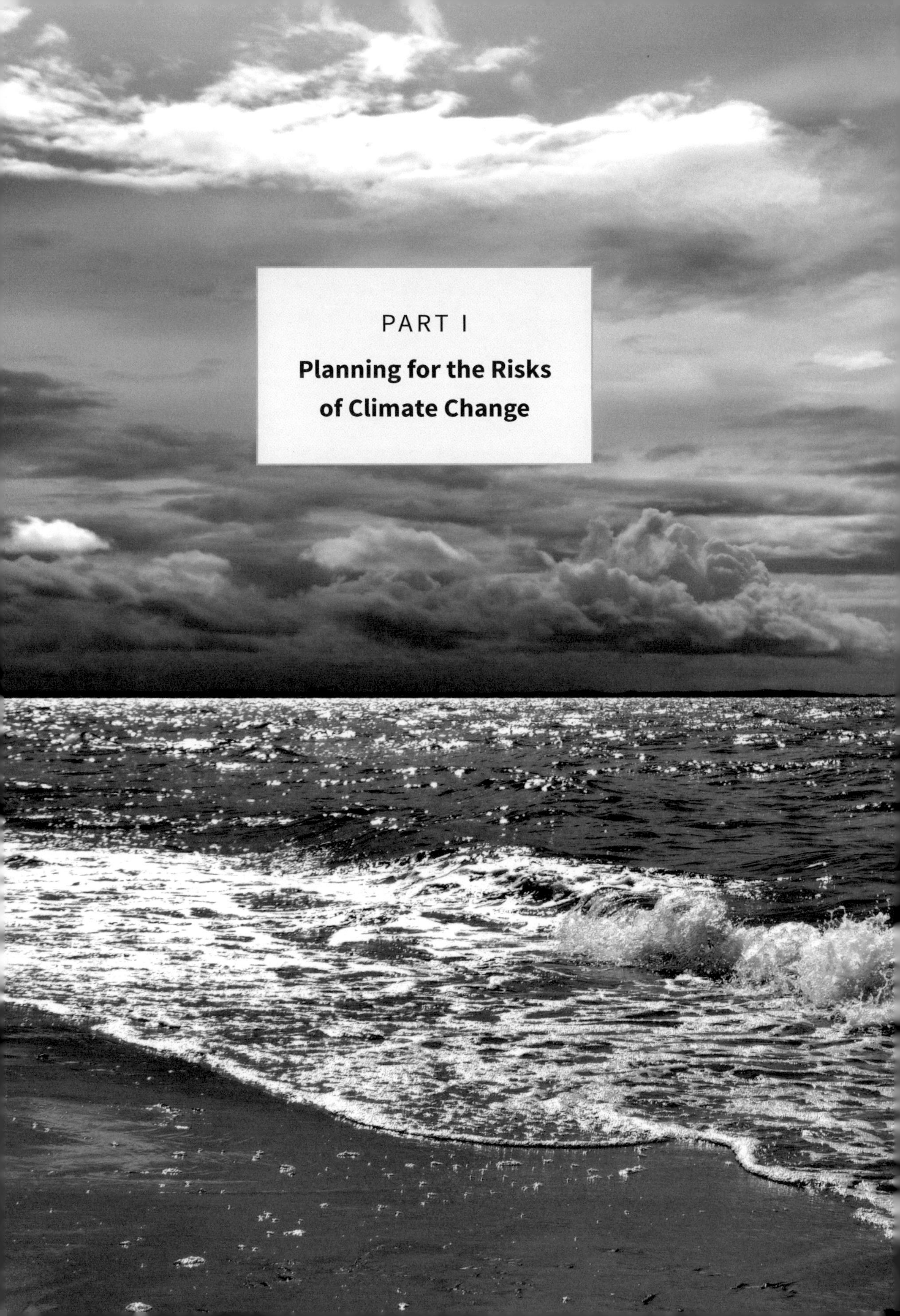

PART I

Planning for the Risks of Climate Change

1

Historical Overview of Climate Change

Patricia B. Gary

1.0 Introduction

The climate has a profound influence over life on our planet. Scientists have warned us for decades that anthropogenic greenhouse gases in the atmosphere are causing man-made climate change, and the risks they predicted are now occurring. The temperature is projected to rise to 1.5°C above pre-industrial levels by 2030, and without transformative policies to reduce CO_2 emissions, temperatures may rise 4.5°C (8°F) or higher by the end of the century. If past trends continue unabated, the impacts of this trajectory will include massive sea-level rise, flooding, infrastructure damage, food and water shortages, ecosystem destruction, and human suffering. Climate change is solvable, but the world must act with urgency to address the climate crisis, phase out fossil-fuel–based technology and CO_2 emissions, and create a path to a more sustainable future.

Design professionals are uniquely positioned to help solve the climate crisis through sustainable and resilient design. There is no universal definition of the metrics for "net zero" but the traditional concept of net zero means a building that achieves a nearly zero balance between energy demand and renewable energy supply. Even under a net zero emissions scenario, global warming from emissions since the pre-industrial period to the present are projected to persist for centuries to millennia, but a slower rate of warming would enhance the ability of people and ecosystems to adapt and become more resilient.

This chapter is a historical overview of climate change. It begins with Earth's natural glacial and interglacial cycles and early human adaptation to climate change, and then presents a timeline of scientific study of climate change from Joseph Fourier's discovery of the greenhouse effect in the 1820s, through the passage of the Clean Air Act in 1970, to our modern understanding of climate change, including the key objective findings by the Intergovernmental Panel on Climate Change, which led to the Paris Climate Agreement of 2015. The chapter concludes with recommendations about the design professional's role in advising clients about their options for reducing carbon emissions, and implementing adaptive designs that are resilient to future climate and weather extremes.

1.1 Climate Change: Definition, Causes, Risks

1.1.1 What Is Climate Change?

The terms "climate" and "climate change" may be defined as follows:[1]

Climate Climate is the long-term local, regional, or global weather pattern over a time span (by convention, 30 years). It is more rigorously described by statistics, such as the means and extremes of temperature, precipitation, wind, and other variables including the frequency and intensity of weather events.

Climate Change Climate change is a statistically significant variation in average weather conditions that persists for multiple decades or longer. The climate changes due to the earth's natural internal processes or external forces, or from anthropogenic changes in the composition of the atmosphere or in land use.

Over long spans of time, from decades to many centuries, the climate system evolves under the influence of its own internal dynamics and due to changes from external factors called forcings that affect climate.[2] Forcings include natural phenomena such as orbital or solar variations or volcanic eruptions; they also include human-induced changes to the atmosphere.[3] Radiation from the sun is the energy that drives the climate system, and there are three fundamental ways to change the solar radiation balance of the Earth, and thereby cause a climate change: "(1) changing the incoming solar radiation (e.g., by changes in the Earth's orbit or in the Sun itself), (2) changing the fraction of solar radiation that is reflected (this fraction is called the albedo—it can be changed, for example, by changes in cloud cover, small particles called aerosols, or land cover), and (3) altering the long-wave energy radiated back to space (e.g., by changes in green-house gas concentrations)."[4]

Earth's global temperature has been rising since the mid-1800s and this change in climate is linked directly to the third factor above: human activities that increase the amount of greenhouse gases, primarily the combustion of fossil fuels. Greenhouse gases (GHG) are gases in the atmosphere such as carbon dioxide (CO_2), methane (CH_4), nitrous oxide (N_2O), and fluorinated gases that absorb infrared radiation, trapping heat in the atmosphere. Global annually averaged surface air temperature increased by about 1.8°F (1.0°C) over the period of 1901–2016, and this period is now the warmest in the history of modern civilization.[5] According to the United Nations Intergovernmental Panel on Climate Change (IPCC), Earth's global mean temperature is currently increasing at 0.3°C–0.7°C per 30 years.[6] The IPCC is clear that at the current rate, global warming will reach 1.5°C above the pre-industrial level between 2030 and 2050.

1.1.2 **The Future Rate of Warming**

Earth's future climate and rate of warming will depend upon how much GHG emissions, and especially CO_2, humans continue to emit into the atmosphere throughout the 21st century. This, in turn, will depend upon the GHG mitigation choices made by society and whether we continue to rely upon fossil fuels or pursue policies that transition to clean energy and a low-carbon economy. Scientists use climate models, which are mathematical representations of Earth's processes, to estimate the potential risks and impacts of climate change and to assess the costs of emissions reductions. For the IPCC's Fifth Assessment Report, the scientific community developed a set of four future GHG concentration pathways, called Representative Concentration Pathways (RCP 8.5, RCP 6.0, RCP 4.5 and RCP 2.0), to support global climate model simulations and research the potential impacts of policy responses to climate change.

Figure 1. **IPCC's Fifth Assessment Report Used RCPs**

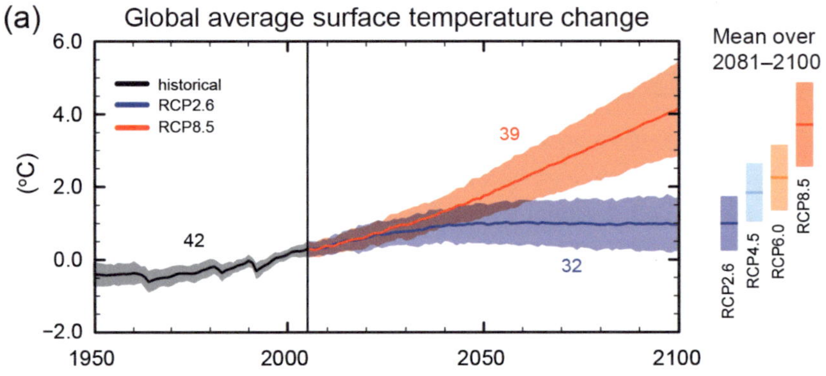

Source: Figure SPM.7, "Summary for Policymakers, The Physical Scientific Basis," *IPCC Fifth Assessment Report,* 2014.

The RCP numbers refer to the amount of radiative forcing produced by greenhouse gases in 2100, relative to preindustrial conditions (i.e., the globally averaged heat-trapping capacity of the atmosphere, +2.6, +4.5, +6.0, and +8.5 watts per square meter or W/m^2).

The RCPs describe possible climate outcomes in the 21st century and are used to test policy decisions on mitigation and adaptation.[7] They help to inform climate change policy by presenting a range of options for governments, policymakers, and others, based on risks and consequences. The IPCC's next report will use new forcing scenarios called Shared Socioeconomic Pathways (SSPs), which link specific policy decisions to projected emissions outcomes.

RCP Scenarios

RCP 8.5 is a baseline, high-emissions scenario consistent with continuation of our fossil-fuel economy, with no significant global efforts to limit GHG, and would lead to about 4.5°C (8°F) temperature increase by the end of the century.

RCP 6.0 and **RCP 4.5** are intermediate GHG scenarios that require mitigation efforts. RCP 4.5 peaks around 2050 and then declines, so that global warming stabilizes at around 2.8°C (5°F) above pre-industrial levels, but still misses the goals of the Paris Agreement to limit global warming to less than 2°C.

RCP 2.6 is a low-emissions scenario, which requires ambitious mitigation efforts, and most closely corresponds to the goals of the Paris Agreement to keep global warming less than 2°C above pre-industrial levels.

Figure 2. **U.S. 4th National Climate Assessment Used RCPs**

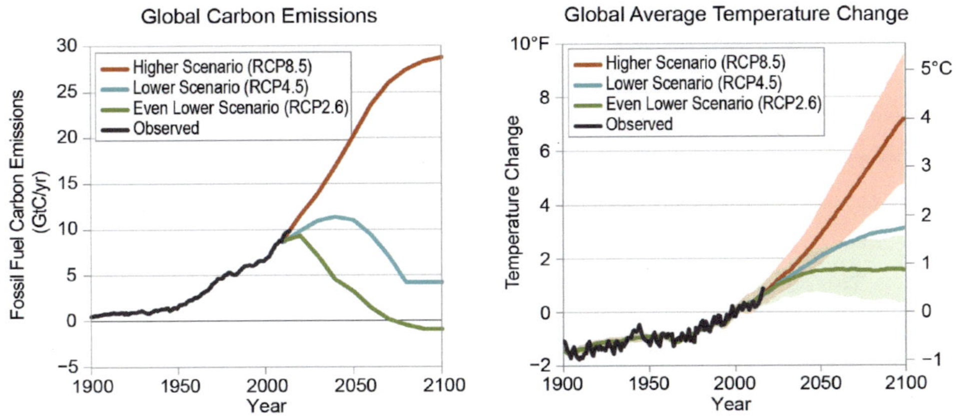

Source: Figure 2.2, "Observed and Projected Changes in Carbon Emissions and Temperature," chapter 2, *4th National Climate Assessment*, 2018.

Who Uses Carbon Emissions Scenarios?

Carbon emissions scenarios are used extensively by scientists, policy makers, and others who study the effects of climate change. They provide a common framework to model and discuss climate change. IPCC's *Fifth Assessment Report* and the *Fourth National Climate Assessment* used RCPs to understand, assess, predict, and evaluate the risks of climate change. New forcing scenarios called Shared Socioeconomic Pathways—SSPs—are being developed and modeled for IPCC's Sixth Assessment Report.

1.1.3 **The Risk of a Tipping Point**

The rise in Earth's temperature has caused profound alterations in Earth's natural systems. We are already experiencing a world-wide increase in the frequency and severity of extreme weather events such as cyclones, heavy precipitation, floods, heatwaves, droughts, and forest fires. As climate change progresses it will cause widespread food and water shortages, loss of ecosystems and biodiversity, crop and species loss, increasing animal- and insect-borne diseases, and other indirect consequences.

The magnitude and rate of climate-related risks will depend on future GHG emission pathways, and the rate of warming. The faster human-induced climate change pushes Earth's climate systems towards warming, the greater the risk of a tipping point, where a threshold is crossed that may lead to irreversible impacts. If climate change is not urgently addressed, once these tipping points are reached, there is a risk of cascading ecosystem collapse, with enormous economic, social and health consequences.

An example of a critical threshold or tipping point would be a change from the current state in which the summer Arctic Ocean is covered with ice, to a new state in which it is ice-free during summer, or year-round. A decline in sea ice decreases the planet's albedo (its reflectivity, or whiteness) which leads to increased Arctic warming and sea ice melt through an amplifying feedback process.

Arctic Sea Ice in September 2019
The red line shows how much it has decreased in the past decade

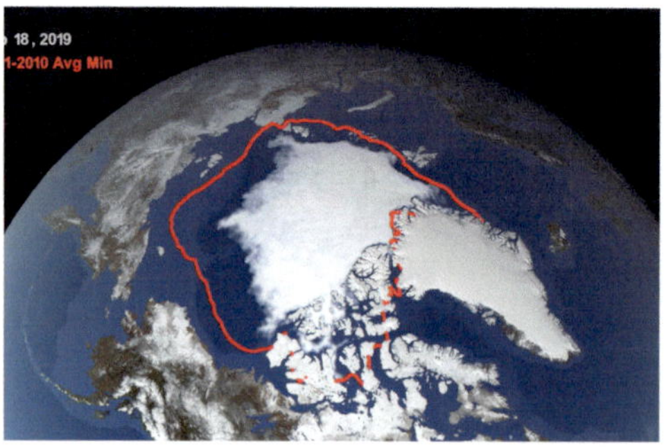

Source: Photo by Maria Jose Vinas, "2019 Arctic Sea Ice Minimum Tied for Second Lowest on Record," NASA Feature, Sept. 23, 2019.

Ice sheets and sea ice have highly reflective white surfaces that reflect solar radiation away from the Earth. When ice melts, it exposes the open sea water, which absorbs, rather than reflects, radiation, and this feedback loop amplifies warming. Long-term feedbacks also occur when thawing permafrost releases carbon dioxide and methane, contributing directly to radiation forcing.

If thresholds are passed, synergistic feedback loops could cause a shift in the atmospheric and oceanic circulation regimes, triggering an acceleration of sea level rise. The rapid deforestation of the Amazon rainforest and other ecosystem destruction is accelerating these shifts. These and other feedbacks could spiral, leading to a "hothouse" world.

A tipping point has already been reached and exceeded for coral reefs, which are vulnerable to heat stress, and reefs around the world are projected to collapse by the end of the century. A temperature spike of 1°–2°C causes bleaching in corals. When sea water temperatures are too warm, corals expel the symbiotic algae (called zooxanthellae) that live in their tissues, provide them with food, and create their beautiful colors. In a phenomenon called bleaching, the corals turn completely white. Only their bright white calcium-carbonate skeletons are visible, and if the water does not cool quickly enough to allow new algae to recolonize the coral, the reef will die. There has been widespread bleaching and die-off of coral reefs in the Florida Keys, Hawaii, other U.S. islands, and globally.

Bleaching of Coral Reefs

Source: Photo from Australian Institute of Marine Science, Australian Government, "Coral Bleaching."

Bleaching from marine heat waves and ocean acidification from dissolved carbon dioxide has caused a massive global die-off of corals, killing approximately 50% of the corals in Australia's iconic 1,500-mile Great Barrier Reef. The demise of coral reefs is a harbinger of radical shifts or tipping points in less sensitive ecosystems.

1.1.4 **The Financial Consequences of Climate Change**

The financial consequences of climate change will be onerous. The National Oceanic and Atmospheric Administration (NOAA) tracks U.S. weather and climate events and reports that U.S. billion-dollar climate-related disasters are increasing:

> Since 1980, the U.S. has sustained 258 weather and climate disasters where the overall damage costs reached or exceeded $1 billion (including adjustments based on the Consumer Price Index, as of January 2020). The cumulative cost for these 258 events exceeds $1.75 trillion.[8]

In the last decade (2010–2019), the U.S. billion-dollar disaster events were historically large, exceeding $800 billion, and included four of the five most costly U.S. billion-dollar disasters in history (Hurricanes Harvey, Irma, Maria, and Sandy). In 2018, Storm Grayson, a nor'easter (called the "Bomb Cyclone") caused a tidal surge and icy flooding from the highest tides in Boston's history, and the Camp Fire was the costliest and deadliest wildfire in California's history. The total cost of California's wildfires for the year 2018 approached $24 billion. NOAA notes that "14 separate U.S. billion-dollar disasters in 2019 represent the fourth highest total number of events (tied with 2018), following the years 2017 (16), 2011 (16) and 2016 (15)."[9] In recognition of the increasing financial risk of climate change, insurers now name climate change as the top risk for their industry.[10]

A recent study by two Environmental Protection Agency (EPA) economists, Jeremy Martinich and Allison Crimmins, published in *Nature Climate Change* in 2019, estimates that total climate damages could reach $500 billion annually by 2090, and that achieving lower emissions through both mitigation and adaptation actions will result in substantial economic benefits.[11] The study quantifies the potential physical and economic damages of climate change to 22 sectors in the U.S. (for example, labor, extreme temperature mortality, infrastructure, coastal property, agriculture, air quality, fish habitat) using a consistent set of climate and socioeconomic assumptions from a coordinated modeling project developed for the Fourth National Climate Assessment. The study assessed two future global warming scenarios: RCP 8.5, the trajectory if past trends continue unabated, with minimal or no change in policy, and RCP 4.5, a pathway that would require much more aggressive mitigation policies than are in place today but would still miss the Paris Climate Agreement targets. The study found that by 2090, impacts on the 22 sectors in the U.S. will cost approximately $220 billion more per year if the U.S. follows the high emissions trajectory.

1.1.5 The Social Consequences of Climate Change

The U.S. Department of Defense (DOD) has warned that climate change is increasing national security threats, including effects on military infrastructure and operations, and the safety of human societies. A DOD report finds that climate change "will have wide-ranging implications for U.S. national security interests over the foreseeable future because it will aggravate existing problems—such as poverty, social tensions, environmental degradation, ineffectual leadership, and weak political institutions—that threaten domestic stability in a number of countries."[12] As examples, the report cites a severe multi-year drought in Syria from 2006–2011, which "contributed to massive agriculture failures and population displacement, . . . [and] massive flooding in Pakistan in 2010 killing more than 2,000 people and affecting 18 million."[13]

The IPCC states with "high confidence" that baseline scenarios without additional mitigation "result in global mean surface temperature increases in 2100 from 3.7°C to 4.8°C compared to pre-industrial levels (range based on median climate response; the range is 2.5°C to 7.8°C when including climate uncertainty)"[14] and that "increasing magnitudes of warming increases the likelihood of severe, pervasive, and irreversible impacts." These impacts on people include:

- Risk of death, injury, and disrupted livelihoods in low-lying coastal zones and small islands due to storm surges, coastal flooding, and sea level rise.
- Risk of severe ill-health and disrupted livelihoods for large urban populations due to inland flooding in some regions.
- Systemic risks due to extreme weather events leading to breakdown of infrastructure networks and critical services such as electricity, water supply, and health and emergency services.
- Risk of mortality during periods of extreme heat.

- Risk of food insecurity and the breakdown of food systems linked to warming, drought, and flooding, particularly for poorer populations in urban and rural settings.
- Risk of loss of rural livelihoods, income, and agricultural productivity.
- Risk of loss of marine and coastal ecosystems, biodiversity, and the ecosystem goods and livelihoods, especially for fishing communities.[15]

All nations will be affected by climate change, but the poorest countries are the most vulnerable and will suffer the most.

1.1.6 **Climate Change is an Ethical Issue**

When federal agencies like the EPA decide whether to implement a new regulation or rule to reduce greenhouse gas emissions, they use a cost-benefit analysis to answer the question: Do the benefits outweigh the costs?[16] The costs of imposing a regulation are weighed against the potential benefits of reducing the harm. But a cost-benefit analysis is a flawed economic framework for climate change policies, because the greatest magnitude of loss will be suffered by future generations. Climate action is not just a matter of science or economics—it is also a moral issue. A cost-benefit analysis (CBA) often trivializes the value of human life, health, and the natural environment because it requires monetary numbers for factors that resist monetization. Economists invent prices for priceless values such as human life and the environment. The economist Frank Ackerman argued that some things don't have a price:

> There are many crucially important values that do not have meaningful prices. This is not a new idea: as Immanuel Kant put it, some things have a price, or relative worth, while other things have a dignity, or inner worth (Kant, 2005 [1785]). No price tag does justice to the dignity of human life or the natural world.[17]

The economists Gernot Wagner and Martin Weitzman make a similar point: "The appropriate price on carbon is one that will make us comfortable enough to know that we will *never* get to anything close to 6°C (11°F) and certain eventual catastrophe."[18]

In 2015, world leaders came together in Paris and 195 nations signed a historic international climate treaty to slow anthropogenic global warming. The aim of the Paris Agreement is to stop runaway climate change by "holding the increase in the global average temperature to well below 2°C above pre-industrial levels and pursuing efforts to limit the temperature increase to 1.5°C above pre-industrial levels." Limiting global warming to 1.5°C will require global human-caused CO_2 emissions to reach net zero by 2050, which in turn will require systemic change and rapid transitions in land use, energy, industry, buildings, transportation, and cities.[19] The existential threat posed by climate change requires leadership and action by governments, businesses, the public—by humans everywhere.

1.2 The Ice Ages and Early Human Adaptation to Climate Change

1.2.1 Climate Shifts from Glacial–Interglacial Cycles

For about the last 2.6 million years, Earth has been alternating between long ice ages and shorter interglacial periods. Earth has experienced seven ice ages in the last 650,000 years—one every 100,000 years. These long time scales, which operate naturally and slowly over tens of thousands and hundreds of thousands of years, are very different than the rapid period of global warming the Earth has experienced since the mid-20th Century due to human activities.

For centuries, the fundamental cause of the timing, or succession, of Quaternary ice ages remained a scientific mystery. In the 1930s a Serbian astrophysicist, Milutin Milankovitch, first proposed that long-term climate change is the result of variations in the Earth's orbital elements, and that Earth's position in relation to the Sun triggers the beginning and end of glaciation periods. He attributed natural glacial–interglacial cycles to astronomical changes in the orientation of the Earth's tilted rotational axis and orbital ellipse.[20] The hypothesis of orbital forcing remained unproven during his lifetime, but in 1976, the journal *Science* published "Variations in the Earth's Orbit, Pacemaker of the Ice Ages,"[21] a study by Hays, Imbrie, and Shackleton, of deep-sea sediment cores that upheld Milankovitch's theory that orbital changes drive the timing of glacial and interglacial periods.

1.2.2 Milankovitch Cycles

Milankovitch cycles have three types of orbital movement: eccentricity (changes to the path of the orbit), in which the orbital shape ranges from more to less elliptical (1 to 5% ellipticity) on a cycle of about 100,000 years; obliquity (tilt of axis), in which the Earth's axis angle decreases from a maximum of 24.5 degrees to a minimum of 22.1 degrees in a cycle that spans about 41,000 years; and precession (how much the Earth wobbles on its axis).[22]

The Earth has an asymmetric distribution of landmasses, with most land masses located in the northern hemisphere. The amount of solar radiation that the high latitudes in the northern hemisphere receive in the summer initiates large scale, natural rhythmic cycles of global warming or cooling, that in the past served as a primary influence and control on climate change.[23]

In the Quaternary, glacial periods have occurred when the Milankovitch factors of obliquity, precession, and eccentricity line up so that the northern hemisphere receives the least amount of summer insolation (solar energy received per surface area). When summer temperatures in the northern hemisphere decline for decades, winter snow and ice persist all year and begin to glaciate, and an ice age is triggered. Eventually the pendulum swings back, and interglacial periods have occurred when the three orbital phases are aligned so that the northern hemisphere has the greatest amount of summer insolation. Cyclical orbital forcing starts interglacial periods with

Figure 3. **Milankovitch Cycles**

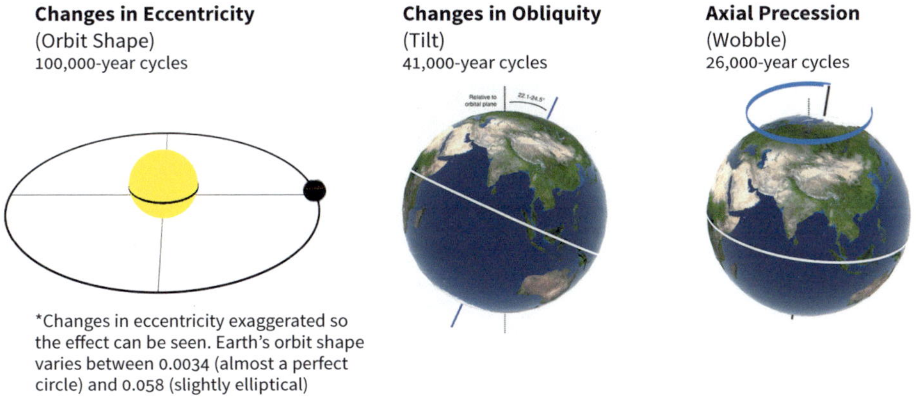

Source: Adapted from Alan Buis, "Milankovitch (Orbital) Cycles and Their Role in Earth's Climate," https://climate.nasa.gov.

small increases of insolation and warming. However, orbital forcings are not the sole cause of the rapid climate transitions that are seen in paleoclimatic records. Positive feedbacks within Earth's climate systems amplify the orbital effects to produce glacial cycles. As the oceans warm, they release carbon dioxide into the atmosphere, and this natural process amplifies the warming in a feedback loop. Glacier retreat and an ice-albedo feedback also contributes to warming. At the end of an interglacial period, a slow process causes the sequential reversal of these conditions.

Earth is currently in the latter half of an interglacial cycle. The Earth's axis is currently tilted at 23.4 degrees (about halfway between its extreme of 24.5 degrees and minimum of 22.1 degrees) and is in the process of decreasing. In approximately 9,800 more years it will reach its minimum tilt. Although we are currently in an interglacial warm period, the Earth should be cooling. As noted by Hays et al., "A model of future climate based on the observed orbital-climate relationships, but ignoring anthropogenic effects, predicts that the long-term trend over the next several thousand years is toward extensive Northern Hemisphere glaciation." Today, NASA scientists note that if there were no human influences on climate, "Earth's current orbital positions within the Milankovitch cycles predict our planet should be cooling, not warming, continuing a long-term cooling trend that began 6,000 years ago."

For the past 800,000 years, until the 20th century, atmospheric concentration of CO_2 ranged from 170 to 300 parts per million (ppm). As can be seen in Figures 4 and 5 (next page), changes in CO_2 concentrations tracked changes in temperature during glacial and interglacial cycles for the past 800,000 years. As Earth emerged from an ice age, temperature and CO_2 levels rose at approximately the same time. The recent increase in atmospheric concentrations of CO_2 caused by human activity is startling because it is unprecedented in the past 800,000 years, and, on a geological scale, is virtually instantaneous. CO_2 levels surpassed 417 ppm in 2020. The last time Earth had a CO_2 concentration of 400 ppm was three million years ago, before humans existed.

Figure 4. **Earth Has Cycled Between Ice Ages and Warm Interglacial Periods**

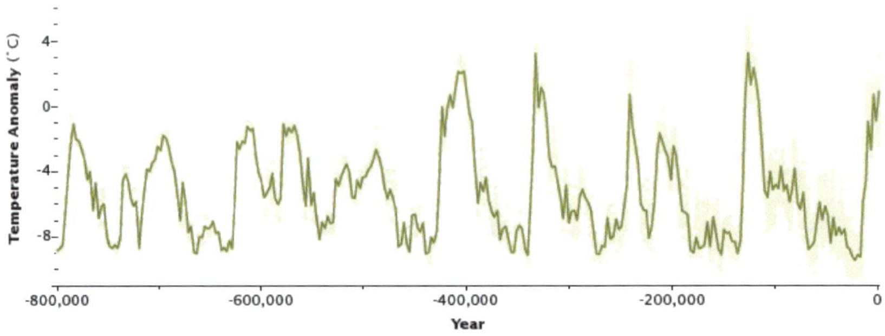

Glacial ice and air bubbles trapped in the ice preserve an 800,000-year record of temperature and carbon dioxide. Earth has cycled between ice ages (low points, large negative anomalies) and warm interglacials (peaks).

Source: "How is Today's Warming Different from the Past?" NASA Earth Observatory, June 3, 2010. Graph by Robert Simmon, based on data from Jouzel et al., 2007.

Figure 5. **CO_2 Levels During Ice Ages and Interglacial Periods, and CO_2 Levels Now**

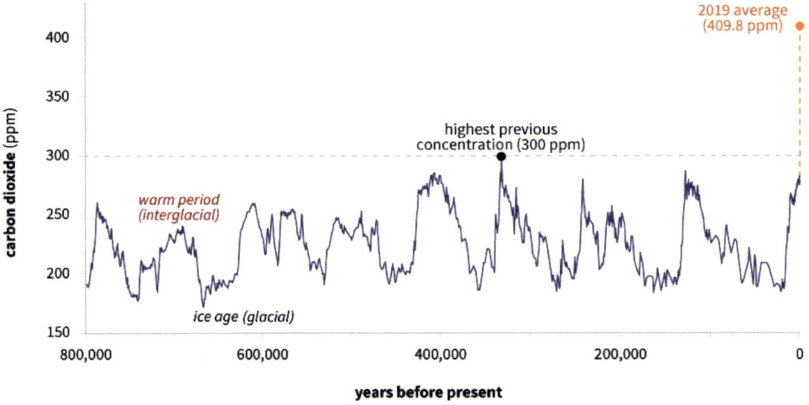

Global atmospheric CO_2 in parts per million (ppm) for the past 800,000 years. The peaks and valleys track ice ages and warmer interglacials. During these cycles, CO_2 was never higher than 300 ppm. On the geologic time scale, the increase (orange dashed line) looks virtually instantaneous.

Source: Rebecca Lindsey, " How do we know the build-up of carbon dioxide in the atmosphere is caused by humans?" NOAA Climate.gov, November 5, 2020. Graph by NOAA Climate.gov based on data from Lüthi, et al., 2008, via NOAA NCEI Paleoclimatology Program.

1.2.3 **Early Human Adaptation to Climate Change**

Since the emergence of our modern human species, *Homo sapiens*,150,000 years ago, human societies have changed adaptively in response to climate-driven environmental change. The paleoclimate history of northeast Africa suggests that warm and wet conditions created green, fertile corridors through the Sahara that facilitated the

Figure 6.

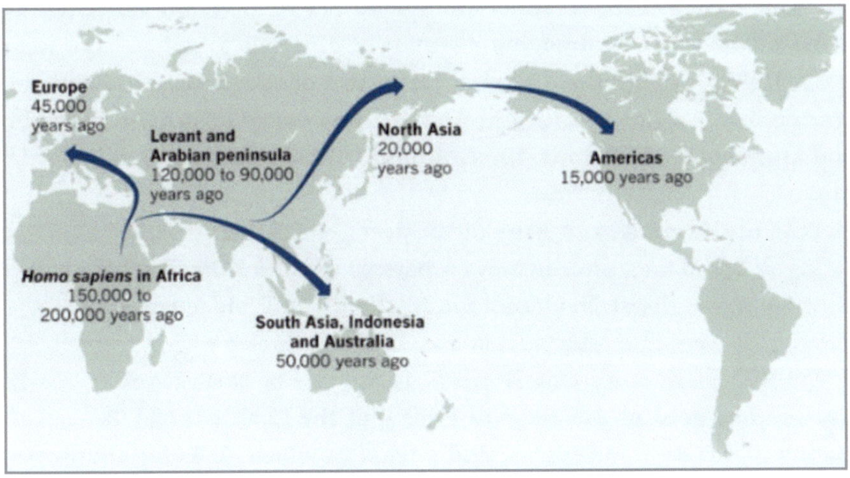

Source: deMenocal, P., Stringer, C., "Climate and the Peopling of the World." *Nature* 538, 49–50 (2016)

Source: Drawing of woolly mammoth by Charles Knight, 1915.

At the close of the last Pleistocene glacial period, 11,000 years ago, the climate became warmer, the tundra turned to forest, and large mammals such as the mammoth and wooly rhinoceros became extinct. Agriculture began in the Holocene, our modern climate era. Glaciers melted, sea levels rose, and Neolithic humans began to farm around the world, allowing populations to grow.

earliest waves of *Homo sapiens* migration out of Africa. However, the primary migration that occurred 60,00–70,000 years ago was driven by a rapidly deteriorating climate at the beginning of the last ice age and a sudden shift to a dry, cold period of extended drought.[26]

Our African ancestors migrated out of Africa and into an ice age, where they became the dominant land animal, and falling sea levels created land bridges that allowed humans to walk from Asia to North America and from New Guinea to Australia, and ultimately to colonize the world.

As civilizations arose in Mesopotamia, China, and South America, legends of a deluge that wiped out humanity appeared in both oral and written traditions. Over the centuries, people have wondered, are these flood stories based on real natural disasters long ago, or are they imaginary?

Written and oral histories from around the world depict the impact that climate had on ancient civilizations. In the Chinese legend of Emperor Yu, a great deluge of the Yellow River engulfed the land over 4,000 years ago, with waters embracing the mountains, submerging the hills, and rising to the heavens. For over twenty years, the floodwaters continued, creating widespread destruction and suffering, until Yu tamed the Yellow River by dredging a system of channels to redirect the waters out to the sea. Unlike western flood stories, the legend of Yu the Great is centered on an engineering feat— a flood control project. Yu was revered for his engineering success and ascended to the throne, thus founding China's first, and semi-mythical, Xia dynasty.

The Epic of Gilgamesh is a Babylonian poem, written in cuneiform on clay tablets dating to 650 B.C.E., and includes a passage about a storm lasting six days and nights, inflicting a global flood and annihilation.[27] Still older versions of the flood story were discovered on tablets from the Tigris-Euphrates region from about 2,000 B.C.E. The Gilgamesh story closely parallels the tale of Noah's Ark in the Book of Genesis, the first book of the Hebrew Bible and the Christian Old Testament, with rain lasting for 40 days and nights, and a flood in which all living things perished, except two of every species.

Photo: Gilgamesh Flood Tablet XI, British Museum.

Andrew George, tr. *The Epic of Gilgamesh*. Harmondsworth: Penguin Books, 2000, pp. 88–99.

Tablet XI

'The stillness of the Storm God passed over the sky,
  and all that was bright then turned into darkness.
[He] charged the land like a bull [on the rampage,]
  he smashed [it] in pieces [like a vessel of clay.]
'For a day the gale [winds flattened the country,]
  quickly they blew, and [then came] the [Deluge.]
Like a battle [the cataclysm] passed over the people.
  One man could not discern another,
  nor people be recognized amid the destruction.
'Even the gods took fright at the Deluge,
  they left and went up to the heaven of Anu.

* * *

'For six days and [seven] nights,
  there blew the wind, the downpour,
  the gale, the Deluge, it flattened the land.

1.2.4 Natural Climate Shifts Have Impacted Different Parts of the World at Different Times

Humanity's ability (or inability) to adapt to changing climate has been a crucial determinant in the rise or fall of civilizations. During abrupt climate fluctuations with prolonged droughts lasting decades or centuries, ancient empires collapsed and entire civilizations vanished. Geological data strongly suggest that the fall of the Akkadians of Mesopotamia, who formed the world's first empire over 4,300 years ago, is related to a 300-year drought that began in 2200 B.C.E.[28] The drought, which may have resulted from cooler sea surface temperatures in the North Atlantic, led to crop failures and famine and also precipitated the collapse of the Old Kingdom of ancient Egypt.

In North America, the Neolithic Anasazi civilization, which arose in 1500 B.C.E., is another example of an ancient civilization that collapsed due to a climate shift that brought drought, food shortages, and starvation. A Navajo word, the term *Anasazi,* is literally translated to mean "enemy ancestors" but has come to mean "the ancient ones."[29]

Cliff Dwellings of the Ancestral Puebloans in Mesa Verde

Source: Cliff Palace, Mesa Verde. National Park Service, 2014.

These Ancestral Puebloans built magnificent, multistoried cliff dwellings high in the canyons of the Four Corners region of southwestern Colorado, southeastern Utah, northwestern New Mexico, and northeastern Arizona. The Anasazi's Mesa Verde settlement in Colorado began at the time of the Roman Empire and flourished for 1,000 years, but within a single generation from 1275–1300 C.E., the Anasazi abandoned their pueblo homes and vanished. Tree ring data indicates that their decline coincided with a prolonged drought. The tree rings show that rain returned in 1300–1340, but the people never did.[30]

Natural climate variability during the past 2,000 years provides an important context for understanding the man-made forcing of Earth's climate since the 1850s. Paleoclimate reconstructions of this time span from proxy data, including historical documents (such as literature, ecclesiastical records, paintings, ship logs, and farmers' logs), tree rings, lake sediments, glacier ice, and corals, indicate there were four natural climate swings during the early part of the Common Era: the Roman Warm Period (in the first centuries C.E.), the Dark Ages Cold Period or Late Antique Little Ice Age (in some regions from 400 to 800 C.E.), the Medieval Climate Anomaly, and the Little Ice Age.

At one time, it was widely believed that the Medieval Climate Anomaly and the Little Ice Age were globally synchronous phenomena. Climate scientists use the term "Medieval Climate Anomaly" to describe a relatively warm preindustrial period from 800 to 1200 C.E. when crops flourished in Europe and the alpine tree limit rose to higher elevations. The term "Little Ice Age" refers to a relatively cold epoch from

1300 to 1850 C.E. associated with an advance of glaciers in the western U.S. and the European Alps.[31] During the past few decades, scientific research has confirmed that these shifts of warming and cooling were not global-scale phenomena, but instead, occurred in different parts of the world at different times, primarily in the Northern Hemisphere, in Europe and North America.[32]

The last written record of the Vikings is a wedding at the Hvalsey Church in 1408

Source: Photo by Ciril Jazbec, from Tim Folger, "Trying to Solve the Mystery of Greenland's Vanished Vikings," *Smithsonian Magazine*, March 2017.

During the warm period, the Norse, or Vikings, sailed from Iceland in 985 and founded a colony in Greenland, which thrived for nearly five centuries, but then suddenly vanished. The Norse abandoned their settlements in Greenland around 1430 C.E., when the climate became consistently colder during the Little Ice Age and farming was unsustainable. Only the native Inuit remained.

Schematic Diagram (Figure 7.1) from IPCC's First Assessment Report

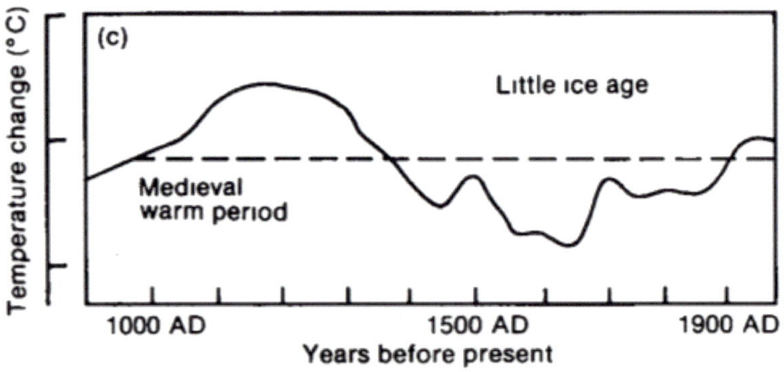

Source: https://www.ipcc.ch/site/assets/uploads/2018/03/ipcc_far_wg_I_chapter_07-1.pdf]

Who Drew Figure 7.1?

This schematic diagram (preceding page) was published as Figure 7.1 in Chapter 7 of the IPCC's *First Assessment Report* (Folland et al., 1990, p. 202) and seems to show that the Medieval Warm Period had higher global temperatures than current times. Yet the curve is hand drawn and the temperature axis has no numbers. The IPCC dropped the drawing in later reports, but it was republished in other literature. In 2008, researchers led by Phillip D. Jones sought to answer the question, "Where did the schematic diagram come from?"

Jones et al. traced the graph's origin to a series of publications from 1965 to 1982 by Hubert H. Lamb, an English meteorologist. In 1965, Lamb coined the term "Medieval Warm Epoch," but his graph is not a global temperature reconstruction. It is based only on data from central England. In their paper, "High-resolution palaeoclimatology of the last millennium: a review of current status and future prospects," Jones et al. note that the graph was "last published (as Figure 30 on p. 84) by Lambç (1982)" and "the Central England curve also appeared in Lamb (1965: Figure 3 and 1977: Figure 13.4)."[33]

Science has progressed, and more recent research, including a 2019 paper by a team of international scientists, continues to verify that climate fluctuations such as the Medieval Climate Anomaly and Little Ice Age were not trends in global or hemispheric mean temperature, but instead were localized events in Europe and neighboring regions that varied in timing.[34] Unlike these natural, pre-industrial climate fluctuations, the unprecedented, rapid anthropogenic (human induced) rise in global average temperatures the Earth is now experiencing has occurred concurrently around the globe and has affected more than 98 percent of the planet.[35] At the end of the 20th century, temperatures were hotter across the Earth than at any time in the previous 2,000 years, and are continuing to rise.

1.3 The Human Origin of Global Warming

1.3.1 Human Activities Have Altered Earth's Natural Cycles

Before the 20th century, the idea that human activity could radically alter the climate on a global level was not a realistic concern. But Earth's average temperature has risen by about 2.1°F (1.2°C) since humans began burning fossil fuels 150 years ago, and the warming rate is accelerating. As noted by NOAA: "[t]he transition from the last ice age to the current interglacial period is estimated to have spanned 5,000 years. Humans could witness the same magnitude of global warming within a span of about 110 years. In other words, if our world warms by as much as 7°F (4.1°C) from 1990 to 2100, as some climate models project could happen, then that warming rate is about 45 times faster than the warming Earth experienced when it emerged from the last ice age."[36]

1.3.2 **The Greenhouse Effect**

Humans have survived climate instability in the past, but the present climate shift is of far greater magnitude than anything the Earth has experienced in the past 2,000 years. Since the Industrial Revolution, human burning of fossil fuels like coal and oil for energy has increased the concentration of atmospheric carbon dioxide (CO_2), thereby causing the atmosphere to trap more heat and exacerbate what is known as a greenhouse effect. Our modern understanding of climate change rests upon the pioneering contributions of scientists in the 1800s such as Joseph Fourier, John Tyndall, Eunice Foote, and Svante Arrhenius, who discovered and studied the greenhouse effect.

1.3.3 **Early Climate Scientists Who Studied the Greenhouse Effect**

Jean Baptiste Joseph Fourier, a French mathematician, is credited as the first scientist to propose that Earth's atmosphere retains heat radiation. In the 1820s, Fourier calculated that the Earth should have an average temperature of −16° Celsius, but since that wasn't the case, he concluded that the atmosphere must be hindering heat from escaping into space.

Jean Baptiste Joseph Fourier (1768–1830)

In 1824, Fourier published a paper, "*Remarques générales sur les températures du globe terrestre et des espaces planétaires*," which compared the Earth with its atmosphere to a box with a glass lid, similar to a hothouse. When sunlight enters, the box's interior heats up and the lid traps the heat. Fournier did not understand the role of carbon dioxide, but his analogy became known as the "greenhouse effect."

In 1856, Eunice Foote, an American scientist, first discovered the ability of carbon dioxide and water vapor to absorb heat.[37] Her paper, "Circumstances Affecting the Heat of the Sun's Rays,"[38] described the warming effect of carbon dioxide in the atmosphere, but Foote's discovery was overlooked, due to her gender.[39] Joseph Henry, the first Secretary of the Smithsonian Institution, presented her paper at the annual meeting of the American Association for the Advancement of Science in Albany, New York, but it remained in obscurity.

Across the Atlantic, John Tyndall, an Irish physicist in Victorian England, was independently investigating the heat-trapping properties of gases.

Tyndall's famous paper, "On the Absorption and Radiation of Heat by Gases and Vapours, and on the Physical Connexion of Radiation, Absorption, and Conduction," explained the scientific basis of the greenhouse effect.[40]

Tyndall's Radiant Heat Apparatus

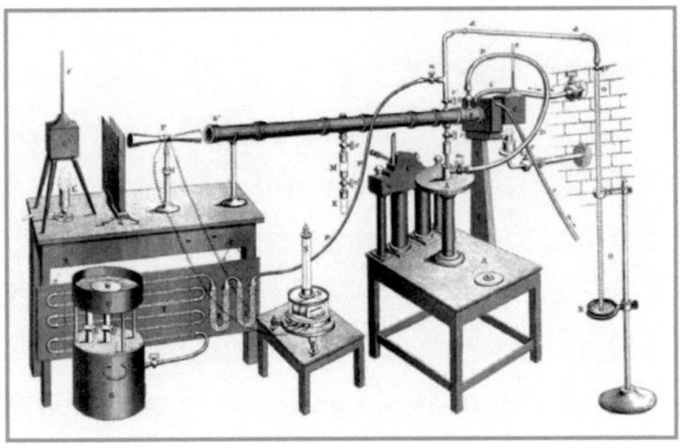

Courtesy the Royal Institution of Great Britain.

Tyndall made technology innovations by designing his own laboratory instruments, a thermopile, and galvanometer apparatus. In 1859, he discovered that small amounts of "perfectly colorless and invisible gases and vapours," including compound gases "carbonic acid" (CO_2) and "water vapour" (H_2O), were able to absorb thermal radiation, but the "simple atoms" of nitrogen and oxygen were inactive.

Following these advances, in 1896, a Swedish scientist, Svante Arrhenius, who later received a Nobel Prize in chemistry, first quantified the effects on the Earth's temperature of increasing and decreasing atmospheric carbon dioxide concentrations.[41] Like Tyndall, his experiments with CO_2 were motivated by an interest in the Quaternary ice cycles. With painstaking numerical calculations, Arrhenius estimated that reducing CO_2 by 50% would cause the climate to cool by 4°–5°C, potentially triggering an ice age, and that doubling CO_2 would cause temperatures to rise by 4°–5°C, leading to rapid warming and glacial retreat. Arrhenius concluded that the industrial burning of coal was adding carbon dioxide to Earth's atmosphere, but in the late 1800s he believed that climate warming would take thousands of years, and have societal benefits, rather than detriments.

1.3.4 Establishing a Modern CO_2 Theory of Climate Change

In 1938, an English steam engineer, Guy Stewart Callendar, published the first evidence directly linking human activities to global warming. His paper, "The Artificial Production of Carbon Dioxide and Its Influence on Temperature,"[42] collected weather data from 147 stations around the world that showed that Earth had warmed about 0.3°C in the previous 50 years, from the burning of coal and fossil fuels.[43] Callendar's estimate of 290 parts per million (ppm) for the atmospheric concentration of carbon dioxide in 1900 is still valid today. Calculating by hand, "he estimated there had been a ten percent increase between 1900 and 1935, which closely matched the amount of fuel burned."[44] Skeptics challenged Callendar's analysis, arguing that natural physical processes, such as the absorption of CO_2 by the oceans, would dissolve all of the harmful CO_2 emissions that humans produce from fossil fuels.

In the early 1900s, the issue of global warming was only being studied by a few academics. Roger Revelle, an expert in oceanography, led the way to our modern understanding by framing the important scientific questions and then securing both the people and resources to answer them.[45]

1.3.5 **Discrediting the Theory that Earth's Oceans Will Absorb All Human Fossil Fuel Emissions**

Revelle served in the U.S. Navy during World War II, and later obtained major funding from the Navy to conduct research in the Pacific Ocean. In 1950, he became the Director of the Scripps Institution of Oceanography in La Jolla, California, and was instrumental in raising scientific and public consciousness about anthropogenically-induced climate change. Revelle discredited a widely-held belief at the time that Earth's systems are self-regulating, and that any excess CO_2 would be absorbed by the vast oceans. He demonstrated that despite the buffer mechanism of sea water, more than half the carbon dioxide produced by carbon combustion and other human activities remains in the atmosphere.[46]

Revelle invited Hans E. Suess, a founder of radiocarbon dating, to collaborate with him at the Scripps Institution. In 1957, Revelle published a now famous article with Suess in *Tellus*, a European meteorology and oceanography journal. In "Carbon Dioxide Exchange Between Atmosphere and Ocean and the Question of an Increase of Atmospheric CO_2 During the Past Decades," he observed that "Human beings are now carrying out a large scale geophysical experiment of a kind that could not have happened in the past nor be reproduced in the future."[47]

1.3.6 **The Keeling Curve**

In 1956, Revelle recruited Charles David Keeling to join the Scripps Institute, and Keeling's targeted research unequivocally proved that atmospheric CO_2 is rising due to human burning of fossil fuels. In 1958, using an infrared (IR) gas analyzer, Keeling began meticulously measuring the amount of CO_2 in the atmosphere from a weather observatory 3,000 meters above sea level on Mauna Loa volcano in Hawaii.

Figure 7. **The Keeling Curve**

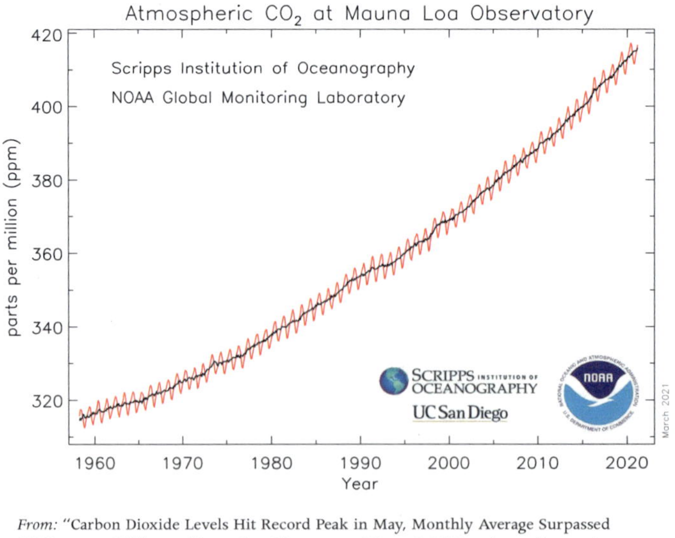

From: "Carbon Dioxide Levels Hit Record Peak in May, Monthly Average Surpassed 414 Parts per Million at Mauna Loa Observatory," June 4, 2019, at https://research.noaa.gov

The first reading on March 29, 1958, measured the atmospheric CO_2 concentration at 313 ppm. Throughout the 1960s, Keeling's measurements were plotted on the now-famous graph known as the Keeling Curve, which showed a steadily increasing annual concentration of CO_2 in the atmosphere since 1958.

Due to seasonal cycling of carbon dioxide, the curve zig-zags in an annual saw-tooth pattern that indicates seasonal variations. For example, during the fall and winter, microbes release CO_2 from decomposing plants, with an annual peak in May, and during the spring and summer, plants absorb CO_2 through photosynthesis, with low levels coming in September, when the growing season ends. In 1974, the National Oceanic and Atmospheric Administration (NOAA) began independent CO_2 measurements at the Mauna Loa observatory.[48]

The graph, with its steady upward curve, is now an iconic symbol of the human impact on the environment and the role of fossil fuels.[49] The Keeling Curve disproved the theory that Earth's vast oceans could absorb all of the excess CO_2 and prevent it from altering the atmosphere.

1.3.7 The Transformative Decade of the 1960s

During the 1960s, when Keeling's measurements for the Scripps CO_2 program were providing the first long-term record of anthropogenic CO_2 emissions, the U.S. was experiencing an environmental crisis. America's cities were trapped in a pall of smog, industrial smokestacks belched out pollution, automotive emissions hung in the air, and waterways were fouled with industrial waste and sewage. Rachel Carson's book, *Silent Spring*, published in 1962, presented frightening evidence that pesticides including DDT not only killed insects, but were toxic to birds and fish, and contaminated the entire food chain, ultimately threatening children. Concern over the impacts of human activity on nature ignited a grassroots environmental movement that demanded aggressive action to clean up the air, water, and land. The first Earth Day in April 1970 was a symbol of environmental consciousness that led to creation of the Environmental Protection Agency (EPA) and passage of the Clean Air Act in 1970 and the Clean Water Act in 1972.

Photo from Operations Excellence, "A Brief History of the Environmental Protection Agency," Sept. 15, 2015, at https://www.opexcel.com/history-of-the-environmental-protection-agency/.

1.4 Modern Awareness of Climate Change

1.4.1 Climate Change Awareness Grew in the 1970s and 1980s

In the 1970s, scientists at the Massachusetts Institute of Technology were beginning to use modeling to study climate change. In 1970, Carroll Wilson of MIT convened a multidisciplinary group of 40 scientists to study humans' impact on the global environment. The next year, he organized a study group in Stockholm of 35 atmospheric scientists from 15 countries. The meeting resulted in a report, "Inadvertent Climate Modification: Report of the Study of Man's Impact on Climate."[50] A dedication in the front of the book, from a Sanskrit prayer, imparts a simple message: "Oh, Mother Earth . . . pardon me for trampling on you."

Throughout the 1970s, as scientific knowledge about global warming grew, the news media also published sensational reports claiming that the Earth was entering a cooling trend. The notion of global cooling was "never more than a minor aspect of the scientific climate change literature of the era,"[51] but in the absence of a consensus about whether the earth was getting warmer or colder, the scientific community continued to opine that more research was needed. Press coverage of climate change was overshadowed by other urgent environmental controversies involving the earth's stratosphere when environmentalists alerted the public that aerosol spray cans were releasing chlorofluorocarbons (CFCs), gases that were destroying the high ozone layer. In 1977, in response to public outrage, Congress restricted use of spray can chemicals in the Clean Air Act.

During the 1980s, two climate scientists, James Hansen and Robert Watson, finally succeeded in focusing a public spotlight on the risks of climate change. They generated enough public concern that the federal government began to sponsor significant climate research. Hansen, a scientist at the National Aeronautics and Space Administration's Goddard Institute for Space Studies in New York, began studying the effect of greenhouse gases on climate. In 1981, Hansen published a paper, "Climate Impact of Increasing Atmospheric Carbon Dioxide," in the journal *Science,* which predicted that burning fossil fuels would increase global temperatures by 4.5° Fahrenheit (2.5° Celsius) by the end of the 21st century.[52]

The topic of climate change soon became political because the government needed to respond to these warnings with policies. In 1981, 1982, and 1984, then-Representative Albert Gore organized House hearings on rising levels of atmospheric carbon dioxide.[53] (In 2006, former Vice President Al Gore's award-winning documentary film, *An Inconvenient Truth,* raised public consciousness, and in 2007, he was awarded the Nobel Peace Prize for his advocacy on behalf of climate change.)

On June 10 and 11, 1986, Robert Watson, a NASA scientist, testified during two days of Senate hearings on the subject of "Ozone Depletion, the Greenhouse Effect, and Climate Change." His testimony—"I believe global warming is inevitable. It is only a question of the magnitude and the timing"[54]—was reported by *The New York Times* and *Washington Post,* and played a key role in alerting the public to the threats posed by climate change. Watson would later go on to chair the United Nations Intergovernmental Panel on Climate Change from 1997 to 2002.

Global warming remained a nonpartisan subject for most of the 1980s, with both Democrats and Republicans calling for more scientific research. In 1987, during the Reagan administration, White House science advisor William Graham and NOAA Administrator Anthony Calio formed a Committee on Earth Sciences, within the Federal Coordinating Council on Science, Engineering and Technology, with the stated purpose of studying global warming and developing policies in response to global change. In 1986, Delaware Senator Joseph R. Biden was the first senator to introduce a bill to establish a task force on the global climate, and in 1987, Congress enacted the Global Climate Protection Act of 1987 (P.L. 100-204).

As the news media's coverage of climate change increased in the late 1980s, public awareness continued to grow. In June 1988, Jim Hansen received national news coverage when he testified before a Senate Committee that he was "99% certain" that global warming was occurring and was caused by human activities.[55] Hansen's testimony warned that computer modeling demonstrated the magnitude of the greenhouse effect, and that it would increase and exacerbate extreme weather events, such as hurricanes, heat waves, and flooding.

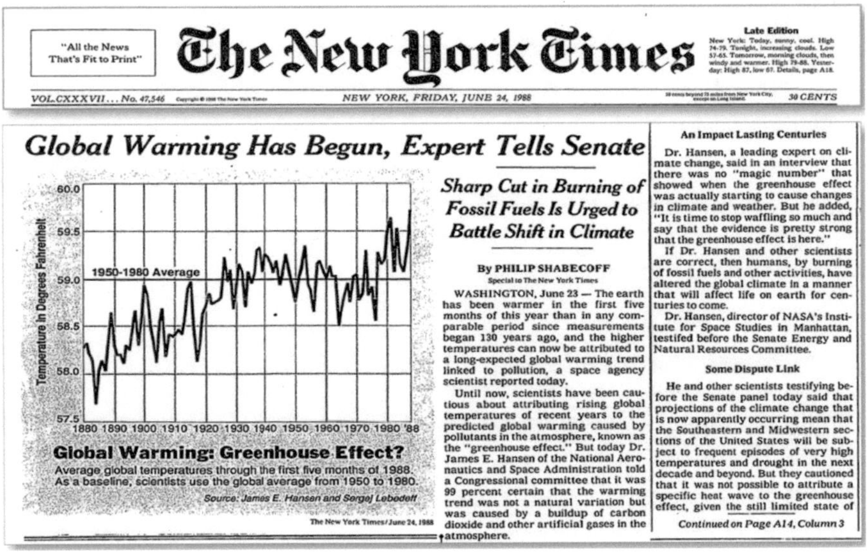

"All the News That's Fit to Print"

The New York Times

Late Edition

New York: Today, sunny, cool. High 74-79. Tonight, increasing clouds. Low 57-65. Tomorrow, morning clouds, then windy and warmer. High 79-88. Yesterday: High 87, low 67. Details, page A18.

VOL. CXXXVII . . . No. 47,546 — NEW YORK, FRIDAY, JUNE 24, 1988 — 30 CENTS

Global Warming Has Begun, Expert Tells Senate

Global Warming: Greenhouse Effect?

Average global temperatures through the first five months of 1988. As a baseline, scientists use the global average from 1950 to 1980.

Source: James E. Hansen and Sergej Lebedeff

The New York Times/June 24, 1988

Sharp Cut in Burning of Fossil Fuels Is Urged to Battle Shift in Climate

By PHILIP SHABECOFF

Special to The New York Times

WASHINGTON, June 23 — The earth has been warmer in the first five months of this year than in any comparable period since measurements began 130 years ago, and the higher temperatures can now be attributed to a long-expected global warming trend linked to pollution, a space agency scientist reported today.

Until now, scientists have been cautious about attributing rising global temperatures of recent years to the predicted global warming caused by pollutants in the atmosphere, known as the "greenhouse effect." But today Dr. James E. Hansen of the National Aeronautics and Space Administration told a Congressional committee that it was 99 percent certain that the warming trend was not a natural variation but was caused by a buildup of carbon dioxide and other artificial gases in the atmosphere.

An Impact Lasting Centuries

Dr. Hansen, a leading expert on climate change, said in an interview that there was no "magic number" that showed when the greenhouse effect was actually starting to cause changes in climate and weather. But he added, "It is time to stop waffling so much and say that the evidence is pretty strong that the greenhouse effect is here."

If Dr. Hansen and other scientists are correct, then humans, by burning of fossil fuels and other activities, have altered the global climate in a manner that will affect life on earth for centuries to come.

Dr. Hansen, director of NASA's Institute for Space Studies in Manhattan, testifed before the Senate Energy and Natural Resources Committee.

Some Dispute Link

He and other scientists testifying before the Senate panel today said that projections of the climate change that is now apparently occurring mean that the Southeastern and Midwestern sections of the United States will be subject to frequent episodes of very high temperatures and drought in the next decade and beyond. But they cautioned that it was not possible to attribute a specific heat wave to the greenhouse effect, given the still limited state of

Continued on Page A14, Column 3

Front page of *The New York Times,* June 23, 1988, from Union of Concerned Scientists, www.uuscusa.org..

Thirty years after Hansen's historic Senate testimony, there is overwhelming evidence that his predictions were accurate—over one-third of the Arctic ice cap has melted, much of California and the American West have burned, lower Manhattan, South Florida, Houston, and New Orleans have flooded, and average temperatures are continuing to climb.[56]

1.4.2 The U.S. Global Change Research Program

The U.S. government's concern with the effects of climate change began more than forty years ago, and the U.S. Global Change Research Program can trace its roots to scientific initiatives by NASA and NOAA.[57] In 1988, the Committee on Earth Sciences was conducting climate change research primarily in three federal agencies—NASA, NOAA, and the National Science Foundation (NSF). During President George H. W. Bush's administration in 1989, the Committee became the U.S. Global Change Research Program (USGCRP) through a Presidential initiative.

President Bush arrived in the White House on a campaign pledge that "Those who think we are powerless to do anything about the greenhouse effect forget about the 'White House effect'; as President, I intend to do something about it."[58] When Bush signed into law the Global Change Research Act (GCRA) of 1990, it established a legal mandate that USGCRP develop and coordinate "a comprehensive and integrated United States research program which will assist the Nation and the world to understand, assess, predict, and respond to human-induced and natural processes of global change."[59] The Global Change Research Act requires USGCRP to conduct an assessment every four years.

Over three decades ago, climate change was already understood to be a matter of increasing societal concern, and the federal government created the USGCRP, a large-scale, interdisciplinary research program, to study it. By the late 1980s, however, the topic of climate change had become politicized. Powerful interest groups such as automobile manufacturers and fossil fuel corporations believed their profits were threatened, and dedicated enormous sums to lobbying against restrictions on emissions, and to blocking political action. Their main rationale was then—and still is over a generation later—to deny the physics of climate change and the fact that human-made global warming is occurring at all.

1.4.3 The Intergovernmental Panel On Climate Change (IPCC)

In 1988, the United Nations Environment Programme (UNEP) and the World Meteorological Organization (WMO) formed the Intergovernmental Panel on Climate Change for the purpose of providing an objective source of scientific information about climate change. As such, the IPCC is the United Nations body responsible for providing the world with a clear view of the current scientific, technical, and socioeconomic state of knowledge relevant to human-induced climate change.[60] It does not conduct new research, but instead bases its assessments on published and peer-reviewed scientific technical literature.

The IPCC presently has 195 member countries, and its deeply vetted assessments provide all levels of governments with scientific information to develop climate policies. It informs policies in international climate negotiations, such as the Conference of the Parties (COP) of the United Nations Framework Convention on Climate Change (UNFCCC). In 2007, the IPCC was awarded the Nobel Peace Prize, together with former U.S. Vice-President Al Gore.

1.4.4 The Hockey Stick Graph Is Published by the IPCC in 2001

In 1998, Michael E. Mann, a professor of atmospheric science, created a graph for a study that became a symbol of human-induced climate change. Mann and his colleagues, Raymond S. Bradley and Malcolm K. Hughes, published "Global-Scale Temperature Patterns and Climate Forcing over the Past Six Centuries" in the journal, *Nature*.[61]

Figure 8. **The Original "Hockey Stick" Graph**

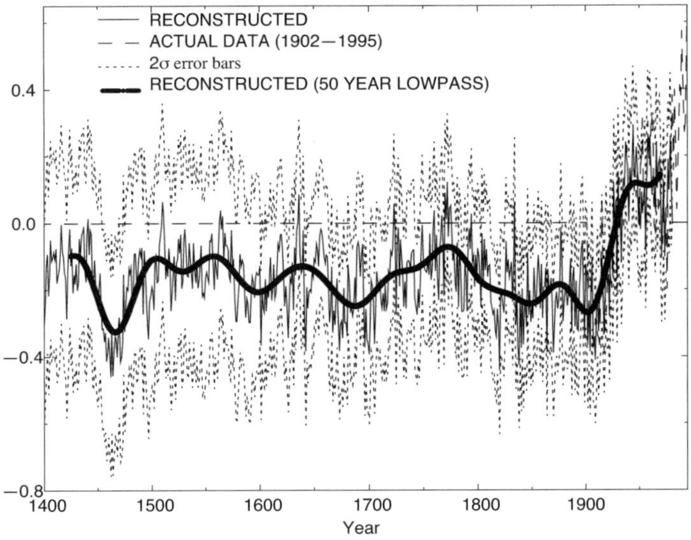

Source:. Michael E. Mann, Raymond S. Bradley, and Malcolm K. Hughes, "Global-Scale Temperature Patterns and Climate Forcing over the Past Six Centuries." *Nature*, vol. 392, April 23, 1998, p. 783.

The study included a visually compelling graph depicting their results, and it became the most politicized graph in the history of science.

The initial research reconstructed temperature for half a millennium, including the centuries before thermometer records were kept, by using proxy data from ice cores, tree rings, corals and lake sediments. The next year, in 1999, Mann et al. published an updated article which included proxy data extending to the year 1000 C.E.[62] Plotted on a graph, the data showed that the dramatic warming of the Northern Hemisphere in the last 50 years was unprecedented in the previous 1,000 years.

Scientists in climate circles recognized that the graph looked like a hockey stick—the temperature remained mostly flat from 1000 C.E. until 1900 (the hockey stick "shaft") and then spiked upward in the 20th century (the hockey stick "blade") due to increases in temperature from the burning of fossil fuels. The graph included wide error bars to the historical temperatures derived from proxy data. Mann's work achieved special prominence in 2001 because he was a contributing author of the IPCC's *Third Assessment Report* (TAR, 2001) and a co-lead author of the chapter, "Observed Climate Variability and Change," which featured the hockey stick graph in Figure 220 (Figure 5 in the Technical Summary).[63]

Figure 9.

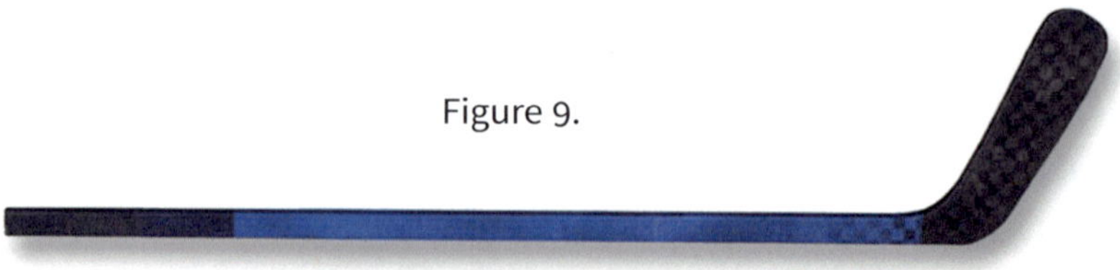

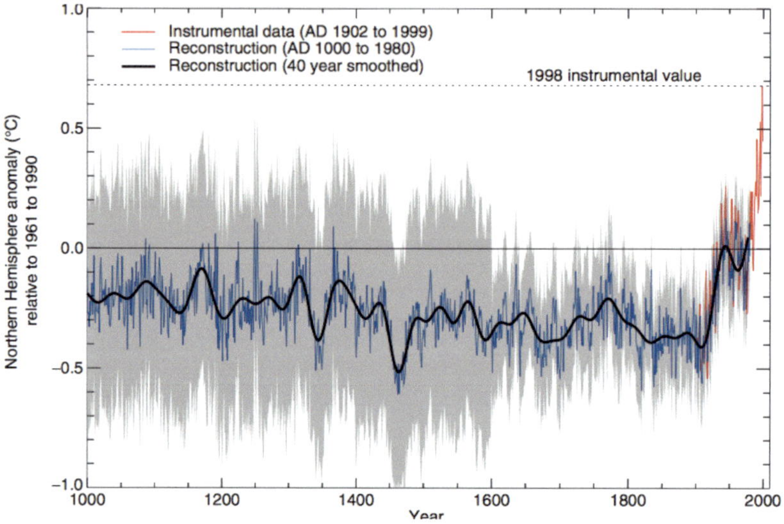

Graph from IPCC's *Third Assessment Report*, 2001, Chapter 2, Figure 2.20 (Figure 5 in the Technical Summary).

The graph appeared multiple times in TAR, including in the *Summary for Policymakers*, where IPCC reported that "[T]he rate and duration of warming of the 20th century has been much greater than in any of the previous nine centuries. Similarly, it is likely that the 1990s have been the warmest decade and 1998 the warmest year of the millennium."[64]

After the IPCC Report was published, Mann's graph became publicly disputed. He was vilified in the media by climate change skeptics, including fossil fuel companies who charged that he falsified the proxy data, and that temperatures during the Medieval Warm Period were higher than those of post-industrial times. Despite the wide error bars on his graph, he was personally attacked and threatened with congressional investigations, which questioned his methodologies. In 2003, Mann was called to testify about his research before the U.S. Congress in hearings sponsored by Oklahoma Senator James M. Inhofe, who called global warming a "hoax."[65]

Subsequently, independent researchers, including the National Center for Atmospheric Research, tested and upheld Mann's work, resulting in many other hockey-stick temperature reconstructions.[66] Congress asked the highest scientific body in the U.S., the National Academy of Sciences, to perform its own investigation, and in a report released on June 22, 2006, the Academy vindicated the work of Mann, Bradley, and Hughes:

> The basic conclusion of Mann et al. (1998, 1999) was that the late 20th century warmth in the Northern Hemisphere was unprecedented during at least the last 1,000 years. This conclusion has subsequently been supported by an array of evidence that includes both additional large-scale surface temperature reconstructions and pronounced changes in a variety of local proxy indicators, such as melting on ice caps and the retreat of glaciers around the world.[67]

Thus, the one hockey stick graph became what Mann calls a "hockey team."[68] "'If you look at the different groups [of researchers] there are literally about two dozen' hockey sticks now," he states.[69] In 2018, the American Association for the Advancement of Science awarded Mann its prestigious prize for Public Engagement with Science, and in 2019, Mann was awarded the Tyler Prize for Environmental Achievement.[70]

New scientific data and research, including a trilogy of papers in 2019 by the Past Global Changes (PAGES) 2k consortium, an international group of scientists, continue to support the conclusions underlying the hockey stick graph.[71] The 2019 studies, published in the journal, *Nature*, used proxy data to develop global maps of temperature change going back 2,000 years. The papers find no evidence for pre-industrial globally synchronous cold and warm periods, and conclude that anthropogenic global warming is unparalleled in the past 2,000 years. The PAGES 2k dataset is archived at the World Data Service (WDS) for Paleoclimatology, hosted by NOAA.[72]

Figure 10. **PAGES 2k Consortium Supports Conclusions of Hockey Stick Graph**

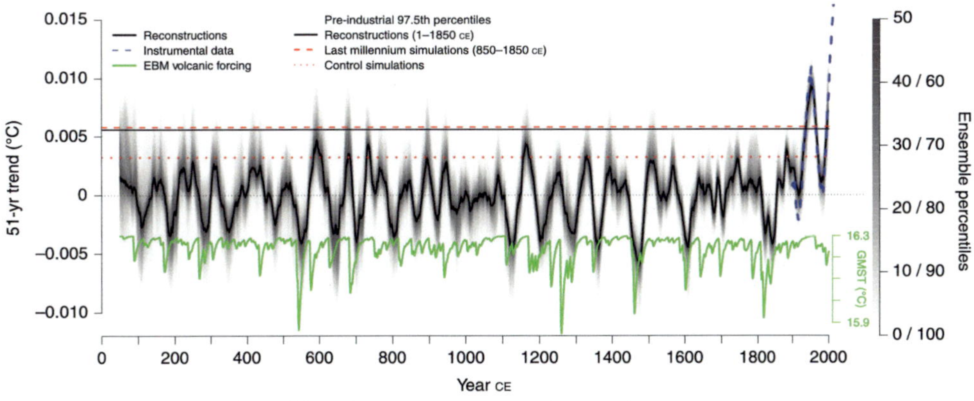

Fig. 4 | Multidecadal temperature trends over the Common Era.

Source: PAGES 2k Consortium, "Consistent Multidecadal Variability in Global Temperature Reconstructions and Simulations over the Common Era," Figure 4, *Nature Geoscience*, July 24, 2019.

Today, NOAA's National Centers for Environmental Information offers paleoclimatic proxy data and paleoclimate reconstructions available to search and download on its website, with over 10,000 data sets derived from natural sources such as tree rings, ice cores, corals, and ocean and lake sediments.[73]

1.4.5 **IPCC's High Degree of Scientific Certainty**

IPCC's *Fourth Assessment Report* (AR4, 2007) recognized the enormous advancements in climate science that were made possible over 40 years by the evolution of scientific methodology and tools, including computer modeling.[74] The advancements allowed "increasingly more definitive statements about human impacts on climate."[75] In the *Fifth Assessment Report* (AR5 2013, 2014), the IPCC finally announced with a high degree of scientific certainty that human activity is the dominant cause of global warming since the mid-1900s, and that urgent action by policy-makers is needed:

> [S]cience now shows with 95 percent certainty that human activity is the dominant cause of observed warming since the mid-20th century. The report confirms that warming in the climate system is unequivocal, with many of the observed changes unprecedented over decades to millennia: warming of the atmosphere and the ocean, diminishing snow and ice, rising sea levels and increasing concentrations of greenhouse gases. Each of the last three decades has been successively warmer at the Earth's surface than any preceding decade since 1850 . . . These and other findings confirm and enhance our scientific understanding of the climate system and the role of greenhouse gas emissions; as such, the report demands the urgent attention of both policy-makers and the general public.[76]

This statement informed negotiations for the Paris Agreement of 2015, in which 195 countries committed to limiting global warming to below 2°C.

1.4.6 **The Paris Climate Agreement**

The Paris Agreement is a historic international treaty that addresses global warming on a world-wide scale. In December 2015, 195 nations joined the Agreement at the 21st Conference of the Parties (COP21) in Paris, France.

The goal of the Paris Agreement is to limit the rise in global average temperature to below 2°C above pre-industrial levels, and to use best efforts to limit the increase to 1.5°C. To reach these goals, "nationally determined contributions" (NDCs) are set by each country. In a speech to world leaders in Paris, on November 30, 2015, President Obama expressed the United States' moral responsibility to the world: "I've come here personally as the leader of the world's largest economy and the second largest emitter [of greenhouse gases] to say that the United States not only recognizes its role in creating this problem, we embrace our responsibility to do something about it."[77]

The Paris Agreement requires each country's NDC to be recorded in a public registry by the secretariat, which is maintained on-line.[78] The U.S. NDC pledges to reduce its greenhouse gas emissions "by 26–28 per cent below its 2005 level in 2025 and to make best efforts to reduce its emissions by 28%."[79] The U.S. could do this

Mal Langsdon, Reuters.

by developing better technologies and infrastructure for renewable energy, such as solar, photovoltaic (PV), and wind, cutting carbon pollution from new and existing power plants, strengthening building and energy codes, implementing more stringent fuel economy standards, and increasing regulation under the Clean Air Act.

Figure 11. **U.S. Paris Agreement Commitment**

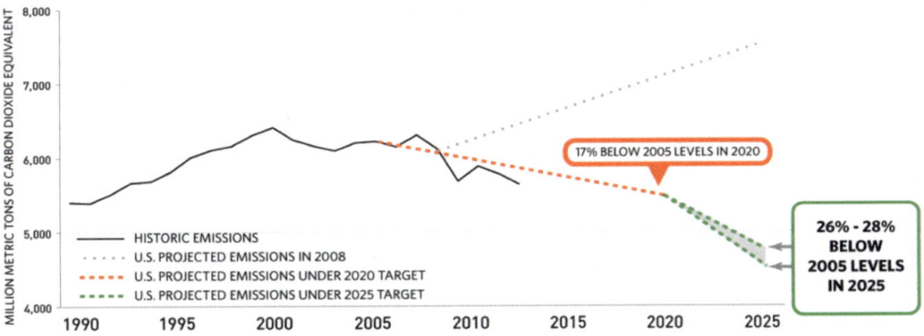

Source: https://www4.unfccc.int.

1.4.7 U.S. Withdrawal from the Paris Agreement

After the November 2016 U.S. presidential election, President Trump announced that the U.S. would reverse its commitment to reduce GHG emissions by withdrawing from the Paris Agreement. On November 4, 2019, the U.S. became the first and only country to attempt to exit the Paris Agreement by submitting formal notification of its withdrawal to the United Nations. Under Article 28, withdrawal is a multi-year process. A party must wait a minimum of three years to give written notification of withdrawal, which does not become effective until one year after notification. Notwithstanding Trump's actions, many companies, states, and local governments vowed their continued support for the Paris Agreement through organizations such as We Are Still In and the Climate Mayors. The U.S. Climate Alliance is a bipartisan organization of 25 governors that is committed to implementing policies that advance the goals of the Paris Agreement, and that aims to reduce greenhouse gas emissions at least 26–28 percent below 2005 levels by 2025.[80]

The U.S. process of withdrawal was not final until the day after the U.S. presidential election, on November 4, 2020. Joe Biden vowed that if elected, he would sign the papers needed to rejoin the Paris Agreement on Inauguration Day, and President Biden made good on his promise on January 20, 2021, by signing an executive order just hours after entering office. The U.S. is again a party to the Paris Accord, and the U.S. will have a seat at the table at the COP26 UN climate change conference, when world leaders meet on November 1–12, 2021 in Glasgow, Scotland to address the climate crisis. Although originally scheduled to be held on November 2020, COP26 was postponed for a year due to the Covid-19 pandemic.

1.4.8 **USGCRP's Fourth National Climate Assessment**

In November 2018, the USGCRP, composed of 13 federal agencies, issued the *Fourth National Climate Assessment* (NCA4), a stark warning about how profoundly human impacts from greenhouse gases have altered our climate. The 1,656-page report begins with an urgent message that "Earth's climate is changing faster than at any point in the history of modern civilization," and warns that "the assumption that current and future climate conditions will resemble the recent past is no longer valid . . . .[81] Chapter 29 summarizes options for mitigation and adaptation and notes that without drastic action, we can expect "substantial [and likely irreversible] damages to the U.S. economy, human health, and the environment," with annual losses growing to hundreds of billions of dollars by 2100.[82]

1.4.9 **Earth's CO_2 Levels Exceed 417 PPM**

In May 2020, the Mauna Loa Observatory recorded a CO_2 average of 417 ppm, the highest that CO_2 levels have ever been in human history.[83]

Air bubbles trapped in ancient glacial ice allow scientists to compare Earth's modern levels of over 400 ppm to those of the past, and ice cores drilled out of the ice sheets in Greenland and Antarctica, provide an 800,000-year-old record of past concentrations of greenhouse gases. Ocean sediment cores provide an even longer baseline, of tens of millions of years. This record of CO_2 from ice and sediment cores indicate that the last time our planet had a CO_2 concentration of 400 ppm was three million years ago, in the middle of the Pliocene Epoch, before humans existed.[84]

In the Pliocene, the average temperature was 3°C to 4°C (5.4°F to 7.2°F) higher, sea levels were 16 to 131 feet higher, there were no ice sheets in the Northern Hemisphere, and forests grew in the arctic.[85] Sediment cores show that the atmosphere

Figure 12.

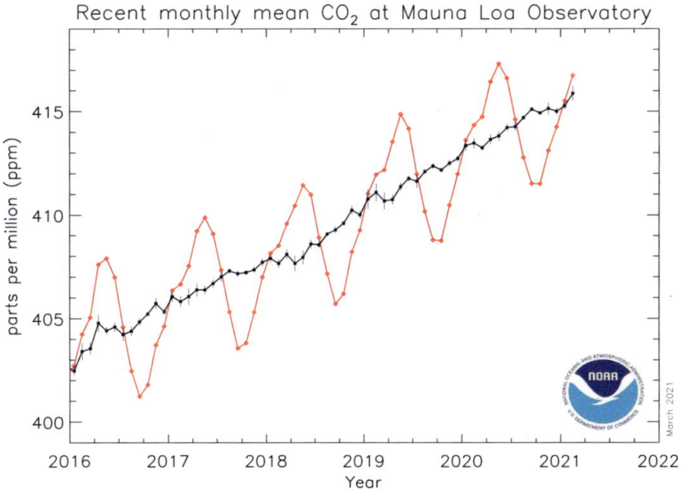

Source: NOAA, "Rise of Carbon Dioxide Unabated, Seasonal Peak Reaches 417 Parts Per Million at Mauna Loa Observatory," June 4, 2020, https://research.noaa.gov/.

has not been consistently above 400 ppm since the Miocene Epoch, 16 million years ago,[86] leading some scientists to suggest that Earth's modern rise of CO_2 warrants classification as a new period, the "Anthropocene," an epoch of rapid climate destabilization where humans have become a major geological force.[87]

1.5 The Climate Emergency

1.5.1 Greta Thunberg Inspires a Youth Climate Movement

In August 2018, sixteen-year-old Swedish climate activist Greta Thunberg began skipping school on Fridays to demand more aggressive climate action by her government. With a hand-painted sign that read "SKOLSTREJK FÖR KLIMATET" (School Strike for Climate), she held solo protests outside of Sweden's Parliament House in Stockholm. Greta inspired a youth movement, Fridays For Future, that went viral on social media under the hashtag #FridaysForFuture, as thousands of students in Europe and around the world held climate strikes to protest their leaders' inaction on climate. Greta's message is simple: the adults in power are betraying young people. Her words at the UN COP24 summit in Poland captured international attention: "You say you love your children above all else—and yet you are stealing their future in front of their very eyes."[88] Greta reframed the conversation around climate change as a moral issue: world governments are failing children by depriving them of their right to a safe climate, because the worst effects of unchecked climate change will be suffered by our youth and future generations.

In 2019, Thunberg addressed multiple international summits and heads of state, including the UK and French parliaments, and demanded that they declare a climate emergency and achieve the emissions targets agreed to in Paris. On Friday, September 20, 2019, she led a massive, coordinated Global Climate Strike in New York City, in which millions of children and adults in 150 countries held parallel strikes.

Photo: European Union.

Greta Thunberg

Greta Thunberg's speech at the UN Climate Action Summit went viral, as she condemned world leaders for betraying young people with business-as-usual policies and reminded them that they are accountable to future generations. Her call to action succeeded, and led many countries and governments to declare a climate emergency.

At the World Economic Forum's annual meeting in Davos, Switzerland, in January 2020, she urged banks, businesses, and governments to stop investing in and subsidizing fossil fuels, and instead, invest in sustainable technologies: "We don't want these things done by 2050, 2030 or even 2021. We want this done now."[89] The challenge confronting us is to ensure that we address the climate emergency with an action plan ambitious enough to dramatically reduce CO_2 emissions and solve the climate crisis.

1.5.2 **The U.S. House Select Committee on the Climate Crisis**

On January 9, 2019, Speaker of the House Nancy Pelosi created the House Select Committee on the Climate Crisis at the start of the 116th Congress. Chaired by Representative Kathy Castor, the bipartisan Select Committee is comprised of nine Democrats and six Republicans and is charged with delivering climate policy recommendations to Congress for future climate legislation.

To inform its policy recommendations, the Select Committee held 17 hearings between April 2019 and February 2020 and consulted with hundreds of experts, including elected officials, tribal leaders, scientists, business representatives, policy experts, public health advocates, and youth activists. On June 30, 2020, the Majority Staff released a 548-page report, titled *Solving the Climate Crisis: The Congressional Action Plan for a Clean Energy Economy and a Healthy and Just America.* Prepared by the Majority Staff, the Climate Crisis Action Plan is a comprehensive outline of policy recommendations that would enable the U.S. to reach 100% clean, net-zero emissions by 2050, in a way that advances justice and equity.

The Climate Crisis Action Plan lays out 12 foundational "pillars" across different economic sectors that would "lead the United States to reach net-zero carbon dioxide emissions before 2050, in line with the IPCC's guidance on emissions reductions needed to limit warming to 1.5°C," and would achieve net negative emissions in the second half of the century. There are hundreds of detailed policy recommendations for Congress. Ending fossil fuel subsidies and production, massive investments in renewable energy, and creation of a national supergrid head up the list.

Under the Trump administration, federal agencies such as the EPA used cost-benefit analyses that hindered federal action to fight climate change; thus, the report recommends changing how climate benefits and costs are evaluated, including the revival of a social cost of carbon (SCC) metric to assess climate policies.

The plan's original March 2020 release date was postponed by the COVID-19 pandemic. As the "Preface" explains, at the time of the report's release in June 2020, "the nation was reeling" from the COVID-19 pandemic, which had already claimed over 120,000 lives, and from an uprising against racial inequality.[90] The plan notes the importance of action on all these fronts: "What we choose to do now shapes the future. What happens next—for racial equality, for public health, for the climate crisis—depends on us."[91]

1.6 The Design Professional's Role

Concern over climate change is growing, and cities and states in the U.S. are beginning to craft visions of carbon neutrality and enact laws with net zero energy targets. Design professionals are a critical component of nationwide efforts to improve energy code compliance and build clean, resilient, and affordable energy systems. As leaders of the design process, architects and engineers have the ability to formulate sustainability goals early in design. The owner makes decisions about design direction, but increasingly, clients are expecting design professionals to advise them about their options to increase energy efficiency and renewable energy use while reducing energy costs and reliance on fossil fuels.

Design professionals are uniquely situated to steer client preferences away from fossil-fuel–based technologies. The challenge will be to find ways to reduce *both* embodied carbon and operational carbon (discussed in chapter 2) within the limits of the project's program, cost, and construction schedule. For design professionals, effective risk management entails presenting the client with a range of options, including the implications of cost and schedule, to fully inform the client's decisions. The design professional should consider including recommendations in scope proposals for the investigation and evaluation of strategies to reduce carbon emissions and meet energy efficiency targets.

Infrastructure must be designed to withstand structural loads, meet the requirements of flood hazard areas, and minimize the impacts of future climate risks, such as increased storm frequencies and intensities, fire hazards, sea-level rise, and storm surge. Standards and regulations usually set only minimum design criteria; thus, effective risk management means that design professionals should inform their clients about the likelihood of greater than specified extremes, and when warranted, recommend more conservative designs, using a low-regret approach.[92]

Pragmatically, however, there may be instances where the client's design decision-making is unduly based upon considerations of cost—there may be a serious disconnect between the design professional's sound and sensible design recommendations and the client's decisions.[93] Ultimately, the owner makes the programmatic decisions that impact project design, and determines the risks they are willing to accept, but design professionals should lead in the process of acculturating their clients about the need to evaluate design options, and implement low-regret strategies that will work under both the current climate and also the uncertain future climate.[94]

Forward-thinking design professionals will counsel their clients to incorporate a climate change risk management system into the organization's asset management program. Such a risk management framework may have features to enable updates over time as conditions change, and may also include a monitoring program to evaluate system performance over time and flexibility to make needed changes.[95] The design professional should document its recommendations and communications with the client, to be able to demonstrate that design options were discussed.

1.7 **Conclusion**

The science is clear that time is running out. We are in a decisive decade, and there is only a small window of time to create a pathway to a sustainable and equitable future. The actions that we take—or do not take—in the near term will determine the future of our planet for centuries to come. We must honor our ethical responsibility to be good stewards of the Earth and safeguard it for future generations.

Notes

1 IPCC, 2012: Glossary of terms. In: *Managing the Risks of Extreme Events and Disasters to Advance Climate Change Adaptation* [Field, C.B., V. Barros, T.F. Stocker, D. Qin, D.J. Dokken, K.L. Ebi, M.D. Mastrandrea, K.J. Mach, G.-K. Plattner, S.K. Allen, M. Tignor, and P.M. Midgley (eds.)]. A Special Report of Working Groups I and II of the Intergovernmental Panel on Climate Change (IPCC). Cambridge University Press, Cambridge, UK, and New York, NY, USA, pp. 555–564.

2 Le Treut, H., R. Somerville, U. Cubasch, Y. Ding, C. Mauritzen, A. Mokssit, T. Peterson, and M. Prather, "Historical Overview of Climate Change," *Climate Change 2007: The Physical Science Basis. Contribution of Working Group I to the Fourth Assessment Report of the Intergovernmental Panel on Climate Change*, Cambridge University Press, Cambridge, UK, and New York, NY, USA, p. 96 (2007).

3 *Ibid.*

4 Jansen, E., J. Overpeck, K.R. Briffa, J.C. Duplessy, F. Joos, V. Masson-Delmotte, D. Olago, B. Otto-Bliesner, W.R. Peltier, S. Rahmstorf, R. Ramesh, D. Raynaud, D. Rind, O. Solomina, R. Villalba, and D. Zhang, "Palaeoclimate," chapter 6 in *Climate Change 2007: The Physical Science Basis. Contribution of Working Group I to the Fourth Assessment*

Report of the Intergovernmental Panel on Climate Change, Cambridge University Press, Cambridge, UK, and New York, NY, USA, p. 449 (2007).

5 Wuebbles, D.J., D.W. Fahey, K.A. Hibbard, B. DeAngelo, S. Doherty, K. Hayhoe, R. Horton, J. P. Kossin, P.C. Taylor, A.M. Waple, and C.P. Weaver, 2017: "Executive Summary," in *Climate Science Special Report: Fourth National Climate Assessment*, Vol. 1 [Wuebbles, D.J., D.W. Fahey, K.A. Hibbard, D.J. Dokken, B.C. Steward, and T.K. Maycokc (eds.)]. U.S. Global Change Research Program, Washington, DC, pp. 12–34.

6 Allen, M.R., O.P. Dube, W. Solecki, F. Aragón-Durand, W. Cramer, S. Humphreys, M. Kainuma, J. Kala, N. Mahowald, Y. Mulugetta, R. Perez, M. Wairiu, and K. Zickfeld, 2018: Framing and Context, Section 1.2.1. In: *Global Warming of 1.5°C. An IPCC Special Report on the impacts of global warming of 1.5°C above pre-industrial levels and related global greenhouse gas emission pathways, in the context of strengthening the global response to the threat of climate change, sustainable development, and efforts to eradicate poverty* [Masson-Delmotte, V., P. Zhai, H.-O. Pörtner, D. Roberts, J. Skea, P.R. Shukla, A. Pirani, W. Moufouma-Okia, C. Péan, R. Pidcock, S. Connors, J.B.R. Matthews, Y. Chen, X. Zhou, M.I. Gomis, E. Lonnoy, T. Maycock, M. Tignor, and T. Waterfield (eds.)]. In Press.

7 Avery, C.W., D.R. Reidmiller, M. Kolian, K.E. Kunkel, D. Herring, R. Sherman, W.V. Sweet, K. Tipton, and C. Weaver, "2018: Data Tools and Scenario Products." In *Impacts, Risks, and Adaptation in the United States: Fourth National Climate Assessment, Volume II* [Reidmiller, D.R., C.W. Avery, D.R. Easterling, K.E. Kunkel, K.L.M. Lewis, T.K. Maycock, and B.C. Stewart (eds.)]. U.S. Global Change Research Program, Washington, DC, pp. 1413–1430.

8 Adam Smith, "2010–2019: A landmark decade of U.S. billion-dollar weather and climate disasters," NOAA, January 8, 2020. Available online at https://www.climate.gov/news-features/blogs/beyond-data/2010-2019-landmark-decade-us-billion-dollar-weather-and-climate.

9 *Ibid.*

10 Rudolph, M. J., Canadian Institute of Actuaries, Casualty Actuarial Society, and Society of Actuaries, 2019, https://www.soa.org/globalassets/assets/files/resources/research-report/2019/12th-emerging-risk-survey.pdf.

11 Jeremy Martinich and Allison Crimmins, "Climate Damages and Adaptation Potential Across Diverse Sectors of the United States," *Nature Climate Change*, 9, 397-404, 2019.

12 Department of Defense, "National Security Implications of Climate-Related Risks and a Changing Climate," July 23, 2015, p. 3, available online at https://archive.defense.gov/pubs/150724-congressional-report-on-national-implications-of-climate-change.pdf.

13 *Ibid*, p. 4.

14 IPCC, 2014, "Summary for Policymakers." In *Climate Change 2014: Mitigation of Climate Change. Contribution of Working Group III to the Fifth Assessment Report of the Intergovernmental Panel on Climate Change*, p. 8.

15 IPCC, 2014. "Summary for Policymakers." In *Climate Change 2014: Impacts, Adaptation, and Vulnerability. Part A: Global and Sectoral Aspects. Contribution of Working Group II to the Fifth Assessment Report of the Intergovernmental Panel on Climate Change*, p. 13.

16 Statement of Solomon Hsiang, Chancellor's Professor of Public Policy, University of California, Berkeley, to the United States House Committee on the Budget, hearing on "The Costs of Climate Change: Risks to the U.S. Economy and the Federal Budget," June 10, 2019.

17 Ackerman, F., "Climate Economics in Four Easy Pieces" *Development*, 51, 325–331, 2008.

18 Gernot Wagner and Martin L. Weitzman, *Climate Shock: The Economic Consequences of a Hotter Planet*, Princeton, NJ: Princeton University Press, 2015, p. 78.

19 IPCC, 2018, "Summary for Policymakers." In: *Global Warming of 1.5°C. An IPCC Special Report on the impacts of global warming of 1.5°C above pre-industrial levels and related global greenhouse gas emission pathways, in the context of strengthening the global response to the threat of climate change, sustainable development, and efforts to eradicate poverty*. World Meteorological Organization, Geneva, Switzerland.

20 Michael Schirber, "Ice Ages Blamed on Tilted Earth," *Live Science*, March 30, 2005. Available online at https://www.livesciencecom/6937-ice-ages-blamed-tilted-earth.html.

21 J. D. Hays, John Imbrie, N.J. Shackleton, "Variations in the Earth's Orbit: Pacemaker of the Ice Ages," *Science*, Vol. 194, No. 4270, December 10, 1976, pp. 1121–1132.

22 Christopher J. Campisano, "Milankovitch Cycles, Paleoclimatic Change, and Hominin Evolution," *Nature Education Knowledge* 4(3):5 (2012); available online at https://www.nature.com/scitable/knowledge/library/milankovitch-cycles-paleoclimatic-change-and-hominin-evolution-68244581/.

23 "Earth's orbital variations and sea ice synch glacial periods," News from Brown, January 25, 2017, available online at https://www.bown.edu/news/2017-01-26/iceages.

24 Chapter 6, "Palaeoclimate, Climate Change 2007: The Physical Science Basis." *Fourth Assessment Report of the Intergovernmental Panel on Climate Change*, p. 449-453.

25 Alan Buis, "Why Milankovitch (Orbital) Cycles Can't Explain Earth's Current Warming," Ask NASA Climate, February 27, 2020, available online at www.climate.nasa.gov.

26 Jessica E. Tierney, Peter B. deMenocal, Paul D. Zander, "A climatic context for the out-of-Africa migration," *Geology*, 2017. DOI: 10.1130/G39457.1

27 ETANA, "The Flood Story in the Gilgamesh Epic," Electronic Tools and Near East Archives, www.etana.org citing Andrew George, *The Epic of Gilgamesh*, p. 88-89 Harmondsworth: Penguin Books, 2000.

28 H. M. Cullen, P. B. deMenocal, S. Hemming, G. Hemming, F. H. Brown, T. Guilderson, F. Sirocko; Climate change and the collapse of the Akkadian empire: Evidence from the deep sea. *Geology*, 28 (4): 379–382. doi: https://doi.org/10.1130/0091-7613(2000)28<379:CCATCO>2.0.CO;2.

29 Owen Jarus, "Mesa Verde: Cliff Dwellings of the Anasazi," *Live Science*, June 15, 2017, https://www.livescience.com/27360-mesa-verde.html

30 George Johnson, "Vanished: A Pueblo Mystery," *The New York Times*, April 8, 2008.

31 National Research Council, "Surface Temperature Reconstructions for the Last 2,000 Years," National Academy Press, Washington, D.C. (2006).

32 Neukom, R., Steiger, N., Gómez-Navarro, J.J., et al. "No Evidence for Globally Coher-

ent Warm and Cold Periods over the Preindustrial Common Era." *Nature* 571, 550–554 (2019). https://doi.org/10.1038/s41586-019-1401-2

33 P.D. Jones, R.K. Briffa, T.J. Osborn, et al., "High-resolution Palaeoclimatology of the Last Millennium: A Review of Current Status and Future Prospects," *The Holocene*, vol. 19, no. 1, pp. 3–49 (2009).

34 *Ibid.*

35 Neukom, R., Barboza, L.A., Erb, M.P., et al. "Consistent Multidecadal Variability in Global Temperature Reconstructions and Simulations over the Common Era." *Nature Geoscience* 12, 643–649 (2019). https://doi.org/10.1038/s41561-019-0400-0.

36 David Herring, "How is the Current Global Warming Trend Different from Previous Warming Periods in Earth's History?" NOAA Climate.gov, July 23, 2014, available online at https://www.climate.gov/news-features/climate-qa/how-current-global-warming-trend-different-previous-warming-periods-earths.

37 Roland Jackson, "Eunice Foote, John Tyndall and a Question of Priority," *The Royal Society Journal of the History of Science*, February 13, 2019.

38 Eunice Foote, "Circumstances Affecting the Heat of the Sun's Rays," *American Journal of Science*, Art. 22, 382–383 (1856).

39 Leila McNeil, This Suffrage-Supporting Scientist Defined the Greenhouse Effect but Didn't Get the Credit, Because Sexism," *Smithsonian Magazine*, December 5, 2016. Available online at https://www.smithsonianmag.com/science-nature/lady-scientist-helped-revolutionize-climate-science-didnt-get-credit-180961291/.

40 John Tyndall, The Bakerian Lecture, "On the absorption and Radiation of Heat by Gases and Vapours, and on the Physical Connexion of Radiation, Absorption, and Conduction," *The Royal Society Publishing*, January 1, 1862, https://royalsocietypublishing.org/doi/10.1098/rspl.1860.0021.

41 Svante Arrhenius, "On the Influence of Carbonic Acid in the Air Upon the Temperature of the Ground," *Philosophical Magazine and Journal of Science*, vol. 41, p. 237–76 (1896).

42 G. Callendar, "The Artificial Production of Carbon Dioxide and Its Influence on Temperature," *Quarterly Journal of the Royal Meteorological Society*, 64, 223–240 (1938).

43 Ed Hawkins and Phil. D. Jones, "On Increasing Global Temperatures: 75 Years after Callendar," *Quarterly Journal of the Royal Meteorological Society*, vol. 139, issue 677 (2013), pp. 1961–1963 . Available online at https://rmets.onlinelibrary.wiley.com/doi/abs/10.1002/qj.2178.

44 James Rodger Fleming, *The Callendar Effect: The Life and Work of Guy Stewart Callendar (1898–1964), The Scientist Who Established the Carbon Dioxide Theory of Climate Change (Historical Monographs)*, Boston: American Meteorological Society, 2007.

45 Walter H. Munk, "Tribute to Roger Revelle and his contributions to studies of carbon dioxide and climate change," Proceedings of the National Academy of Sciences of the United States of America, August 5, 1997.

46 "Roger Revelle, 1986 Batzan Prize for Oceanography/Climatology," International Batzan Prize Foundation (1986), https://www.balzan.org/en/prizewinners/roger-revelle.

47 Roger Revelle and Hans E. Suess, "Carbon Dioxide Exchange Between Atmosphere and

Ocean and the Question of an Increase of Atmospheric CO_2 during the Past Decades," *Tellus* vol. 9, no. 1 (February 1957) pp. 18–27.

48 "Carbon dioxide levels hit record peak in May," NOAA Research News, June 4, 2019, https://research.noaa.gov/article/ArtMID/587/ArticleID/2461/Carbon-dioxide-levels-hit-record-peak-in-May.

49 "The Keeling Curve: Carbon Dioxide Measurements at Mauna Loa," American Chemical Society, available at https://www.acs.org/content/acs/en/education/whatischemistry/landmarks/keeling-curve.html/.

50 William H. Matthews, William H. Kellogg and G. D. Robinson, eds., *Man's Impact on the Climate* (SMIC). Cambridge, MA: M. I. T. Press, 1971. xxiv, 308 pp.

51 Thomas C. Peterson, William M. Connolley, and John Fleck, "The Myth of the 1970s Global Cooling Scientific Consensus," *Bulletin of the American Meteorological Society*, vol. 89, issue 9 (September 2008), p.1331.

52 J. Hansen, D. Johnson, A. Lacis, S. Lebedeff, P. Lee, D. Rind, and G. Russell, "Climate Impact of Increasing Atmospheric Carbon Dioxide," *Science*, vol. 213, no. 4511, August 28, 1981, p. 957.

53 Roger A. Pielke Jr., "Policy history of the US Global Change Research Program: Part I. Administrative development," *Global Environmental Change* 10, p. 16, September 25, 2000.

54 *Ibid*, p. 17.

55 Philip Shabekoff, "Global Warming Has Become Real, Expert Tells Senate," *The New York Times*, June 24, 1988.

56 Elizabeth Kolbert, "Listening to James Hansen on Climate Change, Thirty Years Ago and Now," *The New Yorker*, June 20, 2018.

57 Pielke, Roger A. "Policy History of the US Global Change Research Program: Part I. Administrative Development," *Global Environmental Change,* 10 (2000), p. 16.

58 Opinion, "The White House and the Greenhouse," *The New York Times*, Section A, p. 30, May 9, 1989.

59 Fourth National Climate Assessment, https://www.globalchange.gov/nca-4.

60 World Meteorological Organization, "Summary of WMO Co-Sponsored Programmes," https://www.wmo.int/pages/summary/cosponsored_summary_en.html.

61 Michael E. Mann, Raymond S. Bradley, and Malcolm K. Hughes, "Global-scale temperature patterns and climate forcing over the past six centuries," *Nature*, vol. 392, April 23, 1998.

62 Michael E. Mann, Raymond S. Bradley, and Malcolm K. Hughes, "Northern Hemisphere Temperatures During the Past Millenium: Inferences, Uncertainties, and Limitations." *Geophysical Research Letters*, vol. 26, no. 6, p. 759-762, March 15, 1999.

63 IPCC 2001, Chapter 2: Observed Climate Variability and Change, Figure 2.20, p. 134, Climate Change 2001: The Scientific Basis, Contribution of Working Group I to the *Third Assessment Report of the Intergovernmental Panel on Climate Change*, Cambridge University Press, Cambridge, United Kingdom and New York, NY, USA (2001). Photo of hockey stick from Michael Mann, "Earth Day and the Hockey Stick: A Singular Message," *Scientific American,* April 20, 2018.

64 IPCC 2001, Summary for Policymakers, A Report of Working Group I of the Intergovernmental Panel on Climate Change, Figure 1, p. 3, Climate Change 2001: Synthesis Report, Cambridge University Press, Cambridge, United Kingdom and New York, NY, USA (2001).

65 Naomi Oreskes and Erik M. Conway, "Attack on Climate Scientist Just Latest in Long Line," CNN Opinion, June 9, 2010. Available online at https://www.cnn.com/2010/OPINION/06/07/oreskes.climate.change/index.html.

66 Eugene R. Wahl and Caspar M. Ammann, "Robustness of the Mann, Bradley, Hughes Reconstruction of Northern Hemisphere Surface Temperatures: Examination of Criticisms Based on the Nature and Processing of Proxy Climate Evidence," *Climatic Change*, 85:33–69 (2007); Geoff Brumfiel, "Academy Affirms Hockey-Stick Graph," *Nature*, 441, 1032–1033 (2006). https://doi.org/10.1038/4411032a.

67 National Research Council. 2006. "Overview," *Surface Temperature Reconstructions for the Last 2,000 Years*. Washington, DC: The National Academies Press, p. 20. Available online at https://doi.org/10.17226/11676.

68 Chris Mooney, "The Hockey Stick: The Most Controversial Graph in Science, Explained," *The Atlantic*, May 10, 2013, www. theatlantic.com.

69 *Ibid.*

70 2019 Tyler Laureates, Tyler Prize for Environmental Achievement, www.tylerprize.org

71 Neukom, R., Barboza, L.A., Erb, M.P., et al. "Consistent multidecadal variability in global temperature reconstructions and simulations over the Common Era." *Nature Geoscience*, 12, 643–649 (2019). https://doi.org/10.1038/s41561-019-0400-0.

72 Neukom, R., Steiger, N., Gómez-Navarro, J.J., et al. "No evidence for globally coherent warm and cold periods over the preindustrial Common Era," *Nature*, 571 (2019): 550–554.

73 NOAA, National Centers for Environmental Information, www.ncdc.noaa.gov.

74 Le Treut, H., R. Somerville, U. Cubasch, Y. Ding, C. Mauritzen, A. Mokssit, T. Peterson, and M. Prather, 2007, "Historical Overview of Climate Change," in *Climate Change 2007: The Physical Science Basis. Contribution of Working Group I to the Fourth Assessment Report of the Intergovernmental Panel on Climate Change*, Cambridge University Press, Cambridge, United Kingdom and New York, NY, USA.

75 *Ibid.*, Chapter 1, p. 95.

76 IPCC, 2013: Summary for Policymakers, "Forward, Preface and Dedication," p. v. In: *Climate Change 2013: The Physical Science Basis. Contribution of Working Group I to the Fifth Assessment Report of the Intergovernmental Panel on Climate Change,* [Stocker, T.F., D. Qin, G.-K. Plattner, M. Tignor, S.K. Allen, J. Boschung, A. Nauels, Y. Xia, V. Bex and P.M. Midgley (eds.)]. Cambridge University Press, Cambridge, United Kingdom and New York, NY, USA (2013).

77 The White House, Office of the Press Secretary, "Remarks by President Obama at the First Session of COP21," November 30, 2015, https://obamawhitehouse.archives.gov/the-press-office/2015/11/30/remarks-president-obama-first-session-cop21.

78 Interim NDC Registry, https://www4.unfccc.int/ndcregistry/Pages/Home.aspx.

79 *Ibid.*

80 "U.S. Climate Alliance Governors Oppose Administration's Withdraw from the Climate Agreement" United States Climate Alliance, November 4, 2019, http://www.usclimatealliance.org/publications/pariswithdrawal.

81 *Fourth National Climate Assessment*, Chapter 1: Overview, p. 4, November 2019, https://nca2018.globachange.gov/chapter/1/.

82 Martinich, J., B.J. DeAngelo, D. Diaz, B. Ekwurzel, G. Franco, C. Frisch, J. McFarland, and B. O'Neill, "2018: Reducing Risks Through Emissions Mitigation." In *Impacts, Risks, and Adaptation in the United States: Fourth National Climate Assessment, Volume II* [Reidmiller, D.R., C.W. Avery, D.R. Easterling, K.E. Kunkel, K.L.M. Lewis, T.K. Maycock, and B.C. Stewart (eds.)]. U.S. Global Change Research Program, Washington, DC, USA, p. 1358. doi: 10.7930/NCA4.2018.CH29.

83 Global Monitoring Laboratory, Earth System Research Laboratories, NOAA, http://www.esrl.noaa.gov/gmd/ccgg/trends/weekly.html.

84 Wuebbles, D.J., D.W. Fahey, K.A. Hibbard, B. DeAngelo, S. Doherty, K. Hayhoe, R. Horton, J.P. Kossin, P.C. Taylor, A.M. Waple, and C.P. Weaver, "2017: Executive summary." In: Wuebbles, D.J., D.W. Fahey, K.A. Hibbard, D.J. Dokken, B.C. Steward, and T.K. Maycokc (eds.), *U.S. Global Change Research Program*, Washington, DC, pp. 12–34.

85 Rob Monroe, "What Does 400 ppm Look Like?" The Keeling Curve, Scripps Institution of Oceanography, December 3, 2013, https://sioweb.ucsd.edu/programs/keeling-curve/2013/12/03/what-does-400-ppm-look-like/.

86 Nicola Jones, "How the World Passed a Carbon Threshold and Why it Matters," Yale Environment 360, January 26, 2017, https://e360.yale.edu/features/how-the-world-passed-a-carbon-threshold-400ppm-and-why-it-matters.

87 "Graphic: Carbon Dioxide Hits New High," NASA Global Climate Change, https://climate.nasa.gov/climate_resources/7/graphic-carbon-dioxide-hits-new-high/.

88 John Sutter and Lawrence Davidson, "Teen Tells Climate Negotiators They Aren't Mature Enough," CNN, December 17, 2018. Available online at https://www.cnn.com/2018/12/16/world/greta-thunberg-cop24/index.html.

89 "Greta Thunberg's Remarks at the Davos Economic Forum," *New York Times*, January 21, 2020, available online at https://www.nytimes.com/2020/01/21/climate/greta-thunberg-davos-transcript.html.

90 House Select Committee on the Climate Crisis, "Solving the Climate Crisis: The Congressional Action Plan for a Clean Energy Economy and a Healthy and Just America," Majority Staff Report, 116th Congress, June 2020, p. 1.

91 *Ibid.*

92 J. Rolf Olsen, *Adapting Infrastructure and Civil Engineering Practice to a Changing Climate*, American Society of Civil Engineers, 2015, p. 23.

93 David Hatem and Patricia Gary, "Climate Change, Sustainability and Resiliency Considerations in Design: Professional Liability Risk Issues for Consulting Engineers," *ACEC Insights*, American Council of Engineering Companies/Massachusetts, Fall 2017, p. 1.

94 *Ibid.*

95 "Adaptive Design and Risk Management," ch. 7 in *Climate-Resilient Infrastructure, Adaptive Design and Risk Management*, Bilal M. Ayyub, ed., Reston, VA: American Society of Civil Engineers. 2018, p. 185.

2

Responding to Climate Change Through Mitigation and Adaptation

Patricia B. Gary

2.0 Introduction

Climate change is one of the greatest challenges of our time. In the 2015 Paris Agreement, 195 nations pledged to keep "global temperature rise well below 2°C above pre-industrial levels." The design and construction sector can play a pivotal role in solving the climate crisis, and by aligning its practices to reach net zero, can accelerate a shift to decarbonization of the globe. Mitigation and adaptation actions build resilience and decrease the vulnerability of the built environment to the impacts of climate change. Mitigation and adaptation are complementary aspects of risk management. Each manage different components of climate-related risk, and have generally been treated separately by researchers and practitioners. Capturing their co-benefits in design will mitigate carbon pollution while adapting and responding to climate impacts.

This chapter proceeds in two parts. Part 1 discusses the vision for a new paradigm of infrastructure design that makes deep cuts in carbon emissions and thereby reduces the likelihood of the severest consequences of climate change. Part 2 discusses adaptive design approaches that manage the uncertainty of future climate change by performing well under a range of possible future conditions. Designs that implement both mitigation and adaptation strategies increase resilience synergies. The chapter concludes with recommendations about the design professional's obligations under the professional standard of care.

PART 1

2.1 Mitigation—The Challenge Confronting the World

Atmospheric CO_2 levels have reached 417 ppm, and far exceed both the 280 ppm of pre-industrial times and the 350 ppm considered to be "safe." In October 2018, the IPCC released its "Special Report on Global Warming of 1.5°C," which describes the expected impacts of 1.5°C and 2°C of warming above pre-industrial time.[1]

The report's clear warning that the world has only a small window of time to reach carbon neutrality caused a shift in public awareness and led to widespread protests for immediate government action to address the climate emergency.

Figure 1. **Evolution of Global Mean Surface Temperature (GMST) Over the Period of Instrumental Observations**

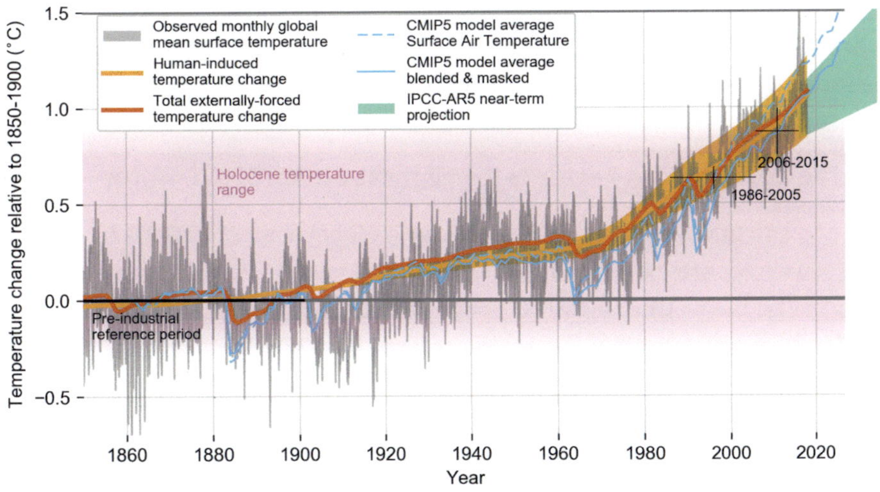

Sourcce: IPCC, *2018 Special Report, Global Warming of 1.5°C*, chapter 1, "Framing and Context," Figure 1.2, p. 57.

The Special Report outlines global greenhouse gas emission reduction pathways that could limit warming to 1.5°C or 2°C. Greenhouse gases (GHG) are gases in the atmosphere such as carbon dioxide (CO_2), methane (CH_4), nitrous oxide (N_2O), and fluorinated gases that absorb infrared radiation, trapping heat in the atmosphere. The Report is unequivocal: the earth is expected to become 1.5 °C warmer as early as 2030 if the current trend continues. To contain warming at 1.5 °C would require the world to make "rapid and far reaching transitions in land, energy, industry, buildings, transport and cities." Global net human-caused emissions "would need to fall by about 45 percent from 2010 levels by 2030 . . . reaching 'net zero' by around 2050."[2]

This means there is only a ten-year window of opportunity to address GHG emissions at the global scale to avoid catastrophic effects of global warming. This ten-year window of time is the essence of the climate emergency confronting the world.

The climate crisis is the challenge of our generation. Unless we take action now, climate change will have devastating and irreversible impacts on life on Earth. The global legacy we leave for future generations depends upon which of the climate model scenarios (discussed in Chapter 1) comes to fruition. We have an ethical obligation to ensure that our children, and their children, and all future generations will live on an inhabitable planet.

2.1.1 Mitigation and Adaptation

There are two main strategies for responding to climate change in the built environment: mitigation and adaptation. Mitigation is the process of reducing and stabilizing the levels of GHG emissions in the atmosphere to limit future climate change. Adaptation is the process of anticipating the actual or expected climate and its adverse effects by taking appropriate action to lessen or avoid harm. Adaptation and mitigation are complementary strategies, and both are necessary because even if mitigation succeeds in reducing all GHG emissions today, adaptation is still needed to address the climate-related changes that have already been set in motion and will continue to affect future generations. Both approaches manage the risks of climate change.

2.1.2 The Construction Sector Is the Highest Global GHG Emitter

The construction sector is the highest single global emitter of GHG by sector, and for that reason, construction is at the forefront of world-wide efforts to achieve carbon neutrality. In the United States, residential and commercial buildings account for 40% of energy consumption. According to data from UN Environment Programme's

Figure 2. **Global Share of Buildings and Construction Final Energy and Emissions**

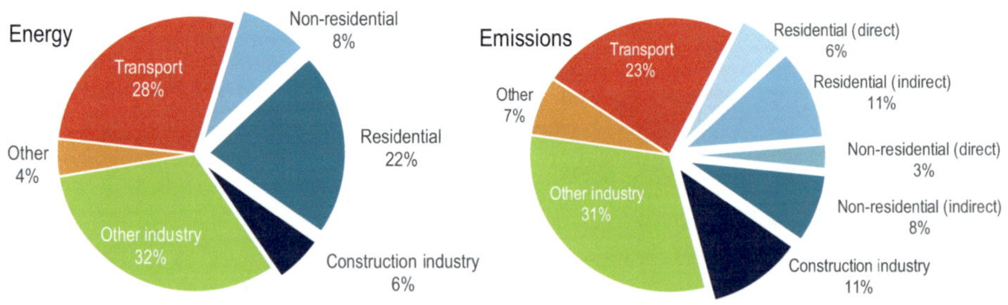

Construction industry is the portion of industry that manufactures building materials such as steel, cement and glass. Indirect emissions are emissions from power generation for electricity and commercial heat.

From UNEP's *2019 Global Status Report for Buildings and Construction Sector*, p. 12.

2019 Global Status Report, the global construction industry accounted for 36% of final energy use and 39% of energy and process-related emissions in 2018. Additionally, 11% of process-related emissions came from manufacturing building materials such as steel, cement, and glass.[3]

Construction sector emissions are at an all-time record high and increased by 2% from 2017 to 2018, while final energy demand rose 1%. These trends indicate that the construction sector is not on track to meet the targets of the Paris Agreement. Even more concerning, UN data projects the world will add more than 2.5 trillion square feet of new buildings in the next 35 years.[4] Decarbonization strategies are vital to addressing the massive amounts of new GHG emissions that will be created. The future of our planet depends upon the U.S. and other countries implementing the mitigation measures in which CO_2 emissions "fall by about 45 percent from 2010 levels by 2030, reaching 'net zero' by around 2050."

2.1.3. **Green Recovery from the Pandemic**

In 2020, COVID-19 caused a world-wide public health crisis, a large-scale loss of life, and an economic shock of epic proportions. According to the World Bank, the severity and breadth of the initial financial collapse from the pandemic was the worst since World War II. The pandemic had a severe impact on the global construction industry, causing construction activities to drop by "10–25% compared to 2019."[5] Globally, in the wake of the pandemic, "more than 25 million jobs (or around 10%) across the [construction] sector" were lost or were at risk of being lost in the short term.[6] In response to the COVID-19 pandemic, the United States and other world governments have used economic stimulus plans to address the impacts of lockdowns and quarantines, and construction of new infrastructure is expected to play an important role in recovery from the pandemic. As the U.N. Environment Programme notes, "The construction sector is essential for an economic recovery after the COVID-19 crisis. In 2015, it accounted for 11%–13% of global GDP and today involves far-reaching value chains of small and large businesses. The sector is a substantial provider of local jobs: 7% of total global employment or 220 million jobs depend on it."[7] A robust recovery plan which stabilizes and then regrows our economy can serve the dual purpose of addressing the climate crisis by prioritizing investment in climate-resilient technology and infrastructure.

The U.S. government's economic stimulus packages present a unique opportunity to support a green recovery from the COVID-19 pandemic at the scale needed to catalyze action to decarbonize the built environment. Infrastructure spending, and stimulus programs for the construction industry, are a proven method of responding to economic crises. They "have large macroeconomic impacts, as many countries have a large need for new buildings and renovation of existing buildings, and because the sector has strong potential for activating local value chains."[8] Long-term recovery strategies can accelerate a just transition away from fossil fuels and toward renewable energy by creating millions of green jobs in clean energy. The International Renewable Energy Agency (IRENA) estimates that renewable energy could employ 40 million people by 2050 if the international community utilizes its full renewable energy potential. Post-pandemic stimulus packages are a historic, once-in-a-lifetime opportunity to build back better, by upgrading electricity grids, investing in renewable energy and green building programs, and creating a shift toward net zero global building stock. By committing resources to sustainable infrastructure, governments can create a pathway to a green recovery from the pandemic.

2.2 Green Building, Sustainability, LEED, and Green Codes

Green building and sustainability are ancient concepts that were understood thousands of years ago. Neolithic Puebloans, the Anasazi, constructed entire villages that still exist in the Southwest and are perfect examples of passive solar design. The Anasazi built their cliff dwellings high up in the ledges of the sandstone canyons of Mesa Verde, Colorado. By using southern-facing cliff overhangs, the spectacular stone dwellings received full sun and solar heat in the winter from the low angle of the winter sun, and the overhanging cliff ledge offered cool shade from the high summer sun.

Contemporary green building has its roots in the environmental movement of the 1960s, when idealistic young people left cities and suburbs and went back to the land to homestead and live self-sufficiently "off the grid" (for example, in a geodesic dome).

Dean and Linda Fleming's dome, Libre, Colorado, 1969. Copyright Roberta Price.

2.2.1 Development of Solar Photovoltaic (PV) Technology

The OPEC oil embargo of 1973 and the subsequent energy crisis created fuel shortages and gasoline rationing, which sent a powerful message about the scarcity of natural resources. The EPA and NASA responded by investing in research and development of cleaner, renewable energies such as solar and wind power, hydroelectric power, and biomass and geothermal energy. There is no defining moment of green building, but

the development of solar photovoltaic (PV) technology and wind turbines had a profound effect. The Solar Heating and Cooling Demonstration Act of 1974 required the installation of solar heating and cooling units in federal buildings by 1977 to acclimate the public to the new technology.[9] President Jimmy Carter had thirty-eight solar panels installed on the White House roof to heat water for the First Family.

2.2.2 **The Concept of Sustainability**

The term "sustainability" arrived on the world political stage in 1983. Norwegian prime minister Gro Harlem Brundtland was chair of the United Nations World Commission on Environment and Development (the Brundtland Commission) and gained international recognition as a leading advocate for the concept of sustainable development. The Brundtland Commission's report, *Our Common Future*, coined the term "sustainable development" in a now-famous quote, and urged the world to take heed: "Sustainable development is that which meets the needs of the present without compromising the ability of future generations to meet their own needs."[10] This led to the convening of the 1992 Earth Summit in Rio de Janeiro, which adopted the United Nations Framework Convention on Climate Change.

Prime Minister Gro Harlem Brundtland of Norway, speaking to the United Nations General Assembly in 1987.

The Brundtland Commission alerted the world to the global character of major environmental problems and the common interest of all countries to pursue policies aimed at environmentally sound development in order to reduce the effects of human activities on the environment for future generations. The report issued a call to action:

> Over the course of this century, the relationship between the human world and the planet that sustains it has undergone a profound change. When the century began, neither human numbers or technology had the power radically to alter planetary systems. As the century closes, not only do vastly increased human numbers and their activities have that power, but major, unintended changes are occurring in the atmosphere, in the soils, in water, among plants and animals, and in the relationships among all of these. The rate of change is outstripping the ability of scientific disciplines and our current capabilities to assess and advise.[11]

Our Common Future opened up a broad international discussion of sustainability. Soon after publication of the Brundtland report, the term "sustainability" came into general use and helped shape the consciousness of the world.

"Green building" refers to practices that make buildings sustainable and resource-efficient throughout their whole life-cycles, from siting to design, construction, operation, maintenance, renovation, and destruction. Green building is seen as a way to reduce man-made GHG and the carbon footprint of buildings. In 1989, the American Institute of Architects (AIA) formed a Committee on the Environment (COTE), one of the oldest organizations in the U.S. dedicated to sustainable design.[12] COTE obtained funding from the EPA to develop an Environmental Resource Guide for building products based on life-cycle analysis.

2.2.3 **LEED, the Green Building Rating System**

The sustainability movement had an enormous impact on the construction industry. In 1993, amidst heightened anxiety about global warming and dwindling supplies of natural resources, the U.S. Green Building Council (USGBC) was formed.[13] A Washington, D.C.–based nonprofit organization committed to sustainability, USGBC's vision of cost-efficient and energy-saving green building led to the development of the Leadership in Energy and Environmental Design (LEED) Green Building Rating System, which is used to measure and evaluate environmental performance of green buildings.

LEED development began in 1994, led by Robert K. Watson of the Natural Resources Defense Council (NRDC), the founding chairman of the LEED Steering Committee. LEED is a voluntary, consensus-based system in which credits are earned for satisfying criteria that address specific environmental impacts inherent in the design, construction, and operation of buildings. The first LEED pilot program (LEED version 1.0) was launched in 1998, and after modifications, Version 2.0 was released in March 2000. LEED equates "green" design with "good" design, with Certified, Silver, Gold, and Platinum award levels. Each level corresponds to the number of credits accrued

in seven green design categories: Sustainable Sites (SS), Water Efficiency (EF), Energy and Atmosphere (EA), Materials and Resources (MR), Indoor Environmental Quality (IEQ), Innovation in Design (ID), and Regional Priority (RP).

LEED's simple, holistic, metrics-based system provided a framework for what it meant to be green, and transformed the building industry. LEED offers proof to clients, stakeholders, and the public at large that environmental goals have been met, and the financial advantages of going green include improved health and indoor air quality, increased worker satisfaction, lowered utility costs, and tax incentives. Policy-makers in government became the key drivers for green building and began to use sustainable design as a statement of their mission. Federal government agencies now insist upon LEED certification in all public projects, and LEED certification is the norm for high-end office buildings and corporate headquarters. USGBC has certified more than 100,000 commercial LEED projects worldwide. There are a proliferation of other green rating and certification programs, including the much more rigorous Living Building Challenge, which requires demonstration of zero energy or net positive energy (generating more energy than consumed and returning it to the grid).

2.2.4 **Green Building Codes**

The years 2007–2009 witnessed cities and states begin to adopt green building codes, often based on LEED. In New York City, the Green City Buildings Act became effective on January 1, 2007, and required all new municipal buildings and renovations in the five boroughs to meet the LEED Silver certification standards.[14] In 2007, Boston enacted Zoning Article 37 which required private developers to build greener by obtaining LEED certification for all construction greater than 50,000 square feet. In 2008, California became the first state in the nation to institute a state-wide green building code, the California Green Building Standards Code (CALGreen). In 2009, the International Code Council (ICC) released the first model green building code, the International Green Construction Code (IgCC), updated in 2018. The IgCC is designed to be incorporated into a jurisdiction's codes and ordinances and function as an overlay to other International Code Council model codes. As discussed below in this chapter, many jurisdictions including New York City, Washington, D.C., and California have now enacted much more aggressive legislation to reduce GHG emissions, which is aimed at meeting the targets of the Paris Agreement.

LEED revolutionized the industry, but it is not enough to solve the climate crisis. The most effective way to regulate GHG emissions from the building sector is to leverage building and energy codes by setting a target of net zero before 2050, with ambitious interim reduction targets. If implemented correctly with net zero requirements, green building codes have the potential to rapidly reduce our carbon footprint and stabilize CO_2 levels in the atmosphere. The climate emergency requires the entire building sector to reach net zero energy by 2050, not just a few aspirational projects. Yet only a tiny fraction —less than .01% of buildings— are considered net zero energy today. As a result, there is an alarming gap between our current action and the scale of action required.

While federal, state, and local governments have critical roles to play in legislating net zero energy as a code requirement, all stakeholders need to radically rethink how infrastructure is constructed, and make a paradigm shift to phase out CO_2 emissions in a meaningful way.

2.3 **GHG Mitigation Strategies**

The carbon footprint of a built asset includes GHG emissions from operational use (i.e., residential and commercial buildings use large quantities of energy for heating, cooling, lighting, and other needs) and emissions from industrial processes. The latter, known as embodied carbon emissions, is the footprint from upfront emissions associated with raw material extraction, fabrication, and installation of building materials. Until recently, assessments of a building's emissions and life-cycle costs have focused primarily on operational energy efficiency. Embodied carbon emissions have historically not been counted in the points assigned by energy performance programs such as LEED, except indirectly through materials reuse, recycled content, or regional materials.

Reducing the ubiquitous embodied carbon in steel, concrete, aluminum, and other materials is one of the most challenging problems of climate change mitigation. The World Green Building Council's (WGBC) report, *Bringing Embodied Carbon Upfront,* notes:

> Carbon emissions are released not only during operational life but also during the manufacturing, transportation, construction and end of life phases of all built assets—buildings and infrastructure. These emissions, commonly referred to as embodied carbon, have largely been overlooked historically but contribute around 11% of all global carbon emissions.[15]

Most definitions of "net zero" include only operational carbon, and in this sense "net zero" is a misnomer. To achieve the aims of the Paris Agreement, the design and construction sector must forge a new path to reduce both operational and embodied carbon emissions.

Design professionals are retained by clients to make recommendations about design. The owner is the project sponsor and makes the programmatic decisions, but as

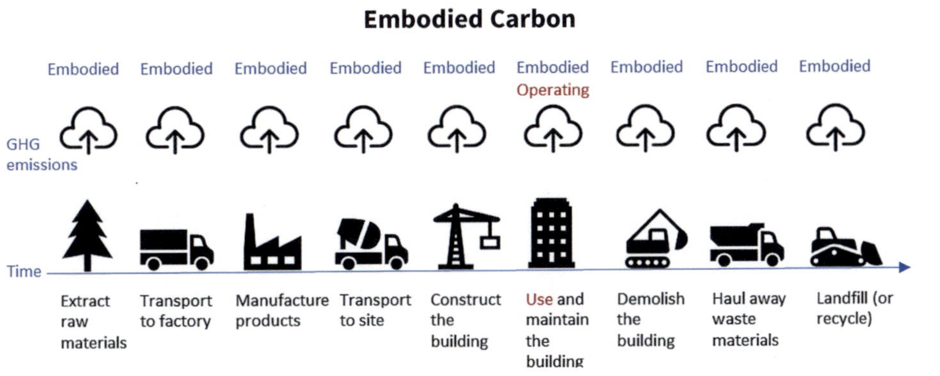

Source: "Embodied Carbon of Buildings and Infrastructure: International Policy Review," Sept. 2017, www.c40knowledgehub.org.

leaders of the design process, architects and engineers have the ability to undertake whole-life carbon thinking early in design. Design professionals are uniquely situated to influence client decisions by identifying low-carbon solutions, advocating for innovative approaches that provide energy savings, and shifting client preferences away from fossil-fuel–based technologies. Material selection during a project's planning process is a way to target carbon emissions, and the key challenge will be to find ways to reduce *both* embodied carbon and operational carbon within the project's program, cost, and construction schedule limits. All project stakeholders should collaborate during the planning process to evaluate emission reduction opportunities. Ultimately, the project owner will make the decisions that impact design, but design professionals should serve a critically important role in influencing client decisions as to design options.

2.4 Reducing Embodied Carbon Emissions

Limiting warming to 1.5°C, or well below 2°C above pre-industrial levels, requires a new vision for reckoning with the industry's most formidable remaining front, embodied carbon. The new vision calls for design professionals to rethink how buildings and infrastructure are constructed using a whole-life-cycle approach that provides opportunities to reduce embodied carbon. Reducing embodied carbon emissions in materials is a previously untapped opportunity that will help to dramatically reduce GHG and meet the timeframe facing the world, of reaching net zero by 2050. For the past decade, scientists and researchers have been working on innovative tools and databases for calculating embodied carbon.

In planning a project, architects can conduct a whole-building life-cycle assessment (WBLCA) to examine the carbon impacts of building materials over their entire life-cycle, from raw material extraction, manufacturing, and construction, to operation and demolition.[17] While this was previously difficult due to the absence of consistent materials data, international standards have been developed—environmental production declarations (EPDs)—which aim to ensure consistency and comparability of outcomes.[18]

Quantifying an Organization's GHG Emissions[16]

An organization's GHG emissions are typically divided into three categories—Scope 1, 2, and 3. Executive Order 13514, signed by President Obama in 2009 (now revoked), applied these scope definitions to encourage sustainable operations in federal agencies.

Scope 1: Emissions resulting from purchased goods and services and their transportation (e.g., boilers, furnaces, turbines, emergency generators, automobiles, ships, aircrafts, and process emissions from the manufacturing or processing of chemicals and materials).

Scope 2: Indirect emissions from purchased energy (i.e., emissions that physically occur at the facility where the electricity, steam, heating, or cooling is generated).

Scope 3: Emissions from sources over which the agency or organization exerts no direct control (e.g., business travel, employee commuting, wastewater treatment).

2.4.1 Life-Cycle Analysis Tools

LCA tools—which enable design professionals in the U.S. to quantify emissions across the whole life-cycle—include Tally, a US-developed plug-in for the design software Revit, and Athena Impact Estimator. Architecture 2030 is introducing the Carbon Smart Materials Palette, a free on-line platform to identify embodied carbon "hotspots" in building materials, and in 2019, Skanska USA released the Embodied Carbon in Construction Calculator (EC3), which is also free; and Autodesk has enabled EC3's integration with BIM 360. WGBC notes that care should be taken when using tools from different countries and regions "as the underlying LCA datasets associated with a particular tool may not be representative, due to differences in the way materials and products are sourced and made."[19]

2.4.2 Concrete and Steel—Hotspots of Embodied Carbon

Structural systems are usually the largest source of embodied carbon. Concrete and steel are two of the most widely used man-made structural building materials in existence but are carbon-intensive because of their manufacturing processes. Transitioning to net zero emissions will depend upon the supply side of industry finding a way to supply structural materials such as steel and cement, or close substitutes, without adding CO_2 to the atmosphere. "Technologies like carbon capture, utilization and storage, or hydrogen-based production of iron for steel making, have been demonstrated but are not commercialized."[20] If engaged early in the design process, a structural engineer can optimize ways to reduce the impacts of materials.

Concrete is made of three basic components: water, aggregate (rock, sand, or gravel) and cement. It is an essential component of the built environment and is used in all types of infrastructure, from buildings, bridges, and roads to sidewalks. Patented 200 years ago, Portland cement's main ingredient, limestone, must be heated at high temperatures (around 1,500°C) and releases massive amounts of emissions

(1.0 ton of CO_2 per ton of cement created) when heated. Production of Portland cement is estimated to be responsible for 5% of total global CO_2 emissions, and is the single largest industrial emitter world-wide, second only to the steel industry at 6.6%.[21] Currently, the primary means of reducing the carbon footprint of concrete is to reduce the amount of cement in the mix. For example, cement can be partially replaced with Supplementary Cementitious Materials (SCMs) such as Ground Granulated Blast furnace Slag (GGBS), Pulverized Fly Ash (PFA), and Silica Fumes (SF).[22]

Steel production also entails a carbon-intensive manufacturing process. During steel manufacture, "carbon (coke from coking coal) is used to reduce iron oxide ore in blast furnaces, producing .6 to 3.1 tons of process CO_2 per ton of crude steel produced. This is in addition to CO_2 emissions from fossil fuels burned to generate the necessary high temperatures (1,100°C to 1,500°C)."[23] Metals like steel can be recycled without loss of physical qualities—thus recycling can contribute to emissions reductions—but recycling will require increased coordination of the demolition and construction phases. Using sustainable mass timber instead of steel is another way to cut GHG emissions. New timber technologies, such as cross-laminated timber (CLT) are highly resilient to fire. Although building structures out of timber may not seem innovative, all-timber construction is a carbon-friendly, sustainable alternative to steel and concrete. Finally, designers frequently call for more material than is structurally necessary; eliminating overengineering and oversizing, and implementing efficiency in material design can reduce embodied carbon by reducing waste.

2.4.3 **Legislation Aimed at Embodied Carbon**

New legislation that regulates embodied carbon will help to accelerate change. Government regulation can influence design and procurement choices, and motivate the supply chain to develop effective solutions.[24] In the U.S., state governments are just beginning to enact laws that will drive the private sector. The state of California's Buy Clean California Act, which became effective on July 1, 2019, is the first legislation in the U.S. to regulate embodied carbon, and targets construction materials used in public works infrastructure projects such as roads, bridges, and public buildings. The law requires EPDs for steel, flat glass, mineral wool, and insulation specified for state projects, and is intended to motivate suppliers to reduce their climate pollution. Marin County, California, has enacted Marin County Code Chapter 19.07, effective in January 2020, which regulates the composition of concrete that "maintains adequate strength and durability for the intended application and at the same time reduces greenhouse gas emissions."

2.4.4 **Voluntary Initiatives to Reduce Embodied Carbon**

Collaborative forums are also setting milestones for reaching embodied carbon solutions. The Structural Engineers 2050 Challenge (SE2050) is a professional organization that has set a goal of net zero embodied carbon in structural systems by 2050, and was inspired by the AIA 2030 Challenge to reduce building operational energy to net zero.[25]

Reducing embodied carbon requires the efforts of all stakeholders across the building and construction value chain, including developers and contractors. Skanska, a global development and construction group, has shown leadership by committing to achieve net zero carbon emissions by 2045, with a 50% reduction by 2030. Skanska USA recently launched the EC3 Calculator, a free tool that calculates embodied carbon emissions within building materials. Skanska developed the EC3 with the University of Washington's Carbon Leadership Forum, Canadian software developer C Change Labs, and Microsoft.

In January 2020, Microsoft announced the company's commitment to be carbon negative by 2030, and by 2050 to remove "all the carbon it has emitted since it was founded in 1975," thus erasing 100% of its carbon footprint. Microsoft is piloting the EC3 in the redevelopment of its Redmond, Washington, headquarters, and set a target of 15% to 30% reduction in embodied carbon, with Skanska as the general contractor.[26]

IPCC's 2018 Special Report includes a poignant quotation: "Pour ce qui est de l'avenir, il ne s'agit pas de le prévoir, mais de le rendre possible." ["As for the future, it is not a question of predicting it, but of making it possible."—Antoine de Saint Exupéry, *Citadelle,* 1948.]

2.5 **Policies and Laws to Achieve Carbon Neutrality**

Government policy can drive private sector choices that impact carbon emissions by improving legislation at the national, state, and city level, including strengthening building codes that influence the design and procurement choices made by other stakeholders. Cities and states in the U.S. have begun to craft visions of carbon neutrality, and many companies and organizations have committed to achieving carbon neutrality. The 2021 International Energy Conservation Code (IECC) will contain a new Zero Energy (ZE) Code appendix that states and local governments may adopt as a base or stretch code.

In 2019, Washington, D.C. and New York aligned with the Paris Agreement by enacting legally binding legislation. The Clean Energy DC Omnibus Amendment Act mandates that Washington, D.C. must be 100% dependent on renewable energy by 2032, and calls for net zero energy building codes by 2026. New York's Climate Leadership and Community Protection Act requires New York State to have a net zero carbon economy by 2050, including 70% renewable energy by 2030 and decarbonization of the electric sector by 2040. In May, 2019, New York City enacted Local Law 97, which establishes an office of building energy and emissions performance, and calls for a 40% reduction in emissions from buildings over 25,000 square feet by 2030 and an 80% reduction by 2050 from 2005 levels. Other states including California, Hawaii, New Mexico, Washington, and Puerto Rico have enacted legislation requiring 100% clean electricity by 2050,[27] while Colorado, Maine, and Nevada have enacted legislation committing to ambitious energy reductions.[28]

In January 2020, the governor of Massachusetts pledged to achieve net zero GHG emissions by 2050, and the Executive Office of Energy and Environmental Affairs (EEA) set the statewide emissions limit to require net zero emissions by 2050.[29] In December 2020, EEA issued a report, Massachusetts 2050 Decarbonization Roadmap, which explores the dynamics of decarbonization by 2050 through an integrated approach by sectors including buildings, transportation, non-energy, and land.[30] Reducing emissions requires a holistic approach, and the EEA report relies heavily on the pillars of decarbonization that have been identified in prior deep decarbonization studies, such as the Deep Decarbonization Pathways Project,[31] both in the U.S. and internationally.

To achieve deep decarbonization in the built environment, the state must expand its clean energy supply. Renewable generation is the key to dramatically reducing carbon emissions, and the availability of a "world-class offshore wind resource off the New England coast" can provide bulk low-cost, carbon-free electricity. The report advises that as of October 2019 "1.6 GW of offshore wind has been procured with another 1.6 GW authorized for procurement and development in the 2020s."[32] In addition to buildout of offshore wind, deployment of solar generation on rooftops and land at scale, together with additional energy storage, and new high-voltage transmission lines are needed to share low-cost clean energy.[33]

Decarbonization strategies for new construction and existing buildings (see Section 2.5.1, "The Concept of Net Zero," below in this chapter)[34] include electrification of end uses and economically advantageous electrification technologies, such as electric heat pump systems for building space conditioning needs, as well as investments in improvements to building envelopes. Implementation of a high-performance net zero emissions building energy code will help to minimize the near-term installation of fossil fuel equipment in buildings.

In March 2021, the Massachusetts Legislature overwhelmingly passed a climate policy bill that codifies the requirement for net zero GHG emissions limits by 2050, and Governor Baker signed that bill into law. The law sets interim emissions limits every five years. Massachusetts will create a local net zero stretch energy code that municipalities have the option to adopt and will eventually be effective statewide. The new law defines environmental justice populations and provides protections for these communities.

As a coastal city, Boston is on the frontline of the climate crisis. In 2016, the City of Boston pledged to achieve carbon neutrality by 2050. In October 2019, Boston issued an updated Climate Action Plan, which outlines strategies to accelerate a shift to net zero new construction by 2030, and retrofit and electrify at least 80% of existing buildings by 2050. To set an example, Boston will adopt a zero net carbon standard for new municipal construction and for city-funded affordable housing. As a first step, Boston recently strengthened its green building review process under Article 37 of the Boston Zoning Code, and Article 80 Large Project Review (50,000 square feet and over) to align with a carbon neutrality policy. Since 2019, private developers of large projects must submit a Zero Carbon Building Assessment that models how the building can achieve a path to zero carbon, zero energy, or energy positive design (ZC/ZE/E+ Design). New zoning requirements are expected to be released in 2021.

A report by the Boston Green Ribbon Commission, *Carbon Free Boston Summary*

Report 2019, notes the experiences of other major cities that voluntary programs do not result in sufficient emissions reduction, and that "emissions-based performance standards for all buildings are the most effective measures for achieving the scale of change required."[35] Boston's Building Energy Reporting Disclosure Ordinance (BERDO) currently has energy reporting requirements for buildings 35,000 square feet or larger. A technical analysis and stakeholder process began in 2020, and BERDO is expected to be amended in 2021 with new emissions performance standards that will require deep carbon cuts in Boston's largest buildings, and deep energy retrofits that achieve at least a 50% reduction in energy use.

The UCLA Luskin Center for Innovation has collected data that shows that "despite [the former] federal policy designed to promote fossil fuels, progress on 100% clean energy commitments has been rapid."[36] The data is encouraging, with 72 cities and counties in seven states already powered by 100% clean electricity sources, including solar and wind. California is the most progressive, with 66 cities and counties at 100%.

2.5.1 **The Concept of "Net Zero"**

There is no universal definition of the metrics for "net zero." The traditional concept of "net zero" focuses on operational emissions, and means a building that achieves a nearly zero balance between energy demand and renewable energy supply. Achieving net zero is usually accomplished through a combination of energy efficiency and renewable energy. In most settings, buildings cannot generate 100% of their energy needs on-site; thus, net zero buildings are still connected to the grid. Voluntary standards and codes typically define net zero as buildings that are highly energy efficient and supply all their energy needs from renewable sources (e.g., on-site and/or off-site solar or wind). Net zero buildings are high performance, which means they implement various energy efficiency measures in order to reduce their energy consumption.

Electrification of end-use services in buildings in conjunction with decarbonization of electricity generation is part of the strategy to achieve net zero operational emissions. The building sector currently uses a significant amount of natural gas, fuel oil, and propane, for space conditioning, water heating, cooking, and other end-use services. By substituting electricity for fossil fuels in end-use services, significant reductions in carbon dioxide emissions can be achieved. On a larger scale, achieving net zero energy emissions by 2050 requires the power sector to meet our nation's electricity demand with low-carbon electricity generation from a clean electric grid.

2.5.2 **Decarbonizing the Grid**

Rapid decarbonization of the grid will require government action at both state and federal levels to transform the power sector. Today, over half of our electricity is produced from burning fossil fuels such as coal and natural gas. Coal is on the decline (550 coal-fired plants have closed since 2010), and renewables such as wind and so-

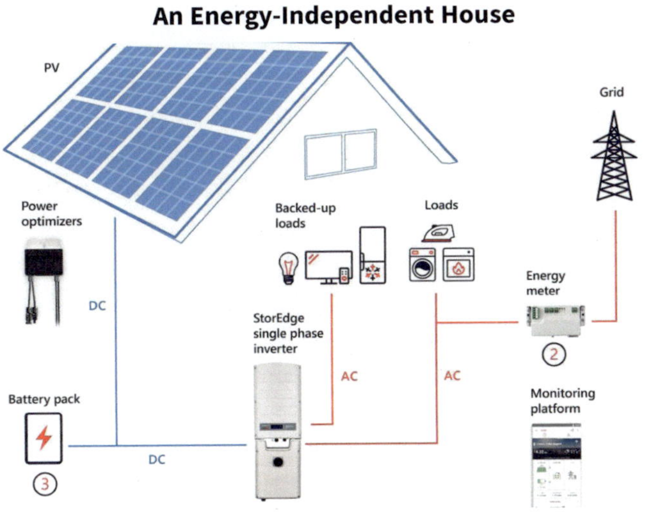

From: "Creating Energy Independence with Solar Panels and Storage Battery Systems in the Home," *Forbes*, Jan. 26, 2020.

lar presently compose approximately 10% of U.S. electricity generation. Electricity is generated at large, central sites and distributed to end users with a network of transmission lines known as the grid. Net zero pathways will need to shift electricity generation from mostly fossil fuels such as gas and coal to 70% wind and solar by 2050.

In June 2020, the House Select Committee on the Climate Crisis issued a report, *Solving the Climate Crisis, the Congressional Action Plan for a Clean Energy Economy and a Healthy, Resilient, and Just America,* which includes recommendations for a fully integrated national transmission grid which it terms a "National Supergrid." The U.S. does not have a national energy grid, and instead has three grids: the Eastern Interconnection, the Electric Reliability Council of Texas (ERCOT), and the Western Interconnection. The National Renewable Energy Laboratory (NREL) is working on a report, *The Interconnection Seam Study*, for the Department of Energy (DOE) that will identify cost-effective options for national, cross-state high voltage direct current (HVDC) transmission lines. Extension and reinforcement of power grids will be essential to getting renewables on line and sharing them over geographical areas.

While some displacement of coal will come from growth of renewable resources, such as solar PV and wind, the DOE recommends that natural gas combined cycle (NGCC) and carbon capture, utilization, and sequestration (CCUS) technologies can play a role. IPCC suggests that natural gas power generation can act as a bridge technology, "with deployment increasing before peaking and falling to below current levels by 2050 and declining further in the second half of the century."[37] Battery energy storage can cost-effectively integrate high levels of solar and wind energy into electricity grids and will be a vital backup to the grid if intermittent weather causes shortfalls from renewables. Investments in advanced battery storage and electrification of transportation will accelerate decarbonization.

2.5.3 **Transitioning to Electricity for End-Use Services**

Assuming a decarbonized power sector, transitioning to electricity for end use services such as heating, cooling, hot water, and cooking, instead of burning natural gas or fuel oil, will greatly reduce a building's emissions. Net zero emissions can also be facilitated by fully utilizing the potential for on-site renewable generation, such as rooftop solar. Design criteria can incorporate energy efficiency technologies including heat pumps, which can move heat from one location to another. For existing buildings, replacing fossil fuel equipment (such as gas and oil furnaces and boilers) with heat pump retrofits during routine home improvements is part of the net zero pathway. Net zero design strategies that optimize energy efficiency and reduce building energy loads include building envelopes, windows, natural lighting, high performance materials, and more efficient heating, ventilating, and air conditioning (HVAC) systems and equipment.

Net zero is now a viable reality, and the key has been a rapid decline in renewable energy costs. IRENA reports: "The fall in the cost of electricity from utility-scale photovoltaic (PV) projects since 2010 has been remarkable—between 2010 and 2018 the global weighted average levelized cost of electricity (LCOE) from solar PV declined by 77%."[38] Plummeting costs for wind and solar energy have made it feasible for the U.S. to almost completely decarbonize the power sector by 2035, according to a report issued in June 2020 by the Goldman School of Public Policy at the University of California, Berkley. The *2035 Report* uses the latest renewable energy and battery cost data to demonstrate the technical and economic feasibility of achieving 90% clean (carbon-free) electricity in the United States by 2035 while injecting $1.7 trillion of private investment into the economy.[39] Robust policy reform is also needed to support rapid electricity sector decarbonization, and the most progress can be made with a federal clean energy standard that complements state policy leadership.

The Unisphere Building

United Therapeutics headquarters in Silver Spring, Maryland, is the largest Net Zero building in the world

Source: Halkin/Mason Photography, LLC, courtesy of United Therapeutics.

PART 2

2.6 Resilience and Adaptation to Climate Change

Resilience is a complex concept that emerged from ecology in the 1970s and is defined and understood differently by different disciplines.[40] The Rockefeller Foundation's 100 Resilient Cities defined resilience as the capacity of individuals, communities, businesses, and systems within communities to survive, adapt, and grow no matter what types of stresses and acute shocks they experience. The concept of resilience is still evolving. Resilience captures the co-benefits and synergies of an integrated approach to climate mitigation and adaptation; i.e., the choices and actions that reduce GHG emissions mitigate the rate and magnitude of dangerous anthropogenic climate change and its impacts (discussed in Part 1), and the adaptive ability of infrastructure to withstand a given range of climate variability, including extreme events, reduces the consequent level of harm (discussed in Part 2). Adaptive capacity allows infrastructure to cope with successively larger climate impacts caused by successively higher levels of global warming, while mitigation lessens exposure to climate-related hazards by reducing the projected upper ranges of global warming. However, even if the world reaches net zero targets, elevated GHG concentrations in the atmosphere will persist for decades or even centuries.[41] Part 2 of this chapter focuses on adaptive design and risk management.

2.7 Risk Management and Adaptive Design

Risk management aims to limit harm in an environment of uncertainty. There is scientific certainty that the climate is changing but there is considerable uncertainty about the magnitude and timing of climate change over the lifecycle of infrastructure. Climate change is exacerbating climate hazards and amplifying the risk of natural disasters.[42] When natural disasters strike, a community's response is often driven by short-term priorities rather than long-term goals of sustainability, adaptation, and resilience. Resilient infrastructure means that risks have been considered and managed to achieve an acceptable level of performance given the available information, and that capacities to withstand and recover from shocks are in place.[43] Measuring the resilience of an infrastructure asset can help stakeholders, including public and private sector owners, to better understand and evaluate risk exposure.

2.7.1 Climate-Resilient Infrastructure

Both qualitative and quantitative techniques are useful in measuring and understanding resilience. "Resilience measurement, like most efforts to measure complex phenomena, requires a multidimensional, multi-method approach. One of the main benefits of qualitative data and subjective data is that they provide a more detailed understanding of the dynamic relationships that explain variations in well-being following exposure to shocks and stressors."[44]

For infrastructure, quantitative resilience metrics are tools that can serve as a framework to analyze new technologies and nature-based design solutions, anticipate their performance, and evaluate strategic investments:

> Resilience can be quantified through computational experiments in which disruptions are triggered, the system performance is analyzed . . . and integrated resilience metrics are computed. By repeating this process, different system design solutions can be ranked with respect to resilience.[45]

The uncertainty associated with climate change is not completely quantifiable, however, and accounting for it in engineering practice requires "an appropriate understanding and treatment of uncertainty, including engineering judgment."[46]

Design teams and owners can evaluate design options by monetizing the costs and benefits of design options based on an analysis of different scenarios with different risk profiles.[47] During the last two decades, researchers have proposed different methods for quantifying the resilience of infrastructure. The Multidisciplinary Center for Earthquake Engineering Research (MCEER) developed the diagram in Figure 3 (below) to illustrate the concept of resilience in an infrastructure asset.[48]

The resilience triangle plots the loss of functionality from a disruption (vertical axis) and the time it takes to return to the initial level of performance (horizontal axis).[49] Figure 3 is used with a suggested framework of resilience called the four "Rs," defined as:

- **Robustness:** the ability of the system and system elements to withstand external shocks without significant degradation or loss of performance
- **Redundancy:** the extent to which systems, system elements, and other elements are substitutable and capable of satisfying and sustaining functional requirements in the event of disturbance;
- **Resourcefulness:** the ability to diagnose and prioritize problems and to initiate solutions by identifying and monitoring all resources;
- **Rapidity:** the ability to recover and restore functionality in a timely way, containing losses and avoiding future disruptions.[50]

Figure 3. **Resilience Triangle**

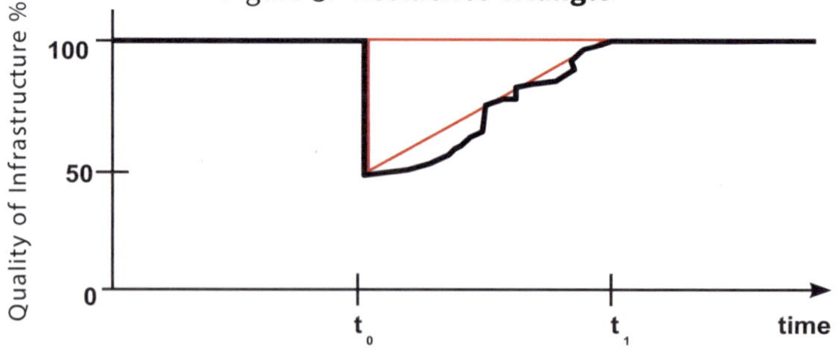

Adapted from Kathleen Tierney and Michel Bruneau, "Conceptualizing and Measuring Resilience: A Key to Disaster Loss Reduction," *TR News 250*, May–June 2007, pp. 14–15.

ASCE's manual of practice, *Climate-Resilient Infrastructure, Adaptive Design and Risk Management*, proposes several model-based approaches that expand upon the resilience concepts in Figure 3.[51]

Figure 4. **Resilience Model Proposed by Ayyub**

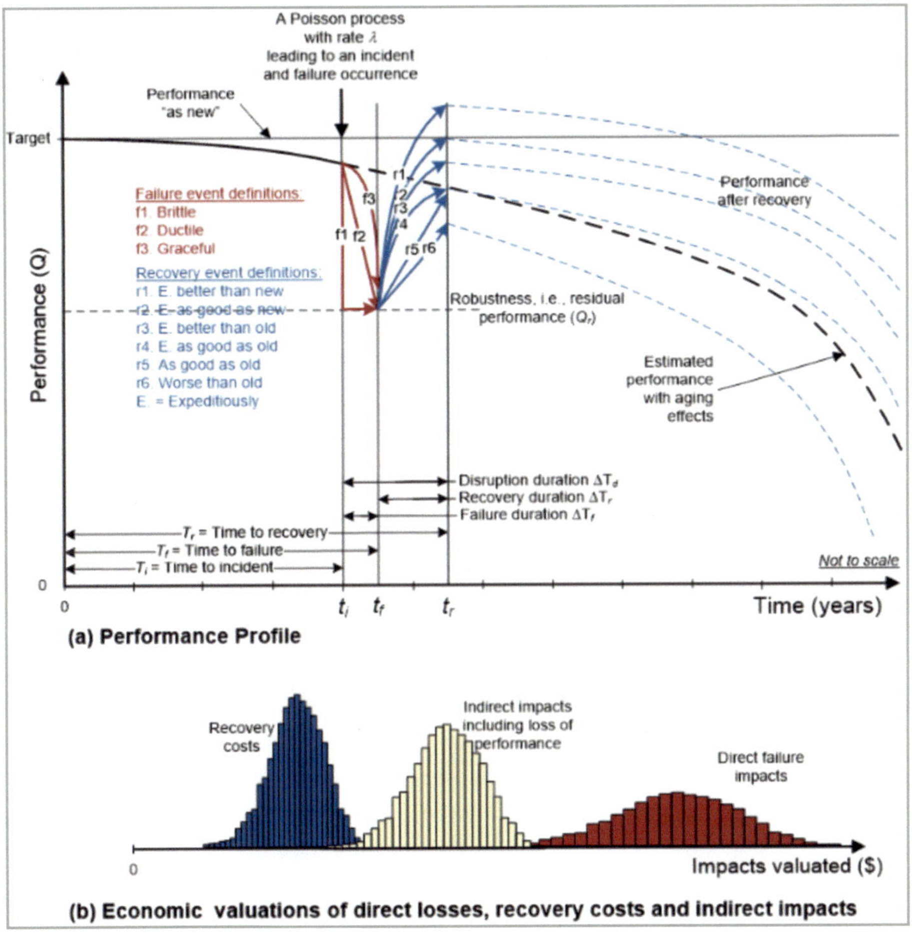

This analysis uses mathematical models to measure resilience. "The failure event is followed by a recovery event with a duration ΔT_r. The recovery event concludes at time tr. The total disruption (D) has a duration of $\Delta T_d = \Delta T_f + \Delta T_r$. The figure shows for illustration purposes three failure events: (1) brittle (*f*1), (2) ductile (*f*2), and (3) graceful (*f*3), and six recovery events: (1) expeditious recovery to better-than-new (*r*1), (2) expeditious recovery to as-good-as-new (*r*2), (3) expeditious recovery to better-than-old (*r*3), (4) expeditious recovery to as-good-as-old (*r*4), (5) recovery to as-good-as-old (*r*5), and (6) recovery to worse-than-old (*r*6). These events define various rates of change of performance of the system."

From: Bilal M. Ayyub, *Practical Resilience Metrics for Coastal Infrastructure Features*, U.S. Army Corps of Engineers, Engineer Research and Development Center, May 2019, pp. 27–28.

The ASCE manual recognizes that the model in Figure 4 is complex and does not lend itself easily to practical applications. Therefore, it suggests several alternative models that are simpler for practitioners such as engineers, city planners, and policy makers.[52]

Still other researchers at the Reliability and Risk Engineering Laboratory, ETH Zurich have proposed a resilience curve for quantifying resilience.

Figure 5. **Resilience Curve**

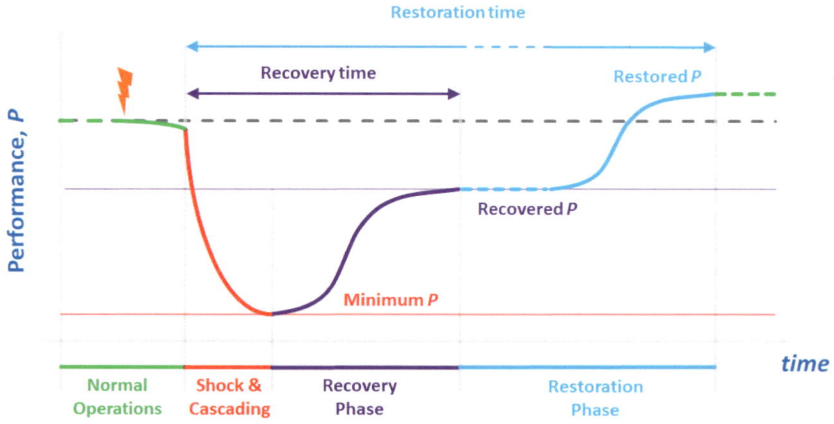

Giovanni Sansavini, Paper 34, "Engineering Resilience in Critical Infrastructures," in Florin, M.-V., & Linkov, I. (eds.), *IRGC Resource Guide on Resilience*. Lausanne: EPFL International Risk Governance Center, 2016. DOI:10.5075/epfl-irgc-228206, p. 207.

Modeling can be used as a decision-making tool to quantify and evaluate risk and the resiliency of infrastructure in a non-stationary environment. ASCE recommends that the mathematical objective is to use low regret strategies and policies "that would work well under both the current climate and the uncertain future climate."[54]

Climate change impacts critical transportation infrastructure such as bridges, which have a long service life, sometimes exceeding 100 years. ASCE's 2021 Report Card for America's Infrastructure gives bridges a grade of C, stating that "7.5% of the nation's bridges, are considered structurally deficient, meaning they are in 'poor' condition." Heatwaves, thermal stresses and precipitation increase corrosion, spalling and cracking in concrete, as well as deck deterioration and deformation. Sea level rise and high runoff from stream flows increases the risk of hydraulic failure, or scour of foundational elements. The clogging of structural elements such as expansion joints with debris can also cause failure. More intense storms, high winds, and storm surge affect slope stability, and climate change is causing die-off of coral reefs which provide natural protection to bridges against the risks of coastal flooding from storms.

ASCE notes that bridge engineers are using innovative, ultra-high performance materials to meet this challenge and increase the resilience of bridges, and are also designing "living bridges" where "sensors are being embedded into new and existing structures to provide continuous feedback on structural conditions."

2.7.2 Structural Health Monitoring

Structural health monitoring (SHM) is innovative technology that uses sensors to gather data. SHM sensor technology can be used to monitor the safety and resilience of critical infrastructure, such as bridges. Current real-time data outputs from sensors are compared with recent and historical sensor outputs to evaluate structural conditions. Monitoring the data allows early detection of damage and prognosis of failure before it happens. In *Structural Health Monitoring, a Tool for Improving Critical Infrastructure Resilience*, Swedish scientists Daniel Honfi and David Lange explain how SHM can be used to measure the 4 Rs of resilience by:

- Developing condition-based maintenance strategies (*robustness*),
- Providing information on alternative possibilities (*redundancy*),
- Prioritizing maintenance and emergency actions for better use of resources (*resourcefulness*),
- Alerting first responders in case of emergency (*rapidity*).

Indian River Inlet Bridge Structural Health Monitoring System

The Indian River Inlet Bridge is a 533-m long cable-stayed bridge in Delaware. The Center for Innovative Bridge Engineering at the University of Delaware worked with the Delaware Department of Transportation (DelDOT) and the design-build team of Skanska and AECOM to plan and install a comprehensive structural health monitoring (SHM) system for long-term in-service monitoring. The SHM system is a fiber-optic based design with more than 120 sensors distributed throughout the bridge to measure the structural response of the bridge under various environmental loads and live load conditions. The system collects data continuously during normal operation, and is also utilized during regularly scheduled controlled diagnostic load tests, to monitor ongoing bridge performance.

Source: Hadi Al-Khateeb, Shenton III, Michael J. Chajes, and Christos Aloupis, "Structural Health Monitoring of a Cable-Stayed Bridge Using Regularly Conducted Diagnostic Load Tests," *Frontiers in Built Environment, Bridge Engineering,* March 29, 2019.

The performance of natural and nature-based (green) infrastructure can also be defined and measured like traditional hard (gray) infrastructure.[53] (Table 1)

Table 1. **Examples of Natural and Nature-Based Features Relevant to Coastal Systems**

Types	Dunes and Beaches	Vegetated Features (e.g., marshes)	Oyster and Coral Reefs	Barrier Islands	Maritime Forests/Shrub Communities
Benefits	Breaking of offshore waves, Attenuation of wave energy, Reduction or prevention of inland water transfer	Breaking of offshore waves, Attenuation of wave energy, Reduction or prevention of inland water transfer, Increased infiltration	Breaking of offshore waves, Attenuation of wave energy, Slowing of inland water transfer	Wave attenuation and/or dissipation, Sediment stabilization	Wave attenuation and/or dissipation, Shoreline stabilization, Soil retention
Performance factors	Berm height and width, Beach slope, Sediment grain size and supply, Dune height, crest, and width, Presence of vegetation	Marsh, wetland, or submerged aquatic vegetation elevation and continuity, Vegetation type and density, Spatial extent	Reef width, elevation, and roughness	Island elevation, length, and width, Land cover, Breach susceptibility, Proximity to mainland shore	Vegetation height and density, Forest dimension, Sediment composition, Platform elevation
Performance measurements suitable for resilience quantification (Examples)	Example: See Section 2.4.1 on beaches after hurricane Sandy	Example: See Section 2.4.1 on beaches after hurricane Sandy			

Bilal M. Ayyub, *Practical Resilience Metrics for Coastal Infrastructure Features,* Table 4, p. 13, U.S. Army Corps of Engineers, 2019. Available online at https://apps.dtic.mil/dtic/tr/fulltext/u2/1073955.pdf.

2.7.3 **Risk Reduction Through Adaptation**

Most infrastructure has a service lifespan of 50–100 years, and needs to remain safe and functional during that time period because it will pose a costly or potentially deadly threat if it fails. All types of infrastructure—roads and bridges, water and wastewater treatment facilities, transportation networks, ports and levees, stormwater systems, hospitals, industrial buildings, and homes—are vulnerable to climate-related hazards. In Boston, where many neighborhoods are built in low-lying coastal areas, approximately $2.4 billion is needed over the next several decades to protect the city from flooding.[55] The EPA estimates that damages to coastal property from sea-level rise and storm surge (SLR/SS) in the contiguous U.S. will be "$5.0 trillion through 2100 (discounted at 3%) in a future without emission reductions," but "[w]hen cost-effective adaptation along the coast is included, the estimated damages are reduced to $810 billion."[56]

Adaptive design aims to strengthen the resilience of critical infrastructure against the risks of climate change, including extreme weather. An adaptation project could involve traditional hard engineering (gray infrastructure), a nature-based solution (green infrastructure), or a gray-green hybrid approach. Examples of green infrastructure include living shorelines, parks, wetland and dune restoration, and green stormwater protection such as green roofs, rain gardens, and bioswales. An adaptation project could also entail soft infrastructure, like development of an early warning system.[57]

2.7.4 Evaluating Projected Climate Risk

A key message of the 4th National Climate Assessment was that climate patterns are expected to deviate from the recent historical record, and that failure to adapt could be costly. The assessment notes: "Adjusting to projected climate risk, rather than relying on interpretations of past impacts, has important implications for the location and design of built human infrastructure [and] engineering standards."[58] To reduce the risk of adverse impacts, information about current and future climate conditions should be incorporated into design standards, policies, and practices. Risk management assessments that are based upon an unchanging or stationary climate should be avoided, but the question is how best to address the risk of extreme events under a changing or non-stationary climate.

Adaptation entails a continuing risk management process. Effective risk management for adaptation will address two timescales: "1) adapting to current variability, which in any particular location may now be different than suggested by the historical record of climate observations, and 2) preparing for future change."[59] The distinction between the two timescales is important for both risk management and cost-benefit analysis. Project decision-makers must appreciate the extent to which the climate has already changed, and also evaluate what the climate will look like in the future. These timescales will call for different types of adaptive action.

2.7.5 FEMA's Flood Maps Are Based on Historical Climate Data

An example of historical, stationary climate data that may not reflect current or future climate conditions are FEMA flood maps, which form the basis of many jurisdictions' building codes. Over fifty years ago, in response to the lack of flood insurance available on the private market, and the escalating costs of disaster relief, Congress enacted the National Flood Insurance Act of 1968, which established the National Flood Insurance Program (NFIP). NFIP is overseen by the Federal Emergency Management Agency (FEMA), and FEMA is responsible for mapping the nation's hazardous flood areas, and determining the size and location of the 1% annual chance of flood zone (100-year flood) and 0.2% annual chance of flood zone (500-year flood).

FEMA's insurance rate maps (FIRMs) have significant influence on the built environment because many jurisdictions have adopted the mapped flood zones as a basis of their regulatory framework. FIRMs are created using historical data, and while they may accurately reflect current or historical information, they do not consider how future climate conditions may influence flooding. This is intentional, because the Code of Federal Regulations, 44 CFR 65.6, provides that FIRM "revisions cannot be made based on the effects of proposed projects or future conditions."[60] Although they were originally intended only to establish flood insurance rates, FIRMs are used across the country as regulatory tools, and are referenced in hundreds and thousands of zoning ordinances and building codes. FEMA is required to "revise and update all floodplain areas identified once during each 5-year period."[61] Yet, that hasn't prevented FEMA flood maps created as long ago as 1983 from regulating the built environment.

After Superstorm Sandy struck New York on October 29, 2012, it became widely known that FEMA's maps for the area were seriously outdated, and that no new data had been collected to update the maps since 1983. The maps vastly understated the dangers and devastation that occurred in New York and New Jersey.[62] Similarly, when Hurricane Harvey inflicted severe damage to Hitchcock, Texas, in August 2017, the FEMA flood maps hadn't been updated since 1983.[63] Although most of Houston's flood maps were updated in 2017, they failed to predict the inundations of Hurricane Harvey, and 40% to 50% of Houston residents who lived outside of FEMA's 100-year and 500-year flood zones had their homes flooded.[64]

2.7.6 **Boston is Vulnerable to Sea Level Rise and Storm Surge**

Boston is another U.S. coastal city that is vulnerable to sea-level rise and storm surge, and whose current flood regulations are based on the historical information reflected in FEMA's flood zones.

The Massachusetts State Building Code is an example of a state building code that regulates on the basis of FEMA flood maps. The current 9th edition of the Building Code, 780 CMR, identifies flood hazard areas subject to a 1% or greater chance of flooding in any given year, based on FIRMs, and includes provisions of ASCE 24-14, which specify the minimum freeboard requirements based upon criticality of infrastructure, with 1–2 feet of freeboard added to the base flood elevation (BFE). Freeboard is a factor of safety which is defined by an amount of building elevation above the BFE. However, accounting for climate change in flood plain management means reconsidering the concept of a 100-year flood under non-stationarity. The statement in our Nation's Fourth Climate Assessment that "the assumption that current and future climate conditions will resemble the recent past is no longer valid" highlights why historical climate data should not be used as a basis for design. FIRMs do not depict future climate conditions, such as sea level rise.

When FEMA's flood maps are the determinant of where the minimum building and land-use codes must be applied, the infrastructure may be vulnerable. As the Earth's temperature rises from greenhouse gas emissions, extreme storms are becoming more common and severe, increasing the frequency of so-called 100-year flood events. Property outside of the delineated 100-year floodplain may be at risk.

Storm Riley flooded Boston in March 2018

> The Massachusetts Building Code, 780 CMR, identifies the 1% annual chance of flood by reference to FEMA's flood maps. Similarly, Boston Zoning Code Article 25, titled "Flood Hazard Districts," defines flood hazard areas as lands in a flood plain, subject to a 1% annual chance of flooding in a given year. Under Article 25, flood hazard districts are a special overlay district. The exact boundaries are determined by 100-year base flood elevations shown on FEMA flood maps. Since FIRMs are based on historical information, there are many areas of future flood risk in Massachusetts and Boston which lie outside these boundaries and are not required to meet minimum standards for flood resistant buildings.

2.7.7 New Tools and Studies Identify Flood Risk

In the face of these concerns, many U.S. coastal communities are looking for better tools to identify climate risk to man-made and environmental assets. The City of Boston is preparing for 40 inches of sea level rise by 2070, and has issued a series of reports, Climate Ready Boston, as a resource for each waterfront neighborhood (East Boston, Charlestown, South Boston, and Dorchester). The reports utilize sea level rise and storm surge (SLR/SS) data and projections from the best available tool, the Boston Harbor Flood Risk Model, to evaluate the community's vulnerability and make recommendations for how Boston can become more resilient to SLR/SS.

Figure 6. **Projections of Mean Sea Level Change Under Three Emissions Scenarios, RCP 2.6, RCP 4.5 and RCP 8.5**

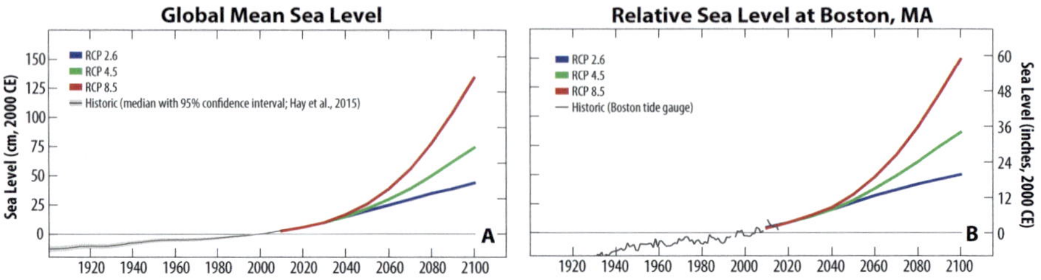

From: Boston Research Advisory Group Report, "Climate Change and Sea Level Rise Projections for Boston," Climate Ready Boston, p. 13, June 1, 2016.

2.7.8 **The Boston Harbor Flood Risk Model**

In 2015, the Boston Harbor Flood Risk Model (BH-FRM) was developed by a project team of scientists, including Kirk Bosma of Woods Hole Group, Inc., Ellen Douglas of UMass Boston, and Paul Kirshen of the University of New Hampshire.[65] The model was developed as part of a vulnerability assessment of Boston's Central Artery/Tunnel Project (CA/T, called the "Big Dig"), which is a critical link in Boston's transportation network. In the event of a disaster, it is an indispensable route for evacuation, emergency responses, and recovery services. The BH-FRM is a state-of-the-art numerical model capable of simulating thousands of potential nor'easters and tropical storms. It uses a geographic information system (GIS) that allows the user to overlay potentially impacted areas. The BH-FRM has been adopted by agencies, municipalities, and private entities as the standard from which to identify the projected SLR/SS from coastal flooding in Massachusetts.

Sea Level Rise Projections for Boston

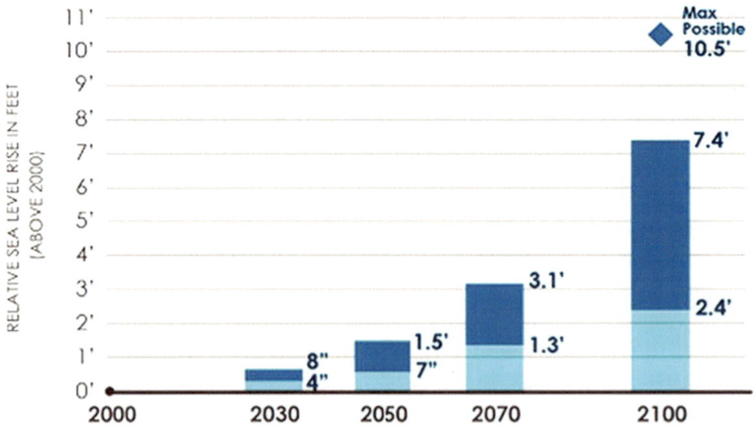

From: BPDA, "Climate Resilience Guidance," 2017.

In October 2017, the Boston Planning and Development Agency (BPDA) updated its policy to require all large projects subject to Boston Zoning Code Article 80 to file a "Resiliency Checklist" that addresses how their building and site design will manage flood risk in the Sea Level Rise-Flood Hazard Area (areas with a projected 1% annual chance of flooding with 40 inches of sea level rise) "as derived from the MassDOT-FHWA Boston Harbor Flood Risk Model (BH-FRM)." The 40 inches of SLR is derived by combining the mean sea level rise (3.2 feet, or 38.2 inches, above 2013 tide levels) plus 2.5 inches to account for local land subsidence.

In September 2019, the BPDA released a draft report, *Coastal Resilience Design Guidelines*, which is intended to help Boston become more resilient to coastal flooding, and provide consistent standards for official project review. The report recognizes that Zoning Code Article 25 does not project future flooding in Boston, because it is based on FEMA flood data, and recommends a proposed Zoning Overlay District that uses BH-FRM mapping.

Boston Neighborhoods with Coastal Flood Resilience Zoning Overlay

Dark shaded area = 1% annual chance of flooding under Zoning Article 25 (FEMA)
Light shaded area = 1% annual chance of flooding in year 2070 with 40 inches of sea level rise

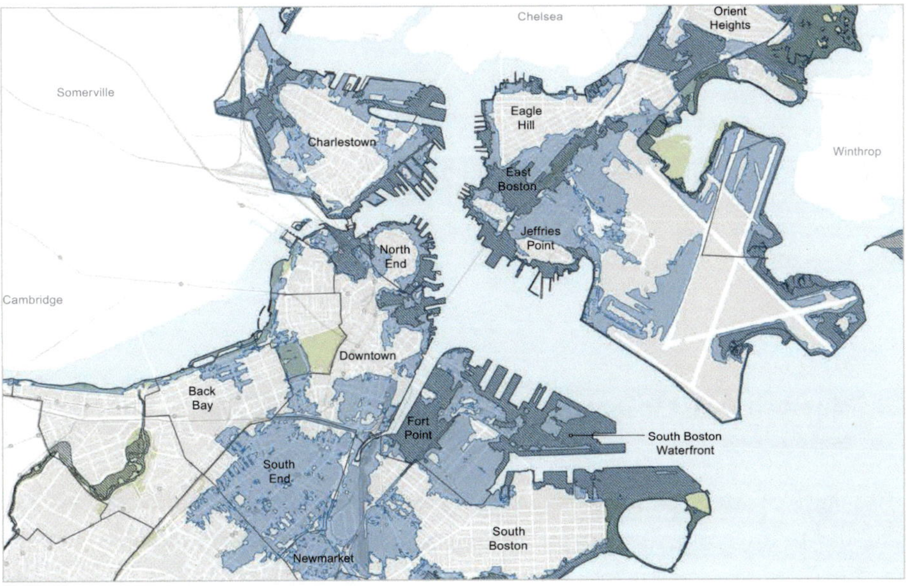

From: BPDA, Coastal Resilience Design Guidelines, Draft September 2019, p. 7.

The spatial boundary of the proposed Zoning Overlay District includes all areas with a projected 1% annual chance of flooding in the year 2070 with 40 inches of sea level rise, as determined by utilizing the BH-FRM.

The City's goal is for the proposed Zoning Overlay and the Design Guidelines to become part of standard development practice for project owners, designers, and builders. BPDA's guidance suggests adaptation strategies, such as raising the elevation of the site and access routes, elevating building ground floors, dry and wet flood proofing, and locating critical building equipment and systems above potential flood heights. In 2018, BPDA also issued Climate Resilient Design Standards and Guidelines for Protection of Public Rights-of-Way (ROW).

The BH-FRM has been extended and updated for the entire coastline of Massachusetts and islands. The expanded model is known as the Massachusetts Coast Flood Risk Model (MC-FRM). Other modeling tools also exist for Massachusetts, and for coastal communities outside of Massachusetts, to assess vulnerability to future sea level rise.

Boston's Coastal Resilience Design Guidelines

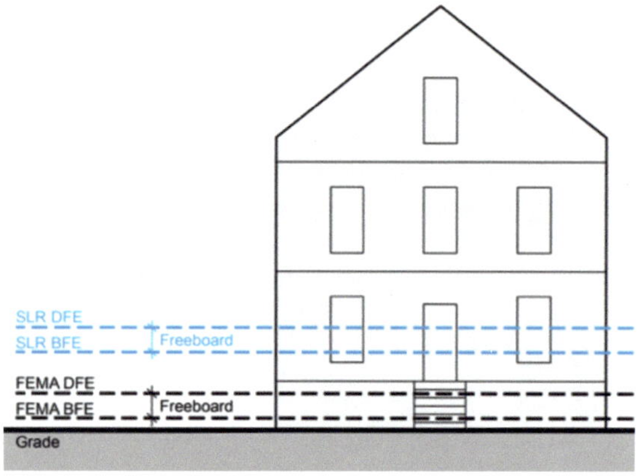

From: BPDA, *Coastal Flood Resilience Design Guidelines*, DRAFT September 2019, p. 19.

Sea Level Rise–Design Flood Elevation (SLR–DFE) is the minimum elevation of the lowest occupiable floor for residential uses, or dry flood-proofing for non-residential uses.

Sea Level Rise–Basic Flood Elevation (SLR–BFE) is the top of the water level, as determined by the 1% annual chance of flood in the 2070 scenario from the Boston Harbor Flood Risk Model, which includes 40 inches of sea level rise (of which 2.5 inches is for land subsidence).

1′–2′ of freeboard is added to the BFE, depending upon use of the building. The SLR-BFE plus freeboard equals the SLR-DFE.

2.7.9. **Massachusetts Is Developing Climate Resilience Design Standards and Guidelines**

The state of Massachusetts is setting a national example by developing design guidelines to integrate climate resilience into the physical assets that are owned and maintained by state agencies. The Resilient Massachusetts Action Team (RMAT), an interagency team led by the Executive Office of Energy and Environmental Affairs (EEA), is tasked with developing the Climate Resilience Design Standards and Guidelines. The goal is to provide a consistent basis-of-design and uniform statewide methodology for climate projections, and then use the data for design criteria to incorporate climate resilience into the State's capital planning process. RMAT is developing a GIS-based web tool for agencies that uses the best available climate science data for the following climate parameters: sea-level rise/storm surge, extreme precipitation (riverine and urban), and extreme heat. The tools are anticipated to be completed in early 2021, and will be accessible at ResilientMA.org.

Users of the Climate Resilience Design Standards Tool will provide project inputs, and the tool will provide outputs. For example, physical assets within a project are assigned three main asset categories: Infrastructure, Natural Resources, and Buildings. After users complete the necessary Project inputs, the tool will provide two outputs.[66] The first output, "Climate Risk Screening," includes preliminary project exposure ratings and preliminary asset risk ratings for each climate parameter. The second output, "Climate Resilience Design Standards," will provide users with standards for each applicable climate parameter; i.e., sea-level rise/storm surge, extreme precipitation (riverine and urban), and extreme heat. The standards output are intended to be used to calculate design criteria while proceeding into the project design phase.[67] For the sea level rise/storm surge parameter, the standards reference the MC-FRM developed by MassDOT and the Woods Hole Group. The MC-FRM is a probabilistic hydrodynamic model that uses values for sea-level rise based upon the RCP 8.5 scenario. The tool outputs are intended to be used by state agencies in requests for proposals (RFPs) for design services.

2.7.10 Sea Level Rise Projections for the United States

Even if anthropogenic CO_2 emissions cease altogether, global mean sea level (GMSL) will still continue to rise for centuries due to global warming that has already occurred. In many locations, the frequency of tidal flooding (typically flooding without a storm present) has increased dramatically over the past few decades, growing from a rare event into a recurrent and disruptive problem.[68] In a 2017 report, *Global and Regional Sea Level Rise Scenarios for the United States,* NOAA developed updated scenarios of GMSL rise, and then regionalized these global scenarios for the entire U.S. coastline, to serve as inputs into assessments of potential vulnerabilities and risks in the coastal environment.

The report's six representative scenarios were developed by starting with estimates of the probability of GMSL change and underlying contributing processes, conditional upon greenhouse-gas emissions pathways, represented by the RCPs (discussed in Chapter 1). The report notes that the collapse of the Antarctic ice sheet may be inevitable as surrounding waters warm, and that "recent modeling of physical feedbacks related to marine ice-cliff instabilities and ice-shelf hydrofracturing (rain and meltwater- enhanced crevassing and calving) used within the physical process models generating GMSL estimates are being incorporated into ice-sheet models."[69] With such feedbacks modeled for Antarctica, additional GMSL rise upwards of 0.6–1.1 m to median estimates under RCP 8.5 are possible, thus the growing evidence of ice loss from Antarctica and Greenland "only strengthens an argument for considering worst-case scenarios in coastal risk management."[70]

Figure 7. **Actual and Projected Average Annual Relative Sea Level (RSL) for Four U.S. Cities**

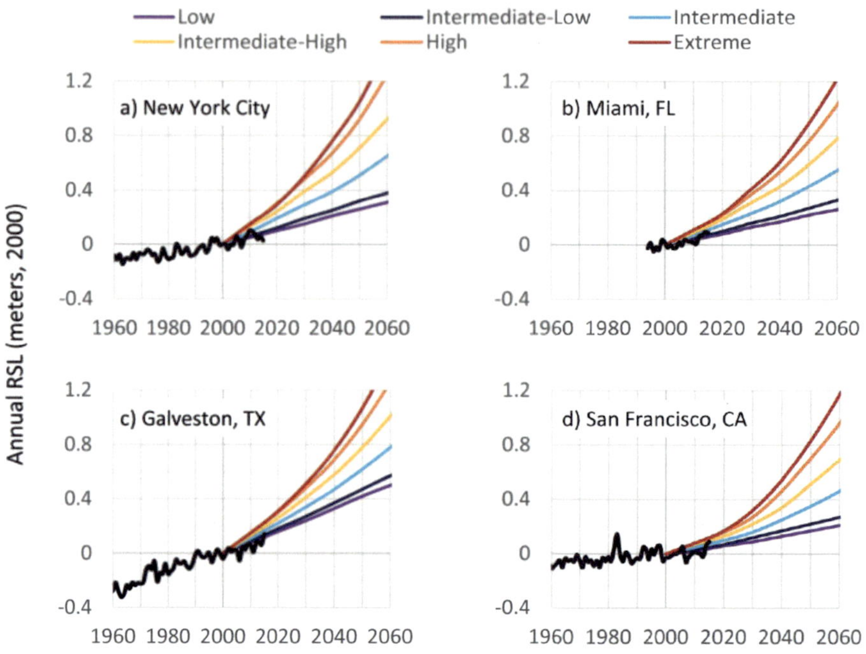

Average annual RSL for New York City (The Battery), Miami (Virginia Key), FL, Galveston, TX, and San Francisco, CA. with their respective (median-value) RSL under the six scenarios. The NOAA RSL observations (tidesandcurrents.noaa.gov/sltrends) are shown relative to the midpoint (year 2000) of the 1991–2009 epoch (1994–2009 at Virginia Key), which is the reference level for the scenarios.

From: NOAA report, *Global and Regional Sea Level Rise Scenarios for the United States*, p. 36.

Figure 7 shows observed annual mean relative sea level (RSL) for four U.S. cities and future RSL under six representative scenarios, which range from a low-end (Low) scenario of 0.3 m to a worst-case (Extreme) scenario of 2.5 m by 2100, as follows:

- Low (0.2m),
- Intermediate-Low (0.5 m),
- Intermediate (1.0 m),
- Intermediate-High (1.5 m),
- High (2.0 m),
- Extreme (2.5 m).

To date, in most circumstances, the range of interannual RSL change/variability since 2000 "has been bounded by the trajectory of the Intermediate-High (1.5-m) scenario."[71] However, it is also important to recognize that "being close to one scenario early in the century does not imply that real-world behavior will follow that scenario throughout the century; for example, rapid ice sheet collapse in Antarctica could conceivably bring the world from the Intermediate scenario early in the century to the Extreme scenario by the end of the century."[72]

Figure 8. **Locations of NOAA National Water Level Observation Network Stations**

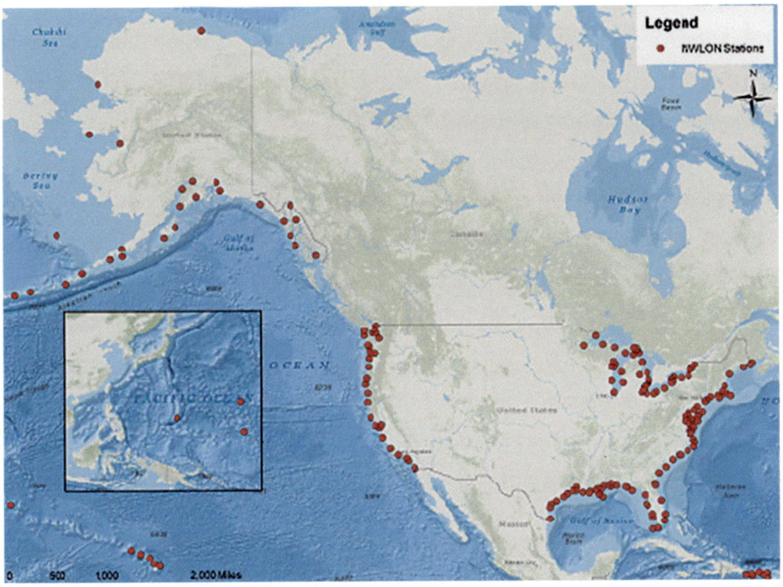

NOAA and its predecessors have gathered oceanographic data along our nation's coasts for over 200 years to protect life, property, and the environment. NOAA monitors the sea levels along the U.S. coasts with the National Water Level Observation Network (NWLON), a permanent observing system of water level stations.

Source: NOAA.

There are a host of simulation models, mapping tools, and analytical techniques that are available to predict potential impacts of sea-level rise. The U.S. Army Corps of Engineers (USACE) Sea Level Change Curve Calculator is a widely accepted web-based modeling tool for SLR predictions that was initially developed to assess the vulnerability of USACE projects. It computes three curves or sea level rise scenarios: low, intermediate, and high. The Sea Level Change Calculator can be used on any locations along the U.S. coast that are part of the NOAA National Water Level Observation Network (NWLON)[73] (Figure 8, above).

The Coastal Storm Modeling System (CoSMoS) is another dynamic modeling system, developed by the United States Geological Survey, which provides detailed predictions of coastal flooding by using sea level rise and storm surge (SLR/SS) data and projections. According to the report, *Rising Seas in California, An Update on Sea-Level Rise Science*, "CoSMoS has now been applied to most of the urbanized California coast (e.g., Southern California and the San Francisco Bay Area) and will soon cover the state's shoreline."[75] In September 2019, a collaboration of state and federal agencies launched *Sea the Future*, a decision support platform for California, to help coastal planners and land managers identify the best sea-level rise visualization tool for their project needs. It is a free online platform that compares sea-level rise modeling tools for the state of California.

ASCE provides the following guidance for engineers considering the use of climate projections:

- Match intended engineering application with appropriate climate model source;
- Recognize that different climate model sources require various amounts of effort for data extraction and preparation;
- Recognize that climate models have various levels of skill at representing historical mean and extreme statistical metrics, and engineers need to understand the major issues and uncertainties involved;
- Create an ensemble and be transparent about assumptions;
- Test robustness of designs to extremes and alternative scenarios;
- Discuss trade-offs and uncertainties in risk, resiliency, performance, and costs with stakeholders; and
- Design for low regret, adaptability, and robustness, and revisit designs when new information is available.[76]

Informed engineering judgment should be used to determine when to use projections of future climate conditions, and engineers should seek to use vetted tools with a documented use history for similar applications.

2.8 What Are a Design Professional's Obligations Under the Standard of Care?

Compliance with the standard of care usually requires more than mere compliance with public law. The design professional's scope of services, other contractual engagement terms, and the exercise of sound and prudent professional judgment are all relevant to compliance with the standard of care.[77] Current trends in climate change are projected to continue into the future, and the frequency, intensity, and duration of extreme weather events is expected to increase. The failure of a project design to account for climate change may result in significant professional liability risk for the design professional and liability risk for its client. If application of the standard of care demonstrates that the design professional's understanding of applicable codes, regulations and data is outdated, an engineer's failure to consider and anticipate future climate impacts in their designs may constitute professional negligence. Project owners expect to be in a position to make informed decisions about resiliency issues and choices, and increasingly, design professionals are expected to be qualified and experienced in their ability to advise clients regarding design options.[78]

Given scientific knowledge that our climate is rapidly changing, design professionals should understand the potential impacts of climate change on a project. Designers can perform vulnerability analyses and risk assessments and inform their clients of conditions that may cause harm, damage, or loss of service during the life cycle of the infrastructure. This may entail hiring climate experts or engineering consultants. Effective risk management strategies include informing clients about climate change risks, and providing design choices and options that will allow the client to make informed decisions that reduce risk, despite uncertain knowledge about the future climate. Ultimately, the owner makes the decisions that impact project design and determines the risks they are willing to accept, but design professionals should lead in the process by recommending low regret strategies.

When building codes, zoning ordinances, regulations, and public data such as flood plain maps have not caught up with the contemporary state of the art and do not reflect the best available data about climate change, design professionals may be uncertain about their obligations, and question what is required under the standard of care. The answer is that a design professional's failure to anticipate and adjust to climate change in their designs could expose them and their clients to legal liability. After Hurricane Harvey flooded tens of thousands of homes in the Houston area in 2017, a number of lawsuits were filed against developers and design professionals.

2.8.1 **The Standard of Care Is Evolving**

The lawsuit, *Alexander v. The Woodlands Development Company*, filed by 487 residents of Timarron Park in Harris County, Texas, illustrates the risks of designing to conform with codes and regulations that provide only minimum standards, because they are based upon an unchanging or stationary climate. Following Hurricane Harvey, several feet of water filled the streets and homes of Timarron Park, a development of single-family homes. The residents sued the development team that planned and constructed their homes and the engineering company that designed the storm water drainage system, alleging claims for negligence, gross negligence, and violations of the Texas Deceptive Trade Practices Act (DTPA).

The tract of land included both a FEMA 500-year floodplain and a FEMA 100-year floodplain. All sections of the development were built outside of the 100-year floodplain, but some portions needed to be filled with dirt to be elevated above the 100-year flood levels. Moreover, the lawsuit alleges that in October 1994, before the defendants designed and developed Timarron Park, a catastrophic rain storm caused the San Jacinto River to rise by 28 feet and Spring Creek, a tributary which borders Timarron Park, to rise by over 10 feet, reaching flood levels approximating a 500-year flood.

The lawsuit, filed in the 215th Judicial District of Harris County, Texas, alleges that the heavy rainfall accompanying Hurricane Harvey in August 2017 was foreseeable, and that the defendants breached the standard of care by building the homes at an unreasonably low elevation relative to its risk of flooding, or by using antiquated data. Additionally, the claims allege that the engineer defendants designed a storm water management system that could not handle anticipated rainfall amounts. The Timarron Park litigation is still ongoing, and exemplifies the type of malpractice claims that can expose design professionals to liability, even if they comply with existing building codes and FEMA flood plain designations. When flooding is foreseeable, design professionals should account for it in their designs.

2.8.2 Risk Communication with Clients

The standard of care is evolving as the scientific knowledge of potential climate change impacts grows. Given the state of knowledge, it will be increasingly difficult to argue that unprecedented weather events were unforeseeable.

Relying upon and complying with existing zoning, building codes and other regulatory standards in a design may be unreasonable if historic climate information is outdated and does not reflect future climate conditions. Design professionals should conduct risk and vulnerability assessments that include steps for anticipating, identifying, evaluating, and prioritizing current and future climate risks and vulnerabilities, and then discuss options with clients for reducing these risks. Risk communication with clients is an important step in managing and reducing risk because it enhances transparency and fosters defensible decision-making. When advising clients about climate risks and the need for adaptive measures, design professionals should document their advice. Ultimately, the higher costs involved in an enhanced, resilient design should be weighed against the heightened risk of liability for failing to do so.

Best-practice designs will incorporate features to increase resilience to extreme weather events and natural disasters. Depending upon locale, this could mean employing wind-resistant, seismic-resistant, or thermal-resistant materials. For example, Miami Beach is planning a $500 million flood control project that will raise the level of its roads to reduce current flooding from sea level rise and high tides, and is grappling with how high the new structures must be, anticipating that sea levels will be even higher in the future.[79]

2.9 Governance and Policy-Making

Despite lofty goals, it is unlikely that the target of net zero building stock by 2030 or even 2050 can be met without stronger governance, new policies, and more stringent building codes and emissions regulation. All construction sector stakeholders have an important role to play in driving the transformation to decarbonization, and delivering more resilient infrastructure. Design professionals can hasten progress to a more resilient building sector by participating in the development of stronger building codes, with mandatory performance standards, as well as initiating and supporting legislative actions and government policies that address climate change.

2.10 Conclusion

Humanity has entered a crucial decade. Our actions now will set a path for the climate that will shape the environment in which our descendants will live for centuries. We have reached a pivotal junction, where all industry stakeholders, including public owners, private developers, architects, all engineering fields, contractors, and suppliers will be needed to steer the building sector to a new paradigm of resilient design and carbon-neutral building systems.

Notes

1 IPCC, 2018: *Global Warming of 1.5°C. An IPCC Special Report on the impacts of global warming of 1.5°C above pre-industrial levels and related global greenhouse gas emission pathways, in the context of strengthening the global response to the threat of climate change, sustainable development, and efforts to eradicate poverty* [Masson-Delmotte, V., P. Zhai, H.-O. Pörtner, D. Roberts, J. Skea, P.R. Shukla, A. Pirani, W. Moufouma-Okia, C. Péan, R. Pidcock, S. Connors, J.B.R. Matthews, Y. Chen, X. Zhou, M.I. Gomis, E. Lonnoy, T. Maycock, M. Tignor, and T. Waterfield (eds.)].

2 *Ibid.*, p. 95.

3 Global Alliance for Buildings and Construction, International Energy Agency and the United Nations Environment Programme (2019), 2019 Global Status Report for Buildings and Construction: Towards A Zero-Emission, Efficient and Resilient Buildings and Construction Sector, pp. 3, 9.

4 *Ibid.*

5 Global Alliance for Buildings and Construction, International Energy Agency and the United Nations Environment Programme (2020): 2020 Global Status Report for Buildings and Construction: : Towards A Zero-Emission, Efficient and Resilient Buildings and Construction Sector, p. 27.

6 *Ibid.*, p. 28.

7 *Ibid.*

8 *Ibid.*

9 Heating and Cooling Demonstration Act of 1974 (P.L. 930-409). Available online at http://www.gpo.gov/fdsys/pkg/STATUTE-88/pdf/STATUTE-88-Pg1069.pdf.

10 World Commission on Environment and Development, *Our Common Future*. New York: Oxford University Press, 1987.

11 *Ibid.*, p. 22.

12 Building, Design & Construction, "White Paper on Sustainability: A Report on the Green Building Movement," November, 2003. Available online at https://www.bdcnetwork.com/sites/bdc/files/BD%2BC%202003%20White%20Paper%20on%20Sustainability.pdf.

13 Patricia B. Gary, "The Future of Megaprojects," ch. 13 in *Megaprojects: Challenges and Recommended Practices*, D. Corkum and D. Hatem, eds., Washington, D. C.: American Council of Engineering Companies, 2010, p. 275.

14 *Ibid.*, p. 254.

15 World Green Building Council, "Bringing Embodied Carbon Upfront: Coordinated Action for the Building and Construction Sector to Tackle Embodied Carbon," September 27, 2019, p. 7. Available online at https://www.worldgbc.org/sites/default/files/WorldGBC_Bringing_Embodied_Carbon_Upfront.pdf.

16 U.S. Environmental Protection Agency, "Greenhouse Gases at EPA," https://www.epa.

gov/greeningepa/greenhouse-gases-epa; see also General Services Administration, Executive Order 13514 Section 13: Recommendations for Vendor and Contractor Emissions, April 2010,

17 Paula Melton, "The Urgency of Embodied Carbon and What You Can Do about It," *Building Green*, vol. 27, issue 9, September 10, 2018.

18 World Green Building Council, "Bringing Embodied Carbon Upfront," p. 27.

19 *Ibid.*, p. 28.

20 *Ibid.*, p. 36.

21 Chelsea Harvey, "Cement Producers Are Developing a Plan to Reduce CO_2 Emissions," E&E News, *Scientific American*, July 9, 2018. Available online at https://www.scientificamerican.com/article/cement-producers-are-developing-a-plan-to-reduce-co2-emissions/.

22 S. Samad and A. Shah, "Role of Binary Cement Including Supplementary Cementitious Material (SCM), in Production of Environmentally Sustainable Concrete: A Critical Review," *International Journal of Sustainable Built Environment*, vol. 6, issue 2, p. 663-674, December 2017. Available online at https://www.sciencedirect.com/science/article/pii/S2212609017300225.

23 Steven Davis, Nathan Lewis, Matthew Shaner, Sonia Aggarwal, Doug Arent, and Ines Azevedo, "Net-zero Emissions Energy Systems," *Science*, vol. 360, issue 6396, eaas9793, June 29, 2018. Available online at https://science.sciencemag.org/content/360/6396/eaas9793.

24 University of Leeds, Arup, and C40 Cities, "Building and Infrastructure Consumption Emissions in Focus," August 2019. Available online at https://c40-production-images.s3.amazonaws.com/other_uploads/images/2390_BIC_Report_FINAL.original.pdf?1570401598.

25 Carbon Leadership Forum and SEI Sustainability Committee, "Structural Engineers 2050 Commitment Initiative." Available online at https://carbonleadershipforum.org/what-we-do/initiatives/se2050/.

26 "Embodied Carbon in Building Materials for Real Estate," Urban Land Institute Greenprint, p. 11, www.americas.uli.org.

27 House Select Committee on the Climate Crisis, 116th Congress, "Solving the Climate Crisis, the Congressional Action Plan for a Clean Energy Economy and a Healthy, Resilient, and Just America," June 2020, p. 23–25 https://climatecrisis.house.gov/sites/climatecrisis.house.gov/files/Climate Crisis Action Plan.pdf.

28 UCLA Luskin Center for Innovation, "Progress Toward 100% Clean Energy in Cities and States Across the U.S.," Nov. 2019, p. 6. Available online at https://innovation.luskin.ucla.edu/wp-content/uploads/2019/11/100-Clean-Energy-Progress-Report-UCLA-2.pdf.

29 Executive Office of Energy and Environmental Affairs, "Massachusetts 2050 Decarbonization Roadmap," December 2020, p.7, n.1. Available online at https://www.mass.gov/info-details/ma-decarbonization-roadmap.

30 Ibid, p.7.

31 Deep Decarbonization Pathways Project, https//www.iddri.org/en/project/deep-decarbonization-pathways-project.

32 "Massachusetts 2050 Decarbonization Roadmap," December 2020, p. 57.

33 Ibid, p. 23.

34 Ibid, p. 44.

35 Boston Green Ribbon Commission, Carbon Free Boston Summary Report 2019, p. 46-47.

36 UCLA Luskin Center for Innovation, "Progress Toward 100% Clean Energy," p. 1.

37 IPCC, 2014: Summary for Policymakers. In: Climate Change 2014: Mitigation of Climate Change. Contribution of Working Group III to the Fifth Assessment Report of the Intergovernmental Panel on Climate Change, p. 21.

38 IRENA, "Future of Solar Photovoltaic: Deployment, investment, technology, grid integration and socio-economic aspects (A Global Energy Transformation: paper), International Renewable Energy Agency, Abu Dhabi (2019).

39 Goldman School of Public Policy, University of California, Berkeley, "The 2035 Report," June 2020,

40 C.S. Holling, "Resilience and Stability of Ecological Systems," Annual Review of Ecology and Systematics, Vol. 4, 1973.

41 Roger N. Jones, Paul Dettmann, Geoff Park, Maureen Rogers, and Terry White, "The relationship between adaptation and mitigation in managing climate change risks: a regional response from North Central Victoria, Australia," Mitig Adapt Strat Glob Change (2007) 12:685–712 DOI 10.1007/s11027-007-9094-5.

42 Bilal M. Ayyub, Robert E. Chapman, Gerald E. Galloway, Richard N. Wright, "Economics of Community Disaster Resilience Workshop Proceedings," NIST Special Publication 1600, National Institute of Standards and Technology, February 2016, p. xxiv.

43 OECD, Climate-resilient Infrastructure, Part 1: "Defining Climate-resilient infrastructure," OECD Environmental Policy Paper No. 14 (2018), http://www.oecd.org/environment/cc/policy-perspectives-climate-resilient-infrastructure.pdf.

44 Maxwell, D., Constas, M., Frankenberger, T., Klaus, D., and Mock, M. *Qualitative Data and Subjective Indicators for Resilience Measurement*. Resilience Measurement Technical Working Group. Technical Series No. 4. Rome: Food Security Information Network, September 2015, p. 5. http://www.fsincop.net/fileadmin/user_upload/fsin/ docs/resources/FSIN_TechnicalSeries_4.pdf.

45 Giovanni Sansavini, Paper 34, "Engineering Resilience in Critical Infrastructures," p. 207. Florin, M.-V., and Linkov, I. (Eds.). (2016). *IRGC Resource Guide on Resilience*. Lausanne: EPFL International Risk Governance Center (IRGC). DOI:10.5075/epfl-irgc-228206 p. 207.

46 Chapter 7, "Adaptive Design and Risk Management," Climate-Resilient Infrastructure and, Bilal M. Ayyub, ed., ASCE 2018, p. 173.

47 Committee on Adaptation to a Changing Climate, Chapter 7, "Adaptive Design and Risk Management," Climate-Resilient Infrastructure, Bilal M. Ayyub, ed., ASCE 2018, p. 176.

48 Michel Bruneau, Stephanie E. Chang, Ronald T. Eguchi, George C. Lee, Thomas D. O'Rourke, Andrei M. Reinhorn, Masanobu Shinozua, Kathleen Tierney William A. Wallace, and Detlof von Winterfeldt, "A Framework to Quantitatively Assess and Enhance the Seismic Resilience of Communities," *Earthquake Spectra*, Volume 19, No. 4, November 2003. p. 735.

49 Kathleen Tierney and Michel Bruneau, "Conceptualizing and Measuring Resilience, A Key to Disaster Loss Reduction," *TR News* 250, May–June 2007, pp. 14–15. https://onlinepubs.trb.org/onlinepubs/trnews/trnews250_p14-17.pdf.

50 Michel Bruneau, et al., "A Framework to Quantitatively Assess and Enhance the Seismic Resilience of Communities," p. 735.

51 Bilal Ayyub, "Practical Resilience Metrics for Planning, Design and Decision Making." *ASCE-ASME J. Risk Uncertainty Eng. Syst., Part A: Civil Eng.* 1(3) (2015), p. 04015008-4, https://doi.org/10.1061/AJRUA6.0000826;

52 Committee on Adaptation to a Changing Climate, Chapter 7, "Adaptive Design and Risk Management," *Climate-Resilient Infrastructure*, Bilal M.Ayyub, ed., ASCE 2018, p. 180-181.

53 Bilal M. Ayyub, "Practical Resilience Metrics for Coastal Infrastructure Features," U.S. Army Corps of Engineers, Engineer Research and Development Center, May 2019, pp. 27–28.

54 Chapter 7, "Adaptive Design and Risk Management," *Climate-Resilient Infrastructure,* Bilal M.Ayyub, ed., ASCE 2018, p. 182.

55 David Levy and Rebecca Herst, Sustainable Solutions Lab, University of Massachusetts Boston, *Financing Climate Resilience*, p. 8, April 2018.

56 EPA, "Climate Change in the United States, Benefits of Global Action," United States Environmental Protection Agency, Office of Atmospheric Programs, p. 7 (2015).

57 S&P Global Ratings, *COP24 Special Edition: Shining a Light on Climate Finance*, December 2018.

58 Chapter 28, "Adaptation, KM2: Climate Change Outpaces Adaptation Planning," *Fourth National Climate Assessment*, November 2019; https://nca2018.globachange.gov/chapter/28/.

59 *Ibid.*

60 Identification and Mapping of Special Hazard Areas, 44 C.F.R. § 65.6 (a) (3). https://www.gpo.gov/fdsys/ granule/CFR-2011-title44-vol1/CFR-2011-title44-vol1-sec65-6

61 U.S. Department of Homeland Security, Office of Inspector General, OIG-17-10, "FEMA Needs to Improve Management of Its Flood Mapping Programs," p. 2, September 27, 2017.

62 Al Shaw, Christie Thompson & Theodoric Meyer, "Federal Flood Maps Left New York Unprepared for Sandy – and FEMA Knew It," Dec. 6, 2013.

63 Michael Keller, Mira Rojanasakul, David Ingold, Christopher Flavelle & Brittany Harris, "Outdated and Unreliable: FEMA's Faulty Flood Maps Put Homeowners At Risk," *Bloomberg*, October 6, 2017.

64 Ramin Skibba, "Why Are FEMA's Flood Maps So Horribly Flawed," *Science*, Sept. 6, 2017.

65 Kirk Bosma, Ellen Douglas, Paul Kirshen, Katherine McArthur, Stephen Miller and Chris Watson, "MassDOT-FHWA Pilot Project Report: Climate Change and Extreme Weather Vulnerability Assessments and Adaptation Options for the Central Artery," June 2015.

66 Climate Resilience Design Standards & Guidelines, Section 1: Project Overview, Draft Document for Public Comment, August 24, 2020, p. 3.

67 Climate Resilience Design Standards & Guidelines, Section 3: Draft Climate Resilience Design Standards Overview, Draft Document for Public Comment Period, August 24, 2020, p. 4.

68 NOAA, "Global and Regional Sea Level Rise Scenarios for the United States," NOAA Technical Report NOS CO-OPS 083, January 2017, p. 1.

69 *Ibid.*, p. 13.

70 *Ibid.*, p. 14.

71 *Ibid.*, p. 35.

72 *Ibid.*, p. 35.

73 *Sea-Level Rise Vulnerability Assessment Tools and Resources: A Guide for Florida's Local Governments*, June 2015.

74 Ashley Miller and Audra Luscher, "NOAA's national water level observation network (NWLON)," *Journal of Operational Oceanography*, November 5, 2018.

75 G. Griggs, J. Árvai, D. Cayan, R. DeConto, J. Fox, H. Fricker, R. Kopp, C. Tebaldi (Working Group). "Rising Seas in California: An Update on Sea-Level Rise Science," California Ocean Science Trust, April 2017, p. 37.

76 Bilal M. Ayyub, "Data and Information Sources," ch. 8 in *Climate-Resilient Infrastructure*, Bilal M. Ayyub, ed., Washington, D.C.: ASCE, 2018, p. 228.

77 David Hatem and Patricia Gary, "Climate Change, Sustainability and Resiliency Considerations in Design: Professional Liability Risk Issues for Consulting Engineers," *ACEC Insights*, American Council of Engineering Companies, Fall 2017.

78 David J. Hatem, "Green and Sustainable Design, Part I: Professional Liability Risk and Insurability Issues for Design Professionals," *Design and Construction Management Professional Reporter*, June 2010.

79 Alex Harris, "Raising Flood-Prone Roads Has Angered Miami Beach Residents. Experts Say They Need to Go Higher," *Miami Herald*, January 22, 2020. Available online at https://www.miamiherald.com/news/local/environment/article239486308.html.

3

Climate Change and Its Relevance to the Design Professional

Lisa L. Churchill

3.0 Introduction

As design professionals, we have become comfortable drawing heavily on past events to inform the work we do today. This is especially true when it comes to the consideration of climate. Until recently, relying on historic meteorological and climate data to inform design decisions was considered best practice. Significant work has been invested to create engineering algorithms, environmental regulations, planning considerations and even policies that incorporate past data into design calculations and decision-making criteria. The assumption had been that there was a predictable consistency to yearly and seasonal temperatures and precipitation amounts, the intensity and frequency of storm events, the probability of particular storm events occurring each year, and even the expected minimal and maximal ranges across a variety of climatic indicators. While there was expected yearly variation, there was no particular "directionality" associated with that variation. When averaged over decades, that variation centered around a predictable mean based on past observations.

Said another way, the 1-in-100 (or 1%) storm would remain the same indefinitely. This was particularly convenient for our industry, as we are often involved in designing assets and systems that will last for 30 years or more. In essence, the assumption was that climate risk would remain constant and quantifiable throughout the life expectancy of a building, bridge, transportation facility—whatever project that was being planned or designed. That allowed climate impacts to be both knowable and predictable over the lifespan of the modern built environment and something that could be easily accounted for and solved in a permanent way.

However, in the past decade, it has become apparent that the degree of climate predictability has been shifting—that in fact what scientists are observing is a shift in those baselines that used to inform our basis of design. What has happened in the past may no longer be a reliable proxy of future conditions—especially when looking several decades out. How do we adapt?

3.1 **The 10,000-year Hall Pass**

For the last 10,000 years, civilizations have evolved during a period of relative climatic stability—a 10,000 year hall pass, if you like.[1] While there have always been yearly variations in weather, the overall trends in temperature and other associated parameters maintained a significant degree of predictability and "normalness"—especially when grouped statistically across decades and centuries. However, that period of predictability is coming to a close. We are currently entering a new phase

Figure 1. **Temperature Variation During the Past 150,000 Years**

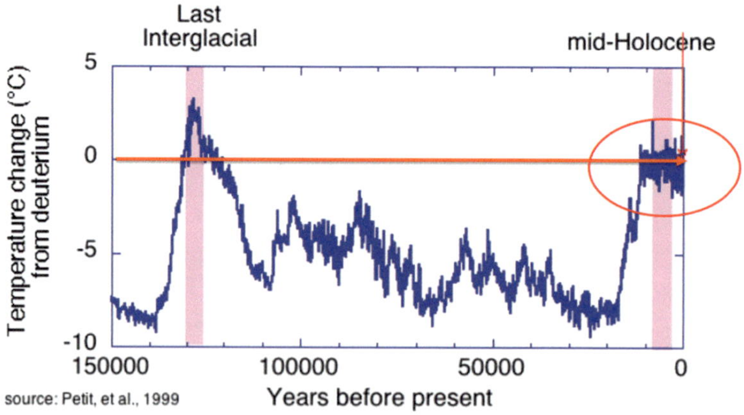

Note the unusual period of climate stability which has occurred during the last 10,000 years. This epoch, known as the Holocene, extends to the present day and is the time over which major civilizations evolved and humans gained dominion over the natural environment.

Source: Petit, Jean-Robert, Jouzel, Jean, Raynaud, Dominique, Barkov, N., Barnola, J.-M, et al. (1999). "Climate and atmospheric history of the past 420,000 years from the Vostok Ice Core." *Nature*. 399. 429. Graph alone available online at http://ossfoundation.us/projects/environment/global-warming/natural-cycle.

where variability is becoming more commonplace and extreme events, once hailed as rare tail-end members, are starting to occur with greater frequency. The challenge becomes how to accommodate those changes within a world where permanence, individualized property rights, and a resource-intensive way of life have become the hallmarks of "modern" society.

During the last 10,000 years, we have seen a shift from nomadic lifestyles to civilizations with permanent settlements and an expectation of a "fixed-in place" style of living with fully renewable resources. We have created significant infrastructure to support this lifestyle and a complexity and dependence on resources that are now being further stressed by increased climatic variability.

Scholars have been able to highlight, within this period of relative global climate stability, localized periods of significant climate stress that both shaped how human cultures evolved[2] and sometimes contributed to the demise of significant civilizations.[3] In fact, the extent of extinction patterns observed today has been compared to other past extinctions, and while the emphasis is on the larger ecosystem, a key culprit that has been identified is climate change.[4] Such a major shift in our natural environment, with its supporting resources and ecosystem services, will translate directly into how we as a society—and more specifically, as experts in the built environment—become more proactive in both projecting and designing for climate resilience.

3.2 Climate Change: No Longer an Irrelevant Externality

The effects of climate change are farther reaching than just their direct physical impacts. Along with accepting the increasing intensity and frequency of extreme weather events (climate shocks), accompanied by shifts in other climate variables such as temperature and sea level rise (climate stressors), there has been the recognition that these changes will also have implications for other aspects of our society, including both immediate financial shocks and longer-term economic ramifications. Many of these impacts are indirect and diffuse and occur at levels and scales that are difficult to track, and, therefore, may show up much later than the actual event itself. What was once a seemingly irrelevant externality is becoming more of a deciding factor in how infrastructure is planned, designed, and funded.

3.3 The Costs of Climate Change

One of the reasons that climate change is becoming more visible is that it is becoming more expensive. Since 1980, NOAA has been tracking the economic impact of extreme weather events for the U.S. At the end of each year, the agency has reported out on the number of events which exceeded a billion dollars in losses. In a recent review,[5] NOAA reported an increase in both the number of severe storms in recent years and the overall financial and economic costs associated with each.

Complicating the increase in extreme weather events is the growing propensity for people and businesses to settle in the areas that are vulnerable to these types of

Figure 2. **Number of Billion-Dollar Natural Disaster Events in the United States, 1980–2020**

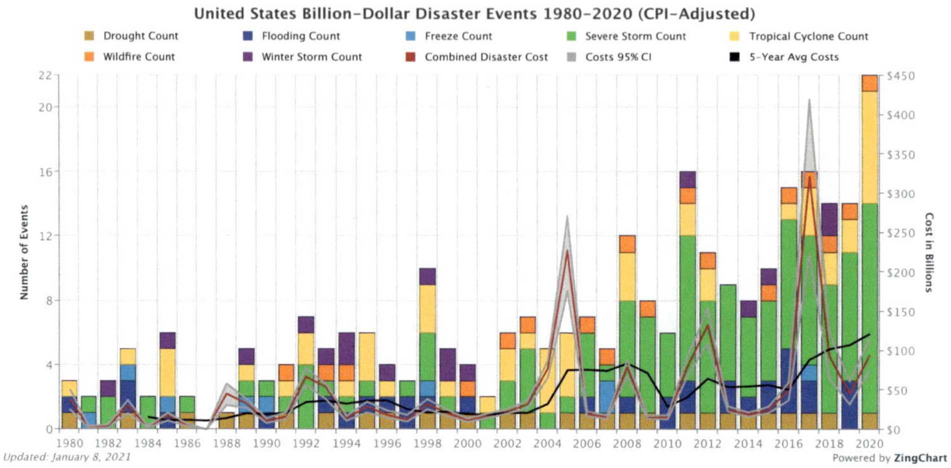

Source: NOAA National Centers for Environmental Information (NCEI) U.S. Billion-Dollar Weather and Climate Disasters (2021). https://www.ncdc.noaa.gov/billions/, DOI: 10.25921/stkw-7w73

Figure 3. **Hurricanes That Caused $10 Billion in Damage in the United States, 2004–Present**

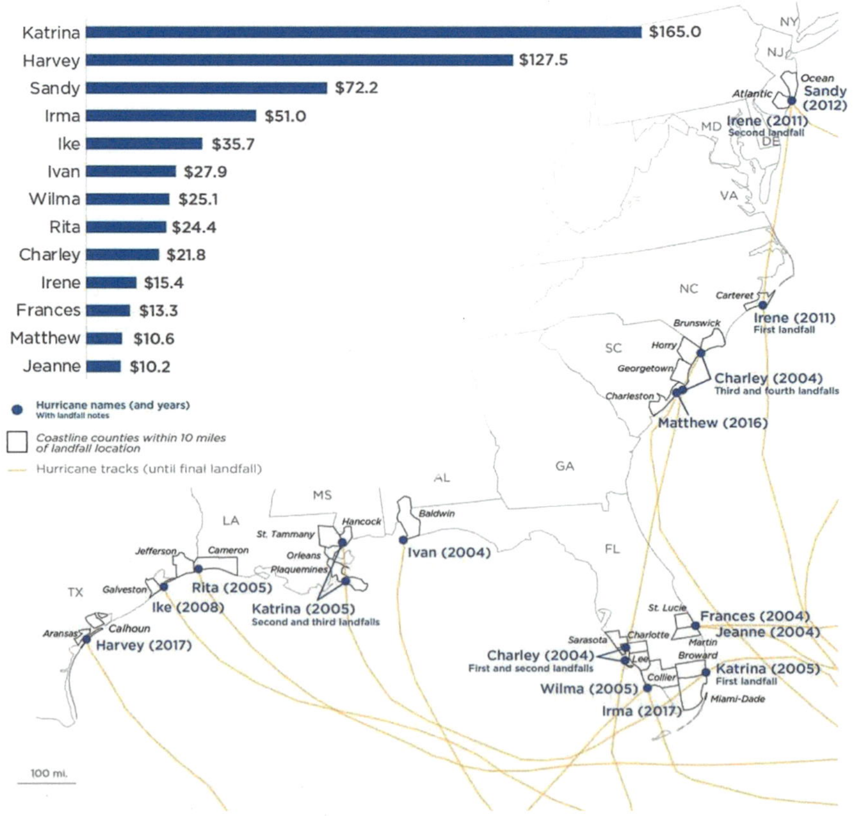

Source: https://www.census.gov/content/dam/Census/library/stories/2019/07/millions-of-americans-live-coastline-regions-graphic-2.jpg

Figure 4. **Projected Migration (Inflow and Outflow) Due to Sea Level Rise**

Top eleven outflow states

Florida	California	North Carolina	South Carolina
Texas	Louisiana	Georgia	Virginia
New York	New Jersey	Massachusetts	

Source: M. Hauer, 2017. *Nature Climate Change* 7(5). DOI: 10.1038/nclimate3271

events, and this is especially true for coastal communities. In the United States, 40% of people now live within coastal communities.[6] The Gulf of Mexico region, which is particularly prone to hurricanes, has seen a 26% growth in population in the last 17 years, compared to the U.S. average of 15% growth during the same period.[7] This means that not only are those areas experiencing a higher frequency of intense storm events, but with more people settling in these areas, there are more people directly affected by each event. As of 2020, there are still insufficient regulatory, underwriting, or financial incentives (or disincentives) to encourage different development patterns. Climate risk is still not being adequately priced—something that will be explored more in Chapter 4, "Climate Change and Insurance: Evolving Trends."

This lack of pricing in current times is creating an ever-increasing economic and financial burden for future valuations. Several studies have come out in recent years to assess the cost of climate impacts to real estate and public holdings. The collaboration between the Union of Concerned Scientist and Zillow resulted in a report that contained a dire outlook for the coastal real estate market. By 2100, the authors projected there could be $1 trillion in properties threatened by sea level rise. Of the roughly 14,000 commercial properties at risk on U.S. coasts within the next 30 years, more than one-third are located in Florida and New Jersey.[8]

There will come a point that the burden of rebuilding and further development in vulnerable spaces is no longer financially or economically feasible. This will likely lead to significant shifts in demographic and economic centers within the foreseeable future. Understanding these trends is key to assessing how the built environment will need to be modified to accommodate what could be significant migration pressures within the coming decades. Figure 4 illustrates one example of the work that is currently being done to project how climate change (in this case, sea level rise) could lead to both intrastate and interstate population relocations within the U.S.

3.4 Who "Owns" the Risk of Climate Change?

There is a well-known disconnect between the planning horizons of infrastructure owners and those of investors. Life expectancies of built assets and systems easily span decades, whereas investment horizons are often a year or less, with a three- to five-year planning horizon being seen as "long term" within the financial sector. Insurers and rating agencies have similarly short horizons, with both entities setting – and resetting – risk calculations year over year. This disconnect has been dubbed the "tragedy of the horizons"[9] and is one of the most fundamental barriers to large-scale investment in resilient infrastructure.

By resetting the risk annually, we fail to capture the true value of resilience. If, instead, that risk were calculated cumulatively over the expected lifecycle of a particular asset, we would see that a 1% (or 1-in-100-year) flood event actually has a 26% chance of happening over 30 years. The probability of that event increases to a 39% chance over 50 years. Describing the risk in this way would likely cause owners and operators to reassess how that risk is managed within their portfolio.

Additionally, by limiting planning horizons to yearly assessments, we also fail to capture a key aspect of the infrastructure budget, namely, the maintenance and operational costs. Commonly, a cumulative tally of those annual costs can easily outstrip the up-front capital costs of the asset or system itself. This further compounds the issue of discounted externalities.

Finally, we also have a very hard time tracking the longer-term cumulative impacts associated with these events. While there is funding available for easily recognizable physical damage from a natural disaster, it is often limited to the immediate repairs that are needed to get the system back in operation. It does not take into account the additional losses associated with a decrease in the life expectancy of an asset as a result of that event.

The markets, investors, underwriters, and even political horizons are often too short to capture these impacts. The true value of resilience is not captured in single, short-lived events but rather through the cumulative impact of those losses, including things like: shortened life expectancies of infrastructure, growing Department of Public Work budgets for repair/additional maintenance, the possibility of losing insurance or of insurance becoming too costly, the potential for federal and state funding for disasters to disappear, outmigration of taxpayers (residents and businesses) who move to less vulnerable areas to live, an increase in energy costs associated with cooling needs, the construction of large-scale mega projects (e.g., seawalls) to allow residents and businesses to remain in place versus moving away, financial disclosure requirements leading to divestment or devaluation, etc.

These are very real and very significant costs, but the current structure of our accounting systems and investment horizons does not allow them to be either adequately captured by or integrated into our larger economic or financial decision-making criteria. It is the combination of these oversights that causes the risk to be significantly underestimated and the true value of resilience to be discounted.

Despite this, there is still a growing awareness that climate change is expensive. And as the impacts and cost increase, there is an ever-growing focus on who actually "owns" that risk. In April of 2014, the greater Chicago area experienced significant flooding after heavy rains. The flooding was severe enough to result in the filing of hefty insurance claims across a number of areas. However, Farmers Insurance filed several class-action suits to challenge those claims. The company argued that the City of Chicago and other local governments were well aware of the potential for this level of flooding as they had recently completed a public climate action plan which stated as much. Farmers argued that the City was aware of those risks and failed to address them in a meaningful way. Based on that, Farmers declined the claims since they felt they could have been avoided had the City addressed the inadequacies in its sewer and storm drain systems. In the end, Farmers withdrew its challenge, but the initial act of filing was dubbed by many as a "shot across the bow" in what is anticipated to become a burgeoning legal field unto itself.[10]

As a whole, the insurance industry is deeply involved in discussions related to the risk allocation and climate risk. FM Global's recent white paper, "Master the Disaster: Why CFOs Must Initiate Natural Catastrophe Preparedness in 2019 and Beyond,"[11] points to CFOs as the responsible parties both for assessing the level of climate risk within their holdings and for allocating adequate capital to reduce the financial impact of natural disasters. Recent statistics released by the QBE Insurance Group noted that there has been both a tripling of extreme weather events since the 1980s and that "the economic costs from natural disasters have exceeded the 30-year average for seven of the last 10 years." Based on this, QBE will now be considering climate change as a material risk, since premiums could become too costly for clients living in highly vulnerable areas.[12]

The financial impacts of inadequate climate risk management are becoming more apparent. Case in point: the PG&E financial failure in 2019 has been labeled as the first climate-change bankruptcy of our time. Wildfires were a discounted risk that became the main driver in forcing a significant utility into bankruptcy. PG&E's bankruptcy is having other unintended consequence as well, including the potential insolvency of major renewable projects that were ultimately funded by PG&E via their ratepayers. It also created a controversy that pit the Federal Emergency Regulatory Commission (FERC) against the bankruptcy courts regarding who had the final say in how PG&E's business model was modified after the bankruptcy. Would it be restructured strictly from a financial perspective, with relation to PG&E's profile, or would it necessarily need to take into account the utility's role in public health and safety, which means it falls more under FERC jurisdiction?[13]

Additional legal challenges have arisen around whether or not the risk allocation is further translated to the level of the individual. In a post–Hurricane Harvey filing against the Arkema Chemical Plant, prosecutors claimed that individuals, not corporations, are responsible for making decisions. Therefore, the culpability for not adequately accounting for climate risks rests with the CEO and CFO, in addition to the company itself.[14] This introduced an interesting nuance to the assessment of climate risk. Would the potential for personal liability result in different climate risk management approaches, and what impacts might that have on the larger business model of an organization and its valuation from an investor perspective?

Legal aspects related to risk transfer and property rights are also being revisited. The ability to use force majeure triggers to cover projected climate impacts is starting to be questioned as it becomes more difficult to assert that these types of events are unpredictable. Likewise, the way that property rights are currently constructed "assumes that environment conditions at a certain location are relatively robust: soil, water levels, climate conditions, air quality, ecosystems are more or less stable ... Most property rights do not foresee amendments caused by environmental changes."[15] This will become a challenge as people are forced to migrate because of climate impacts. Detailed assessments of legal considerations from climate change are addressed in Chapter 5, "Climate Change Lawsuits: Trends and Analysis of Legal Actions," and Chapter 7, "Contracting Practices and Risk Management."

3.5 **Linking Climate Risk to Financial Assessments and Valuations**

In *The Climate Casino: Risk, Uncertainty and Economics for a Warming World,*[16] William Nordhaus, a Nobel Prize–winning economist, studied the economic effects of climate change. It is a revealing piece on why climate risks (and resilience) are not currently included in valuation assessments and the combination of socioeconomic reasons that have led to this. He points out that most people tend to think that climate change is a question for natural scientists to tackle. However, he argues that climate change is better addressed by social scientists and economists. In his view, people are the root of the issue and the source of the solutions. Human constructs, including how we assess value, have led to this crisis.

He also questions the ability of "markets [to] automatically solve the problems they generate" and that the "invisible hand" of markets cannot properly set prices without considering key externalities. It is his view that these externalities will only be seriously considered if they are regulated, and that global externalities (such as climate change) are particularly difficult to solve since there is no "workable or governmental mechanism to deal with them." This is an explanation for why we are still missing a strong market signal for climate resilience.

Barron's published a pithy essay on this topic: "It Still Makes Financial Sense to Burn the Globe."[17] In this, Georg Kell points to four key aspects that need to be changed in order to truly capture the economic and financial impact of climate risk.

1. The current short-term focus of the markets means that long-term risks have little implication for today's decision-making processes;
2. Reliance on past observations to inform current and future growth will not adequately account for the effects of climate change;
3. Current practices do not allow for pricing of future risk. Mark Carney, chair of the Financial Stability Board (2011–2018), has warned of a "Climate Minsky" moment because of this – referring to the work of economist Hyman Minsky whose analysis showed how banks overreached before the 2008 crisis; and,
4. Traditional economic theory, which suggests that price signals reflect all available information, isn't helpful in this case. Current market prices clearly do not account for the cost of carbon or the accompanying economic impacts that will increase as climate change continues.

Even within the public sector, there is an inability to track climate change at a financial level. At the city level, traditional fiscal health indicators used to assess a municipality's financial standing do not record the impact of climate change or extreme weather events. It is not that there are no significant costs incurred during those events, but rather the inability to track those costs is an artifact of how we currently collect and report the data. The costs themselves are not recorded in an easily accessible or standardized method.[18] By failing to capture these types of costs, the actual impacts are greatly underestimated. Without the ability to adequately quantify those losses, we cannot accurately price the value of avoided costs, and therefore it becomes nearly impossible to create a case for large-scale investment in resilience.

However, investors are beginning to integrate climate risk into their financial assessments. One prominent example of that change is the Task Force on Climate Financial Disclosure (TCFD). This effort, spearheaded by Mark Carney, governor of the Bank of England (2013–2020), and Michael Bloomberg, two central figures within the larger financial and investor world, has led to both an awakening and a larger reckoning among investors. It requires that asset owners and investors disclose the *financial* risks associated with the *physical* impacts of climate change. Physical climate impacts can no longer be treated as discounted externalities; they must be seriously considered within the larger assessment of financial health for any institution in both the private and public sectors. The disclosure requirement was originally aimed at asset owners and operators and boldly stated the potential for insolvency if climate risks were not addressed in a meaningful and immediate fashion. A recent report, "Task Force on Climate-related Financial Disclosures,"[19] noted that by 2100, there will be up to $43 trillion of value at risk from climate change. While such disclosure it is still voluntary, it is quickly becoming an industry standard and expectation. It is anticipated that it will develop into a standardized reporting protocol with similar fiduciary responsibilities as are currently in place with pension fund liability reporting requirements.

In addition to TCFD, there are other disclosure criteria, such as the Global Real Estate Sustainability Benchmark (GRESB), Environment, Social, and Corporate Governance (ESG), and even aspects of credit ratings, that are also looking to create more direct indicators between climate risk and the financial health and valuation of assets and systems. And the markets are starting to reflect that interest. In 2019, ESGs covered nearly $20 trillion in investments, the green bond market was valued at $100 trillion and there was $160 investment in climate bonds. And, on March 15, 2021, the U.S. Securities and Exchange Commission announced their interest in standardizing climate disclosure for publicly-owned companies, requesting public comment on the topic.

3.6 **Resilience and Uncertainty**

As experts of the built environment, our community, our clients, and the public at large expect us to fully the understand the larger environment in which we are designing. To best serve their needs, we must have a full appreciation of socioeconomic conditions and the larger natural world in which all of this exists. We need to be comfortable working at the intersection of all of these spheres and being ready for further change.

Climate change has interjected a significant degree of uncertainty into what, until very recently, was a more static and predictable natural ordering. Up until this time, there has been the assumption of stationarity with respect to climate—that while there is an expected variability in daily, monthly, and even annual climate phenomena (with a particular focus on temperature and precipitation), there would be no fundamental shifts in the patterns when viewed at decadal or longer time periods. In other words, while we might expect extreme weather events during any year, when we looked back at the overall trend, those extremes would average out and there would be no statistically significant shift. However, this is no longer the case; the recent past has shown a definite trend away from those historic means. This has forced us to reassess our past assumptions and create new methods that maintain the critical functions of a given asset, system, community, and to then design around that.

With the recognition of growing uncertainty within our systems, the industry transitioned from solely focusing on risk to centering our approach around resilience. The questions moved from what are the hazards that could negatively impact this system, to what are the key functions of the system and what are the potential disruptors that we might plan for (recognizing that some of those disruptors could present new opportunities in addition to challenges).

There has been significant past and ongoing research around resilience within the built environment, which has greatly advanced our knowledge and practices as design professionals. The Cities Resilience Index and Resilience Shift are two representative and comprehensive treatments on the subject.[20] For this work, we will adopt a slightly modified version of the Rockefeller Foundation's 100 Resilient Cities definition of resilience, viewing it through the lens of climate risk:

> Resilience is the capacity of individuals, communities, institutions, businesses and systems to survive, adapt and thrive in the midst of climate shocks and stressors.

Resilience is not strictly about managing risk. Instead the focus is on an understanding of the critical functionality of an entity or system, the larger "ecosystem" (built, natural, economic, and social) in which it operates, and an appreciation of its dependence on and influence over other systems. Resilience is about maintaining the functionality of critical operations in the midst of a variety of disruptors – of which climate change is only one.

3.7 **Becoming Comfortable with Uncertainty**

Our pivot towards resilience has come about with the recognition that the world itself is becoming less predictable. Climate and the increasing frequency and intensity of severe weather events is one of the major changes that is fueling uncertainty. As design professionals, we have to become comfortable with uncertainty and be able to train ourselves to consider events that may have been more readily discounted in the past.

In 2010, Nicholas Taleb published a book on the Black Swan theory, outlining the impact of highly improbable events on historical outcomes, as well as their potential to significantly shape future outcomes.[21] The Black Swan theory is especially applicable to climate change, where there is the considerable uncertainty and a lack of recent historic analogues by which to predict potential outcomes and behaviors.

Black swan

The opposite of a Black Swan is a phenomenon known as the Gray Rhino, a concept introduced by Michele Wucker (2016).[22] A "gray rhino" event is a highly probable, high-impact, but largely ignored threat. They are not random surprises, but rather, events that are quite knowable, occurring after a series of warnings and in the face of ample evidence. They are akin, in some ways, to the elephant in the room—something that everyone knows is there but for various reasons does not want to consciously address.

Rhinoceros

In truth, climate change can be thought of as both. It has elements of the Black Swan, in that there is a high degree of uncertainty regarding how far systems can be stressed,[23] and an inability to fully predict human behavior and political will necessary to change course. However, it is also a Gray Rhino. We have ample evidence of climate change and its recent impacts and enough data to identify trends, but we still refuse to address it in meaningful and transformational ways. Why is that?

To answer that, we need to invoke a third animal analogy. In 2017, Robert Meyer and Howard Kunreuther published *The Ostrich Paradox*.[24] They argued that people fail to act on key information and warning signs based on common behavioral short-cuts we have developed to process the complex environments in which we live. The tendencies include:

Ostrich

Myopia – focusing on short horizons
Amnesia – forgetting lessons of the past
Optimism – underestimating the likelihood of extreme events
Inertia – maintaining the status quo
Simplification – inability to focus on all the relevant facts
Herding – tending to follow our neighbors

The ability to think past these habits requires us to actively short-circuit what have otherwise become familiar and convenient thought pathways for us. Dealing with uncertainty means that we have to become comfortable with scenario planning to supplement more traditional probabilistic work, to bend a bit more towards possible futures, while also remaining grounded in current needs. It will require us to accept changing boundary conditions between the natural and built environments, and the resultant socioeconomic needs that will likewise be addressed by our work.

3.8 **Moving Forward**

Climate change has required us to rethink our previous assumptions of the natural world. As a profession, we are pivoting away from the expectation of stasis and predictability and towards more uncertain and dynamic conditions. It is an exciting time to be a design professional. We are being asked to create something new and improved—to design for a future that will be informed by sustainable and resilient practices.

As design professionals, our roles have always been more than just responding to specific proposal requests. As stewards of the built environment, we work closely with our partners to frame the larger picture, to identify key drivers of change, to hem in uncertainty, and to produce an integrated approach that is creative, responsive, and transformational. We have an obligation to make that practical and achievable.

We are reminded that this will call for more diverse partnerships. Viable solutions will require a deeper understanding and appreciation of the natural world, keen insights into governance and policy making, familiarity with market and investment drivers, and a commitment to equity and inclusion. This will require us to expand our traditional cadre of design professionals to include scientists, policy makers, legal experts, economists, finance specialists, social scientists, and others in deeper and more meaningful ways. There also needs to be innovation with our governance structures to allow for more regional and meaningful interventions, where risk and solutions are considered along natural and social boundaries (e.g., watersheds and communities), not arbitrary jurisdictional boundaries.

Climate change has revealed the lack of resilience in our current systems. The traditional business model of the built environment—how it is developed, funded, and maintained—is simply unsustainable. A backlog of deferred maintenance in every sector of infrastructure is evidence of that.[25] Climate change is adding further stress to that system. It is also providing us with an impetus and urgency to think more broadly, be more creative, and build something today that will shape a better world for this generation and the following ones.

Notes

1 Lisa Dickson, "The 10,000-Year Hall Pass," October 8, 2019. Available online at https://newcities.org/the-big-picture-the-10000-year-hall-pass/.

2 *See* Lewis Dartnell, *Origins: How Earth's History Shaped Human History*, New York Basic books, 2020 (in which the author describes the deep imprint of earth's physical history on how the human culture evolved).

3 *See* Elizbeth Kolbert, *Field Notes from a Catastrophe*, New York: Bloomsbury, 2015; and Kyle Harper, *The Fate of Rome*, Princeton, NJ: Princeton University Press, 2019 (on the role that climate change played in bringing about the fall of the Roman Empire).

4 *See* Elizabeth Kolbert, *The Sixth Extinction: An Unnatural History*, New York: Henry Holt and Company, 2014.

5 NOAA National Centers for Environmental Information, "U.S. Billion-Dollar Weather and Climate Disasters: Overview," available online at https://www.ncdc.noaa.gov/billions/.

6 NOAA Office of Coastal Management, "Fast Facts: Economics and Demographics," available online at https://coast.noaa.gov/states/fast-facts/economics-and-demographics.html.

7 Darryl Cohen, "About 60.2M Live in Areas Most Vulnerable to Hurricanes," United States Census Bureau, America Counts (blog), July 15, 2019, available online at https://www.census.gov/library/stories/2019/07/millions-of-americans-live-coastline-regions.html.

8 Kristy Dahl, "Hurricane Michael Threatens Gulf Coast Homes and Military Bases," Union of Concerned Scientists (blog), October 9, 2018, available online at https://blog.ucsusa.org/kristy-dahl/hurricane-michael-threat-to-military-and-gulf-coast.

9 Carney, M. "Breaking the Tragedy of the Horizon: Climate Change and Financial Stability." Speech by Mark Carney, Governor of the Bank of England and Chairman of the Financial Stability Board, at Lloyd's of London, London, September 29, 2015.

10 Alexis Petru, "Farmers Insurance Drops Climate Change Lawsuits Against Chicago-Area Cities," *TriplePundit*, June 10, 2014 available online at https://www.triplepundit.com/story/2014/farmers-insurance-drops-climate-change-lawsuits-against-chicago-area-cities/43196.

11 "Master the Disaster" is available as a PDF via a link on this page: https://www.fmglobal.mobi/insights-and-impacts/2019/master-the-disaster?csrt=12575847481551753443.

12 "Climate change could make premiums unaffordable: QBE Insurance," Reuters,

February 17, 2020, available online at https://www.reuters.com/article/us-climate-change-qbe-ins-grp/climate-change-could-make-premiums-unaffordable-qbe-insurance-idUSKBN20B0DA.

13 Russell Gold, "PG&E: The First Climate-Change Bankruptcy, Probably Not the Last," *Wall Street Journal,* January 18, 2019, available online at https://www.wsj.com/articles/pg-e-wildfires-and-the-first-climate-change-bankruptcy-11547820006.

14 Christopher Mele, "Chemical Maker and Its Chief Indicted for Explosions During Hurricane Harvey," New York Times, August 3, 2018, available online at https://www.nytimes.com/2018/08/03/business/arkema-chemical-plant-explosion-texas.html.

15 Fennie van Straalan, Thomas Hartmann, and John Sheehan, eds., *Property Rights and Climate Change: Land Use Under Changing Environmental Conditions,* New York: Routledge, 2018.

16 William Nordhaus, *The Climate Casino: Risk, Uncertainty and Economics for a Warming World,* New Haven, CT: Yale University Press, 2013.

17 Georg Kell, "It Still Makes Financial Sense to Burn the Globe," *Barrons*, March 21, 2019, available online at www.barrons.com/articles/climate-change-is-still-too-cheap-esg-investing-51553104880.

18 Urban Land Institute, *Climate Risk and Real Estate: Emerging Practices for Market Assessment*, Washington, D.C.: Urban Land Institute, 2020.

19 "Task Force on Climate-related Financial Dislosures," available online at https://www.fsb-tcfd.org/wp-content/uploads/2020/03/TCFD_Booklet_FNL_Digital_March-2020.pdf.

20 Cities Resilience Index, online at https://www.cityresilienceindex.org/#/ , and Resilience Shift, online at https://www.resilienceshift.org/.

21 Taleb, N., *The Black Swan: The Impact of the Highly Improbable*, New York: Random House Trade, 2007.

22 Michele Wucker, *The Gray Rhino: How to Recognize and Act on the Obvious Dangers We Ignore*, New York: St. Martin's Press, 2016.

23 For more on this topic, *see* "Planetary Boundaries Research," Stockholm Resilience Centre, Stockholm University," available online at www.stockholmresilience.org/research/planetary-boundaries.html.

24 Robert Meyer and Howard Kunreuther, *The Ostrich Paradox: Why We Underprepare for Disasters*, Philadelphia: Wharton School Press, 2016.

25 ASCE, "2021, Report Card for America's Infrastructure," available online at https://infrastructurereportcard.org/.

4

Climate Change and Insurance: Evolving Trends

Nancy M. Rigassio

4.0 Introduction

An underwriter once described to me the difficulty in pricing insurance based on historical data. This underwriter likened the task to driving forward in the car while looking only in the rearview mirror.

Insurance is an important financial recovery tool—a necessary resource for rebuilding after a natural disaster. We all look to insurance when something goes wrong. Insurance may be a homeowner's last hope when disaster strikes. It may be a business owner's ability to recover lost income when a covered loss interrupts the business operations. Insurance is a vital resource and especially important when government funding for assistance becomes strained. The accelerating effects of climate change make the role of insurance more important than ever.

At present, the entire insurance sector is experiencing challenges with continued low interest rates and increased competition from new insurers entering all sectors of the insurance market. Worsening natural disasters, intensified by global warming, impact insurance premiums and coverage. Personal lines insurance, which includes property and casualty policies, experiences the biggest losses from natural disasters. These losses involve property damage to automobiles, homes, and commercial property; business interruption; and logistics and supply-chain disruption.

In parallel, the need for insurance is growing, driven by behavioral changes in society and customer expectations for insuring risks. Insurers have reacted to the evolution of technology and created new products to respond to an increased demand for cyber coverage. Climate change, made evident by intensified natural disasters like fires and floods, has become a long-term risk with complicated impacts. Insurers are responding and using their expertise to provide solutions to losses that have increased as the result of climate change.

Insurers traditionally look at historical data to develop models for assessing natural disaster risks. They use modeling for ratemaking, risk transfer arrangements, and measuring cumulative exposures and capital adequacy. For example, gathering and examining years' worth of loss histories from flood events enables insurers to estimate what damage an insured might be exposed to the next time water reaches a certain height. But modeling and governmental agency mapping may not be as accurate for predicting the risks associated with climate change because past losses no longer provide a complete picture. These examples reveal shortcomings in the ability to predict the complete damage picture caused by natural disaster risks:

- During the first half of 2019, natural catastrophes in Germany caused insured losses of more than €1.3 billion due to the severe thunderstorms and violent hailstorms combined with a heatwave.[1]
- Hailstorms also caused widespread losses in Italy and Greece in 2019, while wildfires burned through areas of Spain after an intense period of hot weather. In 2018, Paris experienced its second wettest winter since 1900, with rainfall of double the annual average.
- In the Fall of 2018, the U.S. National Oceanic and Atmospheric Administration (NOAA) concluded that significantly higher rainfall in parts of Texas may redefine the 100-year events as 25-year events. This won't come as a surprise to Houstonians, who have seen three "500-year" floods in three years.

4.1 Insurance Modeling

In the 1800s, residential property insurers hung maps on walls and used pins to map the concentrations of exposure to fire and lightning risks. The pins-in-the-maps modeling ended in the 1960s. The manual pin pushing became too cumbersome and time consuming because of the numbers of pins.[2]

Economist Karen Clark developed the first catastrophe modeling service in 1987 with a company called Applied Insurance Research, now known as AIR Worldwide. The two other leading catastrophe modeling companies are Risk Management Solutions and EQECAT. Clark recognized the usefulness of modeling tropical storms using computers to estimate damage and loss, applying this technique from the academic study of storms.[3]

Catastrophe modeling ("cat modeling") became widely adopted in the insurance industry after Hurricane Andrew in 1992. The models take input from meteorologists, seismologists, hydrologists, statisticians, engineers, actuaries, and computer scientists, and attempt to predict the losses from insured perils. Advancements in technology and higher resolution studies in the cat models provide more accurate risk exposure and enable insurers to better quantify the financial impact for a range of potential future disasters. Ironically, events that reveal deficiencies in a cat model frequently drive advancements to the models.[4]

Cat modeling software can predict the geographical location of a future event and the intensity of the event. Based on the estimated probability of loss, the cat model user can predict the direct, indirect, and residual losses expected from the event.

The term "direct loss" refers to losses from damage to physical structures and contents, deaths, and injuries. The term "indirect loss" refers to loss of use, additional living expenses, and business interruption. The term "residual loss" pertains to demand surge, labor delays, and inflation in the cost of materials. When the event is a hurricane, residual loss includes evacuation, sociological risks, and expenses due to political influences.

Hurricanes in the Atlantic region during the 2004 and 2005 seasons created a challenge to the widely accepted assumptions on cat modeling and introduced the concept of "secondary losses" to insurers. These two consecutive years of hurricane activity brought record losses. They also brought a new focus on the impact of aggregate losses from multiple hurricanes. Cat modelers for Hurricane Katrina did not anticipate the levees in New Orleans would fail, exposing the inability of models to capture the effects of nonmodeled perils like levee failures.[5]

The losses from Hurricane Katrina in 2005 included greater losses from secondary flooding than the original wind-generated catastrophe, prompting cat modelers to include the impact of secondary losses in the models.[6]

Figure 1 shows an example of a modeled surge depth for Hurricane Sandy in 2012. The 9.4-foot storm surge, plus a full moon high tide, created a record 14-foot storm tide above Mean Lower Low Water at Battery Park, at the southern tip of Manhattan.[7]

Figure 1. **Model of Surge Depth of Hurricane Sandy in 2012**

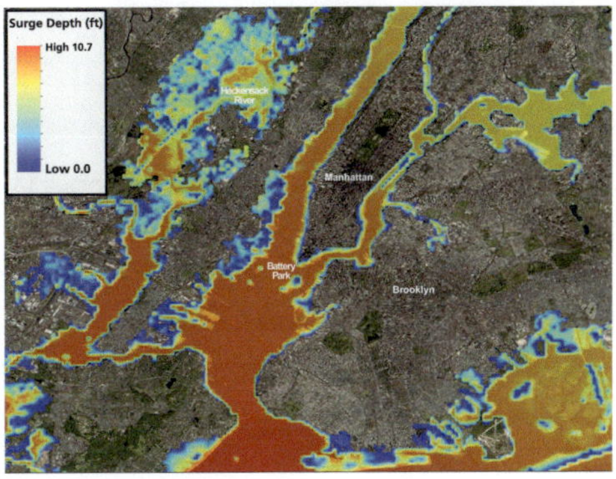

Source: AIR Worldwide.

Cat models provide more than a projection of the estimated damages and losses from an event. Insurers rely on the models to determine rate, premiums, risk selection, mitigation, and risk transfer arrangements. Risk transfer agreements include ceding to reinsurance, allocating a portion of the risk to the policy holder by asking for higher deductibles, excluding specific risks, or imposing sub-limits for certain events. Cumulative exposures are measured according to geographic location, building size, occupancy, construction type, and age of the insured premises. To determine the premium, the insurers need to calculate cover for average annual loss, expenses, and risk load. The cat models allow insurers to assess the adequacy of the insurers' capital reserves.

Cat models, regardless of the modeled event, contain four basic *modules*: (1) probability of the event occurring, (2) intensity of the catastrophe, (3) vulnerability, and (4) allocation of loss amongst the impacted entities. Based on historical data, the first module, an "event" module, generates possible catastrophic scenarios and produces probabilities of the event occurring. The event module addresses the chance that a specific event will occur. The second module, an "intensity" module, considers the intensity of the occurrence based on the level of physical hazard specific to the geographical location. The third module, a "vulnerability" model, quantifies the impact of the event by estimating how badly the structure (and related property, i.e., building contents) will be damaged. Given exposure characteristics and event intensity, the vulnerability module predicts monetary loss to the property. The fourth module, financial loss, measures monetary loss amongst various parties: the policy holder, the insurer, the reinsurer. The financial loss module takes into consideration the parties' insurance policy conditions (i.e., deductible, attachment points, the primary policy limit) and reinsurance treaties.[8]

Like other tools, cat models have limitations. In considering the rapid changes in our society, environment, and climate, applying the traditional cat modeling to predict the effects of climate change may fall short in accurately assessing frequency and severity. Relying on cat modeling to predict losses from enhanced catastrophes from climate change reminds me again of the underwriter's example of driving forward while looking only in the rear-view mirror. Climate change has become an important issue in the insurance industry because it has the potential to change the risks associated with certain hazards. Recognizing these limitations, insurers will continue to rely on cat modeling to predict the potential impacts of climate change, anticipating increased hazards from bigger storm surges, higher sea levels, and higher temperatures. Cat modeling may also be used to quantify the adaptive and mitigating measures we take today to address climate change and predict how those measures will provide benefits by reducing losses from weather-related events. In this way, cat modeling may assist insurers in providing leadership in helping to reduce the risk, not just transfer of the risk.[9]

4.2 Wildfires Are the Wild Card

Between January and August of 2020, fires swept through an estimated 7,861 square miles in the west of Brazil bordering Paraguay and Bolivia, creating a jarring and unprecedented loss in the Pantanal, the tropical wetlands that provide a delicate ecosystem and biologically diverse habitats. To get an idea of the size of the devastation, the burned area is larger than the state of New Jersey. This year's fires in the Pantanal are separate from the estimated 13,200 square miles of tree cover that have burned this year in the Amazon.

We humans haven't anticipated the severity or the frequency of this peril. For me, wildfires confirm the accelerating impact of climate change. Wildfires are treated like other perils (hurricane, flood, and earthquake), but the increasing preference for

Figure 2. **Evaluation of Frequency of Four Recent Wildfires in the United States**

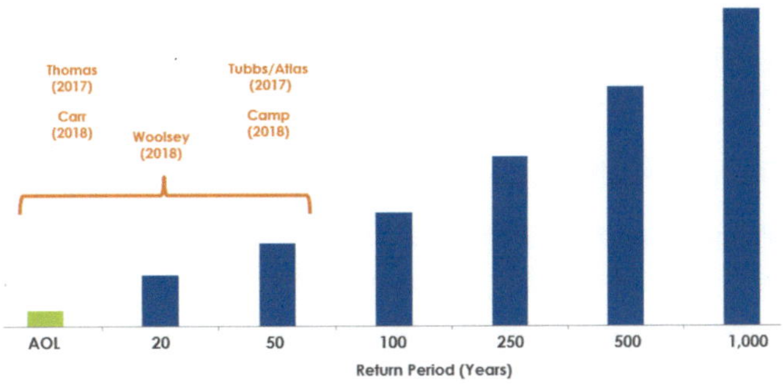

Source: AIR Worldwide.

development in the wildland–urban interface, the clearing of land for ranching and farming, exacerbated by more frequent and severe droughts that create abundant combustible material, make the damage resulting from wildfires a worrisome risk. That is why I refer to wildfires as the wild card.

Losses from only one month, the November 2018 fires in Northern and Southern California, are estimated at $6.8 billion. Compare that one-month number to the losses from the wildfires in the entire year 2017, estimated at $11.79 billion in insured losses.

California wildfires in 2018 caused at least $12.5 billion in insured losses. Just three fires were responsible for $11.4 billion of the losses (the Camp, Woolsey, and Hill fires). For 2019, the *Insurance Journal* estimated that the insured losses from California wildfires amounts to $25.4 billion, but that estimate goes only through Oct. 29, 2019.[10]

Data reported from the National Interagency Fire Center[11] confirm that the number of wildfires in 2020 did not break the record, but the size of the areas burned set a new record. The size of the area that burned exceeded 10.27 million acres. The smoke from the wildfires in 2020 negatively impacted the air quality and caused health concerns for the population in the western United States. Images shared by NASA's Earth Observatory ("A Wall of Smoke on the U.S. West Coast"[12] and "East Troublesome Fire Spread"[13]) demonstrate the extent and intensity of those plumes.

AIR Worldwide modeled gross losses for significant 2017 and 2018 U.S. wildfire events using 2017 industry exposures. This was done in response to the question asked about where losses from extreme events fall on the exceedance probability curve. The modeling concluded that *none* of the four referenced U.S. wildfires (see Figure 2) are considered rare events with high losses. The conclusion gleaned from this model reflects increasing U.S. wildfire risk, in large part due to the choice for living in the wildland–urban interface.[14] This chart places the Thomas and Carr fires in the average occurrence loss (AOL) period. The Woolsey fire may be placed in the period characterized as the 10-year return period. The Tubbs/Atlas and Camp fire events may be characterized as between a 30- and 60-year return period. It's helpful information even though it makes an alarming point. The wildfires referenced in the model are not considered rare events with high losses. Instead, they have low return periods, which reflects an increase in wildfire risk in the United States.[15]

4.3 Historical Lessons

Prudential Property and Casualty had concentrated personal lines insurance policies in south Florida, and experienced catastrophic losses of $1.2 billion when Hurricane Andrew impacted the coast in August 1992. That loss represented 10 percent of Prudential's capital and resulted in a downgrade by the rating agencies. By 1993, throughout Florida, 16 insurance companies announced hundreds of thousands of non-renewals of homeowner policies and rate hikes.[16]

Hurricane Andrew, a category 5 storm, was the costliest disaster in U.S. history when it struck Florida and then Louisiana in 1992 with overall damage at the time of $15.5 billion, according to the Insurance Information Institute (I.I.I).[17] Before Andrew, insurers writing in Florida were not looking closely at their exposure in hurricane prone regions. It was the first cat 5 hurricane in Florida since the Labor Day Hurricane of 1935, a hiatus that gave everyone who lived there and insurance companies a false sense of security.

Allstate, the largest insurer in Florida that canceled policies in response to the losses from Hurricane Andrew, planned in May 1993 to terminate coverage for 300,000 of its 1.1 million Florida customers. In June of 1993, PruPAC planned to terminate 25,000 of its 130,000 Florida policies. The insurers described these actions as a "survival decision." The Florida Insurance Commissioner responded by imposing a six-month moratorium on cancellation of insurance policies, at least until the end of the hurricane season. After Hurricane Andrew, the state formed the Florida Hurricane Catastrophe Fund to protect and maintain insurance industry capacity. It was founded in 1993 during a special legislative session and still provides reimbursements to insurers for a portion of their hurricane losses.

The U.S. wildfires of 2017 and 2018 eliminated 25 years of the industry's profits;[18] 2018 was the second year in a row that wildfire losses were ceded to reinsurance layers to absorb losses from wildfire events. The risk to an insurer for issuing policies in geographic concentration can lead to insolvency of the insurer, as evidenced by the regulatory seizure of Merced Insurance Company, a small regionally concentrated homeowners' insurer that the California state regulators declared insolvent. The court allowed the state agency to liquidate Merced's assets after net incurred losses exceeded its ability to pay claims from California's Camp Fire event.

Regulation imposed by the state of California in December 2019 prevented insurance companies from cancelling homeowners' policies for customers in or alongside ZIP codes struck by recent wildfires. Regulators' efforts to protect the consumers could make the insurance challenge more difficult in geographical areas that experience extreme climate changes, specifically wildfire risk. While attempting to protect the homeowners from cancellations, the regulations pose a new factor in insurers' need to determine rate, premiums, risk selection, mitigation, and risk transfer arrangements.

4.4 New Tools: Dynamic Metrics

Insurers' active involvement in a global effort to limit the earth's temperature rise may be the most important step in minimizing the effects of climate change. What new tools are insurers implementing as preventative medicine to lessen the impact of the insured loss or adapt to climate change?

- An artificial intelligence tool that provides underwriting assistance and new, predictive claim modeling with more usable, meaningful data. This isn't a substitute for underwriting decisions, but a tool to assist in predicting large losses.
- For inland marine insurance, the use of a supply-chain platform to offer practical risk management recommendations. Sensors placed on the cargo monitor the quality of the conditions and send real-time information to shippers, clients, and insurers as the goods move through the supply chain.
- With reliance on supply chains, businesses need to know the extent and nature of their risks – in what geographical areas do supply chains overlap? Several years ago, the automotive industry relied on spare parts suppliers in areas in Thailand that had been subject to flooding. The automotive industry found that the flooding event in Thailand affected the factories in Europe and Japan.

These adaptive tools respond the same way insurers are responding to health care, with the emphasis on prevention through health services—living a healthy life and minimizing the need for medicine—rather than the expensive treatment. How can we do the same to mitigate climate change?

4.5 Paris-aligned Underwriting Approach

In December 2015, 195 countries met in Paris to negotiate the Paris Agreement. Countries that ratified the agreement legally bound themselves to collectively hold global warming below 2° C, as compared to pre-industrial levels, and pursue best efforts to limit warming to 1.5° C by 2100.

Through underwriting decisions, insurers can show the risks that society is taking and foster preventative actions to mitigate those risks. Through investments, insurers can send the right signals to the investment community and the specific companies the insurers choose to place their investments. For my employer, I can report that in 2015, AXA legitimized coal disinvestment by making underwriting restrictions to effect a total exit from the coal industry. In 2017, AXA restricted oil sands in its insurance business.

There is more. Insurers have the responsibility to share knowledge about new risks. Insurers can evaluate the entities they consider insuring and the opportunities for investing according to the degree of Paris-alignment that the potential insureds and investment portfolios reveal. Insurers can use the "investment temperature concept" when assessing portfolios to consider the investments' alignments with the objectives of the Paris Agreement. Impact investing means making green investments and divesting from coal and oil sands. Using metrics, insurers may analyze the carbon-neutral strategies developed by companies and steer toward low-temperature businesses and away from those that fail to align with the Paris Agreement.

Insurers can implement underwriting restrictions, reduce their own environmental footprints, and create insurance products with added environmental value, climate resilience, and adaptation solutions. This will require undertaking more risk modeling, rallying shareholder engagement, investing in academic research and outreach, and participating in thought leadership.

Achieving long-term investment in climate neutrality includes net-zero stock picking and gradual sector reallocations. Reaching this ambition requires real-world changes and requires pressure from policy makers and regulators (the catalyst to business efforts), markets and consumers (reorienting supply and demand), and insurers and investors (providing capital and business incentives toward net-zero models). There has never been a better time than now for resolute climate action – investing in and insuring a world we can leave better for future generations.[19]

Notes

1 Gesamtverband der Deutschen Versicherungswirtschaft e.V. (German Insurance Association) https://www.en.gdv.de/en/issues/our-news/insurance-in-figures-52050.

2 Patricia Grossi and Cheryl TeHennepe, "Catastrophe Modelling Fundamentals," *The Review: A Guide to Catastrophe Modelling*, 2008. London: Informa UK Ltd.

3 Zarah Hirji, "Risky Business: Modeling Catastrophes," *Earth*, October 2012. Available online at https://www.earthmagazine.org/article/risky-business-modeling-catastrophes.

4 National Association of Insurance Commissioners (NAIC), Center for Insurance Policy and Research https://content.naic.org/cipr_home.htm.

5 Hirji, "Risky Business: Modeling Catastrophes."

6 NAIC Center for Insurance Policy and Research, https://content.naic.org/cipr_home.htm. https://www.actuaries.digital/2015/09/22/catastrophe-modeling-in-a-post-katrina-world/

7 AIR Worldwide, "AIR Hurricane Model for the United States," 2020, p. 4. Available online at https://www.air-worldwide.com/siteassets/Publications/Brochures/documents/AIR-Hurricane-Model-for-the-United-States-Brochure

8 American Academy of Actuaries Extreme Events and Property Lines Committee, *Uses of Catastrophe Model Output*, July 2018. Available online at https://www.actuary.org/sites/default/files/files/publications/Catastrophe_Modeling_Monograph_07.25.2018.pdf.

9 Celine Herweijer, director of the climate change practice at RMS.

10 Nic Querolo and Brian K. Sullivan, "California Wildfire Damage at $25.4B and Rising," *Insurance Journal*, October 29, 2019. Available online at https://www.insurancejournal.com/news/west/2019/10/29/546878.htm.

11 https://www.nifc.gov

12 https://earthobservatory.nasa.gov/images/147261/a-wall-of-smoke-on-the-us-west-coast.

13 https://earthobservatory.nasa.gov/images/147452/east-troublesome-fire-spreads-to-the-rockies.

14 Tammy Viggato and Philip Camp, "Is Your Wildfire Risk Management Going Up in Smoke?" *AIR Currents* (blog), AIR Worldwide, February 22, 2018. Available online at https://www.air-worldwide.com/Publications/AIR-Currents/2018/Is-Your-Wildfire-Risk-Management-Going-Up-in-Smoke-/.

15 Tammy Viggato, "How Modeled and Industry U.S. Wildfires Differ," *In Focus* (blog), AIR Worldwide, August 8, 2019. Available online at https://www.air-worldwide.com/blog/posts/2019/8/how-modeled-and-industry-u-s—wildfire-losses-differ/.

16 Christina Sherry, "Florida Homeowners Feel Pinch as Insurance Companies Bail Out," *The Washington Post*, June 13, 1993. Available online at https://www.washingtonpost.com/archive/politics/1993/06/13/florida-homeowners-feel-pinch-as-insurance-companies-bail-out/7cbe1b22-de98-47b9-97d2-e81781f96ff0/.

17 Insurance Information Institute, "25 Years Later, Hurricane Andrew's Lessons Still Resonate," August 11, 2017. Available online at https://www.iii.org/press-release/25-years-later-hurricane-andrews-lessons-still-resonate-081117.

18 Christopher Flavelle and Brad Plumer, "California Bans Insurers From Dropping Policies Made Riskier by Climate Change," *The New York Times*, December 5, 2019. Available online at https://www.nytimes.com/2019/12/05/climate/california-fire-insurance-climate.html.

19 AXA Group, "2020 Climate Report: Renewed Action in a Time of Crisis," June 2020. Available online at https://www-axa-com.cdn.axa-contento-118412.eu/www-axa-com%2F3800520b-ce0f-4aa7-908d-3ec367b21d39_2020_climate_report_axa.pdf.

PART II

Climate Change: Legal Issues and Considerations

5

Climate Change Lawsuits: Trends and Analysis of Legal Actions

Elena Mihaly and Deanna Moran

5.0 Introduction

Litigation has long been recognized as an important tool for compelling change in individual behaviors and industry standards, often serving as a stopgap where existing regulations are inadequate. Nowhere is this dynamic more evident than in cases of product liability, particularly the wave of successful tobacco litigation in the 1980s and 90s brought by state attorneys general. The litigation played a significant role in changing both the public perception of smoking and instituting culpability on the part of tobacco companies. Other notable examples include automobiles, firearms, and asbestos. Similarly, litigation played a crucial role in the regulation and clean-up of hazardous waste sites through the Comprehensive Environmental Response, Compensation, and Liability Act (CERCLA), extending liability not only to those who generate, dispose, or transport hazardous waste, but also to those who finance or design such operations.

This tactic of using legal intervention to shape behavior and standards has evolved over time with climate change litigation emerging as one of the newest areas of application. Although climate change case law has, to date, primarily revolved around mitigation efforts to reduce greenhouse gas (GHG) emissions, a more recent area of litigation is developing around climate adaptation and the liability of various actors in the built environment sector for their "failure to adapt." This term refers to the failure of actors to undertake climate adaptation measures to protect against foreseeable harms from the increasingly severe weather events that are becoming "the new normal." In light of this trend, a growing number of people are asking the question of who owns the liability for non-resilient assets and systems.

This shift from mitigation to adaptation claims may be, in part, due to the barriers that plaintiffs face for holding certain actors liable for their role in contributing to climate change-related impacts. For instance, cases against large GHG emitters—such as oil and gas companies or utilities—seeking mitigation-based remedies have not fared well.[1] The result has been more targeted cases at certain players, including design professionals, for their role in failing to adapt to changing climate conditions. There has been a marked increase in the number of these cases as the effects of climate change have become more severe and frequent.

However, despite the new wave of cases with specific allegations related to climate change, liability cases dealing with the impacts of extreme weather, including flooding and high winds, are not new. Examining and dissecting the fact-specific circumstances of these cases helps us understand the components of emerging climate change-based claims, and provides a preview of how courts may decide future claims.

There are a number of legal avenues by which a design professional may be held liable for failure to adapt to climate change. This chapter focuses on claims arising out of common law (specifically negligence), as well as claims deriving from a design professional's duties contained in statute, regulation, and contract. To understand the implications of climate change on these areas of law, we look at several case studies—both new and old—and highlight the ways in which design professionals may be susceptible to liability.

5.1 Common Law Tort Claims

Tort law is the body of law governing our duties to others and the damages that may be due if those duties are violated. The most frequently pursued tort claim is a negligence action (discussed in detail below), but a viable claim could also arise in a climate change context under theories of nuisance or trespass.

Negligence claims reflect and enforce the generally accepted principle that everyone should act in a reasonable way so as not to injure those around them. To establish a negligence claim in court, an injured party must first establish that the person or entity causing the harm had an obligation or "duty" to behave in such a way as to avoid the harm. This duty is often called the "standard of care," and professionals such as doctors, teachers, etc. must meet this standard of care when acting in their professional capacity. Average citizens are also held to a certain duty of care.

In addition to establishing a "duty," an injured party must also prove that the person or entity breached (i.e., violated) the duty, and the breach was the cause of the harm to the injured party. Courts will look at a number of factors in assessing whether the duty was breached, including: (1) the standards in any underlying contract; (2) the defendant's knowledge of the risk; (3) applicable codes and regulations; (4) relevant industry customs; and (5) the foreseeability of the harm.

Below, we discuss how our expanding awareness of climate change and increasing sophistication to predict its impacts affect each of these factors. A similar analysis to the one below would be undertaken by a court in reviewing an extreme weather-related negligence claim.

5.1.1 **Standards in a Contract**

A contract may contain clear, written standards to which a design professional must adhere, and failure to do so could result in a breach of professional duty. In the climate change context, the inclusion of certain contractual standards may also give rise to a need to incorporate reasonably foreseeable data on climate impacts. For example, a contract may require hurricane straps on a building, or that infrastructure like a bridge should be constructed to a twenty-five-year design life. These contractual terms establish a legal duty which a design professional must follow to avoid being vulnerable to negligence liability. We discuss contract liability, as apart from common law liability, further in section 5.3, below.

5.1.2 **Knowledge of the Risks Posed by Climate Change**

A design professional's actual or potential knowledge of the risks posed by climate change will also factor into the breach analysis. For example, if publicly available storm surge maps indicate that a structure is likely to flood during the lifespan of a building, a design professional could be liable for failing to build the project such that it could withstand that flooding. This could be true whether or not the design professional had actual knowledge of the flood risk if those maps were readily available. Moreover, a design professional could be found liable if harm results from designing a structure based on floodplain or other maps that the professional knew or should have known misrepresented risks, given climate projections.[2]

Thus, when planning a project, design professionals should discuss with the owner the level of due diligence required to determine the appropriate design standards for climate resilience. It may be necessary to conduct research on publicly available weather data and projections as well as climate impact maps and models to determine if the building should be designed to guard against certain known or highly likely future climate impacts. Also, as a matter of good practice, design professionals should consult with appropriate professionals to determine the accuracy of climate data, including flood or storm surge maps, and regional climate vulnerability studies.

5.1.3 **Applicable Codes and Regulations**

If design professionals do not comply with applicable codes and regulations for design or construction, this could result in a finding of *per se* (i.e., automatic) negligence.[3] Even where applicable statutes, codes, and regulations or the permits issued under them make no explicit reference to climate or weather risk, they may include narrative requirements (such as a duty to take reasonable care), general duties (e.g., to use "best engineering practice"), or references to privately developed codes, any of which may import a duty to identify and address risks, like climate risk, that are well understood and reasonably quantifiable for purposes of design and construction. Failure to address those risks could be seen by a court as a breach of the standard of care.

Importantly, compliance with explicit regulatory requirements does not necessarily shield a design professional from liability, since many building and design regulations may not incorporate climate changes that have occurred or become evident since enactment or adoption of the statute or regulation, or that are anticipated during the expected life of the project or permit time frame. Some countries have recognized this shortcoming. For example, the Canadian government recently announced an initiative to reform its National Building Code to incorporate climate change into the design parameters. A briefing note prepared for the Canadian Deputy Minister of Infrastructure in 2018 said the code rewrite will be the "first substantive" introduction of climate change considerations for buildings and crucial infrastructure in Canada.[4] The Australian Building Codes Board similarly acknowledges its building code is "likely to be deficient in some areas" in the event of "climate changes in accordance with high emissions scenarios."[5] The joint Australia-New Zealand Design Standards expressly concedes that, regarding wind speed calculations, "[n]o account has been taken of any possible future trend in wind speeds due to climate change."[6]

In the United States, most states have adopted the International Building Code (IBC), either in full or with amendments. As happened in the building codes of the countries mentioned above, the IBC fails to account for sea level rise or climate change more generally.[7] Thus, while a defendant's adherence to zoning or building codes and other regulations can help determine if a breach of duty occurred, compliance alone with these laws will not shield a design professional from liability for damages resulting from failure to account for climatic changes not considered or evident at the time the code or rule was adopted.

While this has not yet become the target of litigation in the U.S., some cities are developing their own building guidelines for climate-resilient construction and have acknowledged the potential deficiency of existing codes and standards. For example, New York City's new Climate Resiliency Design Guidelines identify climate change risks and appropriate resiliency interventions for city projects—such as raising machinery when building is in a potential floodplain.[8] Similarly, the City of Boston also established its own Climate Resilient Design Standards and Guidelines for design professionals to use when designing flood barriers to protect the public rights-of-way.[9] The Boston guidance document explicitly acknowledges that current standards do not contemplate climate change, noting that the guidance is "meant to augment existing City and State design standards by considering climate impacts. . . ."[10]

5.1.4 **Industry Custom**

Similarly, while the prevailing industry custom may offer courts a useful guide to establish whether or not a defendant breached the appropriate standard of care, even compliance with a pervasive practice may not necessarily equate with meeting the relevant standard of care if there is a relatively low-cost, ascertainable means of eliminating a hazard. In a landmark admiralty case called *T.J. Hooper*, a tugboat owner sought to limit his liability after losing the cargo of two barges in a storm.[11] The tugboat at issue in the case, like most tugboats at the time, lacked a functioning radio that could have received the daily weather report warning of the impending storm.[12] The court nevertheless rejected mere compliance with industry custom as a defense to liability. As the famous federal Appeals Court Judge Learned Hand wrote, "[A] whole calling may have unduly lagged in the adoption of new and available devices. . . . Courts must in the end say what is required; there are precautions so imperative that even their universal disregard will not excuse their omission."[13] Thus, adherence to common industry practice does not foreclose liability, especially when there are known methods for avoiding the risk of injury.[14] In the climate change context, it may not be sufficient for a design professional to point to existing industry customs that lack meaningful consideration of future risks, even if that has been the accepted practice.

5.1.5 **Foreseeability of the Harm**

Whether the harm that occurred was foreseeable is another important factor in determining negligence liability. Foreseeability is one of the most complex concepts in negligence, but distilled to its core, it is the ability of the party being blamed for the injury to have anticipated that such an injury could occur. It is not a question of whether the party actually did foresee the injury, but whether they should have seen that such an injury could occur. For instance, in *Schlitz v. Cullen-Schlitz and Associates, Inc.*, the Iowa Supreme Court upheld a jury verdict finding an engineering firm negligent in designing a sewage treatment facility that included a dike surrounding the facility.[15] The facility was constructed in a floodplain and the dike was insufficient to keep out floodwaters that ultimately damaged the facility. In upholding the jury verdict, the court cited testimony from witnesses who testified that a flood the previous year was much more severe than the one that ultimately damaged the facility, and therefore the engineer should have designed the facility to safely accommodate that size flood.[16]

Of particular note for climate change risk, an event need not have actually occurred in the past for a court to find that it was foreseeable that it could occur in the future. For example, an Illinois Appeals Court held an engineer liable for negligently designing a concrete pylon that collapsed due to high winds onto a customer as she entered a shopping mall, despite the fact that such high wind speeds had never been recorded at the site.[17] The court decided the engineers could have predicted the high winds with existing technology, and that the engineers "failed to exercise that degree of care in the performance of professional duties imposed upon them as members of a licensed profession which exists in large part to prevent harm to the public from structurally unsafe buildings."[18]

Another important point about foreseeability is that design professionals may be expected to design around circumstances that are not identified in a particular contractual agreement. That is, even when a design professional fulfills their contractual duties, they still have a duty to account for reasonably foreseeable events.[19] Thus, merely fulfilling the terms of a contract may not be enough to shield a design professional from negligence liability when reasonably foreseeable risks are ignored.

While design professionals' professional liability insurance generally covers negligent performance of professional services—and typically does not exclude coverage of climate-related damages—the structure of these insurance policies provides cold comfort that climate-related negligence claims will always be covered. Professional liability insurance is "claims-based" (as opposed to "occurrence-based," like automobile or home insurance). This means a design professional is covered based on when the claim is filed, not when the negligence occurred. There could be a scenario where an insurance company does not exclude negligence claims related to climate change *when the design decision was made*, but by the time three to four years passes when the building is flooded or some other climate-related harm becomes evident, the insurance company has chosen to adopt a blanket exclusion for anything related to climate, and the design professional would not be covered. In other words, coverage could change in the future for things design professionals are doing today. It is likely that, at some point, insurance companies will either have to begin raising premiums to accommodate the growing number of climate change-related claims, or they will stop covering damages associated with particular climate-related impacts.

Given the rapidly accumulating and sophisticated body of evidence concerning climate change and its effects, it is becoming increasingly difficult to make the case that because certain events were unprecedented or not mentioned in a contract, they were not foreseeable. The case discussed below provides a good example of this issue playing out in court today.

In *Lijo Abraham et al. v. Costello, Inc.*, Cause No.-201822747 (Dist. Ct. Texas 2018), a group of homeowners in Fort Bend County, Texas, sued Costello Engineering, Inc. for negligently engineering and designing their neighborhood's levee improvement district (LID), after their homes were destroyed by serious flooding in Hurricane Harvey.[20] The LID for their neighborhood was designed by Costello Engineering with a dual purpose—to keep water out of the neighborhood from the nearby Brazos River, and to evacuate rainwater from the area during heavy rainstorms. Yet, during Hurricane Harvey, the LID did not accommodate the rainfall and the plaintiffs' properties were severely flooded. Defendant Costello argues the rainfall in and around plaintiffs' housing development during Hurricane Harvey was "the result of an Act of God."[21] The plaintiffs meanwhile contend the rainfall during Harvey was foreseeable, given the history of rainfall in the Texas Gulf Coast, and that defendants were negligent in failing to engineer the LID to accommodate it. At the time of writing this chapter, parties were in the midst of the discovery phase of litigation, and the Texas District Court had scheduled a trial on the merits in the fall of 2021.

The plaintiffs brought several causes of action against Costello, including for negligence, asserting that Costello breached its duties and was negligent in one or more of the following ways, among others:

- Failing to use ordinary care in the operation, design, engineering, and maintenance of its pump, levee, drainage, and stormwater management systems in the plaintiffs' neighborhood;
- Failing to use ordinary care in developing rules, plans, policies, and procedures to protect against foreseeable risks of harm; and
- Failing to adequately warn or disclose the risk of flooding to the plaintiffs.

The plaintiffs also brought a cause of action for negligence *per se*, nuisance, and gross negligence. On the latter cause of action, the plaintiffs' assert that "Defendant's conduct, when viewed from the standpoint of the actors at the time of the occurrence, involved an extreme degree of risk, considering the probability and magnitude of the potential harm to others."[22] The case is pending in Texas District Court and will likely come down to the question of whether or not the rainfall during Harvey was foreseeable or an unknowable Act of God.

This "Act of God" defense also arose in a criminal negligence suit filed by Harris County, Texas, against Arkema, Inc. and two of the chemical plant operators for alleged harms resulting from chemical releases due to foreseeable flooding during Harvey.[23] In the Arkema case, the chemical company described the Harris County Prosecutor's charges as "doubling down on an unprecedented and outrageous attempt to criminalize a natural disaster" and accused the County of "trying to prosecute a company for the Act of God that was Hurricane Harvey."[24] The case against Arkema was described as "the highest profile criminal case ever brought against a company and its employees for allegedly failing to adequately prepare for the effects of climate change."[25] On October 1, 2020, a Texas state judge threw out these criminal charges against Arkema and the former manager of the plant, cutting short a trial and finding that no evidence in the record supported the felony charges.[26] If this case had resulted in a conviction, it would have sent a strong signal to the petrochemical industry and other industrial plant operators about how they need to prepare for disasters. However, University of Houston environmental law Professor Tracy Hester said the fact that the case fell apart does not mean the dialogue is over. "There's a lot of attention to the risks posed to Harris County and the city of Houston by industrial incidents, explosions, fires releases," he said.[27] "And as a result, I don't think that the failure to reach a verdict, in this case, means that that examination, the risk of future criminal prosecutions, is over."[28]

5.1.6 **Tort Claims and the Act of God Defense**

The "Act of God" defense originates in the longstanding notion that weather and its consequences are divine and are therefore beyond the control of humankind.[29] Thus, a party should not be liable for harm caused by an extraordinary natural event, an Act of God, which is both unforeseeable and impossible to prevent.[30] However, climate change has begun to erode two of the foundational concepts underlying the Act of God defense: the idea that human and nature are separate, and the idea that natural events are unforeseeable.[31] Thus, with the realization of humans' role in driving climate change, the question of whether extreme climatic and weather events are really unforeseeable Acts of God has come to the fore.[32]

While the exact definition of an Act of God varies depending on the jurisdiction, the definitions can be grouped into three general themes: forces of nature, lack of foreseeability, and sole proximate cause.[33] Under the forces of nature approach, the natural event must generally be "extraordinary," "unusual," or "unprecedented" to qualify as an Act of God, as determined by the intensity of the event, characteristics of the area, and climatic history.[34]

Under the lack of foreseeability approach, the court focuses on whether the defendant could have reasonably foreseen the event and prevented the injury.[35] Because foreseeability underlies the traditional tort analysis, some commentators have suggested that the Act of God defense is no longer necessary given that the analysis is simply the application of traditional negligence theory. Under this view, if an event is truly so unforeseeable as to constitute an Act of God, then there is no duty under traditional negligence analysis, and the Act of God defense can therefore be viewed as a no-duty defense.

The third approach to the Act of God defense in tort law focuses on whether the force of nature is the sole proximate cause of the injuries at issue.[36] Under this strict approach, if some element of human activity, intervention, or inaction contributed in some way to the injury, then the Act of God defense will not allow the defendant to escape liability.[37] For example, in *McFarland v. Entergy Mississippi, Inc.*, the court held that the Act of God defense failed in a case involving an ice storm that downed power lines because the power company negligently failed to satisfy its duty to protect the public against downed lines after the storm struck.[38]

Legal scholars critique the Act of God defense as antiquated and increasingly at odds with traditional negligence analysis on several grounds. First, as climate-related data and modeling have improved, extreme natural events have become more foreseeable.[39] Therefore, it becomes more difficult for the defendant to show that he or she took reasonable precautions in light of the increasingly foreseeable risks, thereby undermining the principle of unforeseeability that underlies the defense and creating tension with traditional negligence analysis.[40] Second, as climate science has developed a stronger understanding of the risks posed by climate change, failure to take reasonable measures to mitigate these risks can be considered acts of people, and not of God or the sole forces of nature.[41]

5.2 **Statutory and Regulatory Claims**

Design professionals may be held liable for failing to adequately consider climate change in infrastructure design plans based on requirements in federal or state statutes or their implementing regulations. Statutes and regulations often contain mandatory duties that can be used to compel design professionals to adopt climate adaptation measures. For example, a duty to prevent the release of hazardous material or to maintain air and water quality could be affected by increasingly frequent and severe weather events. Failure to anticipate and account for foreseeable climate conditions like extreme heat and precipitation, storm surge, wind, and sea level rise could trigger a breach of these duties.

In September of 2016 the nonprofit environmental advocacy organization Conservation Law Foundation, Inc. ("CLF") sued ExxonMobil Corporation et al. under the Clean Water Act and the Resource Conservation and Recovery Act.[42] The Clean Water Act counts allege, among other violations, that ExxonMobil failed to comply with regulations and permit language requiring that its oil production and storage facilities be built, maintained, and inspected "in accordance with good engineering practices."[43] CLF contends, in part, that because ExxonMobil failed to consider current or imminent increases in intense precipitation, more intense storms, rising seas, or other extreme weather in its management of the Everett, Massachusetts, oil terminal facility, it has not met the regulatory standard of good engineering practice.

In March 2019, the United States District Court for the District of Massachusetts partially granted ExxonMobil's motion to dismiss with respect to alleged injuries that are unlikely to occur until far in the future. In particular, the court ruled that CLF "does not have standing for injuries that allegedly will result from rises in sea level or increases in the severity and frequency of storms and flooding that will occur in the far future, such as in 2050 or 2100."[44] However, the judge's order also noted that CLF stated "a plausible claim that there is a 'substantial risk' that severe weather events, such as storm surges, heavy rainfall, or flooding, will cause the terminal to discharge pollutants into [nearby communities] in the near future and while the [Clean Water Act] Permit is in effect."[45] The court also allowed CLF's Resource

Conservation and Recovery Act claim to proceed, except to the extent that it was based on discharges from point sources covered by the permit, because CLF plausibly alleged an imminent threat of harm.

ExxonMobil's primary argument for dismissal of the climate change counts was that they have no continuing duty under the permit to evaluate the Everett Terminal's preparedness for severe weather events. The court found that ExxonMobil's interpretation of its permit violated the principles of construction. It ruled "the permit does not impose static or only static requirements."[46] Rather, it requires ExxonMobil to regularly review and update its practices to reflect or address any materially changing circumstances. Specifically, this obligation requires ExxonMobil to consider foreseeable severe weather events, including any climate change–induced weather events.[47] In April 2019, ExxonMobil filed a motion to stay the case, which the court granted in March 2020. CLF appealed this decision to the First Circuit and as of the writing of this chapter, the case is still pending.[48]

CLF also has a similar case pending before the Rhode Island Federal District Court against Shell Oil entities and Motiva Enterprises, LLC, alleging that those companies had failed to comply with the Clean Water Act as well as the Resource Conservation and Recovery Act at their bulk storage and fuel terminal in Providence, Rhode Island (Providence Terminal).[49] CLF alleged that the Providence Terminal was "at risk from coastal flooding caused by sea level rise, increased and/or more intense precipitation, increased magnitude and frequency of storm events, and increased magnitude and frequency of storm surges—all of which will become, and are becoming, worse as a result of climate change."[50] CLF also alleged that the terminal's location, elevation, and lack of preventative infrastructure made it "especially vulnerable to these risks" and that defendants had not taken action to address these vulnerabilities at the Providence Terminal, despite having "long been well aware" of climate change's impacts and risks and having incorporated such risks in "ongoing company investments," including projects off the coast of Nova Scotia and in the North Sea.[51] CLF asserted that Shell's "knowing disregard of the imminent risks" of climate change and failure to fortify the Providence Terminal against such risks constituted violations of the Clean Water Act and Resource Conservation and Recovery Act.[52] On September 28, 2020, the District Judge denied defendants' motions to dismiss several of CLF's claims for lack of legal standing, concluding that "[t]he Complaint makes clear that a major weather event, magnified by the effects of climate change, could happen at virtually any time, resulting in the catastrophic release of pollutants due to Defendants' alleged failure to adapt the Terminal to address those impending effects. While it might not occur for many years, the fact that it is certainly impending is enough to meet the standard [for legal standing]."[53] At the time of writing this chapter, this lawsuit was headed towards discovery and potentially a trial.

In addition to encountering mandatory duties that exist in statute and regulation, design professionals and others may be subject to evolving standards for climate-related risks that are not yet written into statute or regulation. For example, in April of 2018, the Virginia Beach City Council voted unanimously to deny developer Argos Properties' rezoning application to develop 32 single-family residential homes on 50 acres of low-lying swamp land.[54] The Virginia Beach City Council denied the application based on concerns over ingress and egress issues from occasional flooding, and the developer's failure to analyze a 1.5-foot sea level rise scenario in its stormwater analysis. Argos claimed that the increase in the stormwater management features required to be on-site by the city council's "ad hoc" criteria would reduce the buildable area and thus, there would be a lower residential density allowed at the property.[55]

Argos challenged the city council's decision in a complaint filed in Virginia state court, alleging that the application was unlawfully denied. Argos alleged that the city council violated (1) the Administrative Procedure Act because its decision was arbitrary and capricious; (2) the U.S. Constitution's Equal Protection Clause because there was no rational basis for the council's recent approval of a rezoning application of a neighboring developer; and (3) the decision was *ultra vires* (beyond the council's power) because the city was required to follow a strict protocol to gain authority to issue more stringent stormwater requirements than those already in statute and regulation and failed to do so.[56]

Notably, Argos did not challenge the underlying rationale for the city council to incorporate consideration of climate change impacts in permitting; rather it focused its challenge on the process. Argos claimed that it satisfied all of the city's submittal requirements that existed at the time of its application and was subsequently informed that it would need to provide additional engineering criteria, including items like average elevation, soil analysis, and stormwater management facility outfall design and location. These criteria were not required by the city's Stormwater Management Ordinance or the state's controlling regulations.

Argos argued that the city council's denial was based on the company's failure to adhere to criteria that were not legally required and that were not developed until after the application was filed. Further, it stated that, "its failure to assess higher water elevations resulting from an anticipated sea level rise does not protect public health, safety or the general welfare, because the site would not have been engineered to this standard even if it had been approved."[57]

In April 2019, the Virginia trial court ruled that the city council had properly denied the developer's application to build a residential development in an area prone to flooding.[58] This case signals that—at least in Virginia— a court was willing to grant a government entity the flexibility to require developers and designers to include consideration of climate risks in permit or rezoning applications without formally amending the requirements of existing statutes, regulations, or policies.

5.2.1 **Statutory Claims and the Act of God Defense**

The Act of God defense appears in the Clean Water Act (CWA), the Comprehensive Environmental Response, Compensation, and Liability Act (CERCLA), and the Oil Pollution Act.[59] Courts have construed the Act of God defense arising in these statutes narrowly, pointing to clear legislative history as support. For example, a House report on the legislative history of CERCLA emphasized that the natural event must be "unanticipated" and noted that "a major hurricane may be an 'act of God,' but in an area (and at a time), where a hurricane should not be unexpected, it would not qualify" under the CERCLA definition of an act of God.[60] The legislative history of the CWA also makes clear that the use of this defense should be limited.[61] A 1970 Conference Report prior to the passage of the CWA defined Acts of God as "only those acts about which the owner could have had no foreknowledge, could have made no plans to avoid, or could not predict would be included."[62]

As a result of this narrow interpretation of the term Act of God, courts have rejected this defense under federal environmental laws in situations involving cold weather,[63] high-flow conditions on a river related to rain and runoff from melted snow,[64] storms and heavy rainfall,[65] and hurricanes,[66] on the grounds that the weather events did not satisfy the definition.[67] Judicial opinions make clear that the Act of God defense is limited to uncontrollable and unforeseen natural events.[68] For example, in *United States v. Stringfellow,* the court found that heavy rains did not qualify as an Act of God since "[t]he rains were foreseeable based on normal climatic conditions and any harm caused by the rain could have been prevented through design of proper drainage channels."[69] Thus, courts readily reject the Act of God defense to environmental liability when the natural phenomena are foreseeable or when human conduct plays a role in the environmental damage.[70]

The narrow judicial interpretation of the Act of God defense in environmental law has led to its infrequent use in recent years.[71] For example, Murphy Oil chose not to contest its liability under OPA for the Murphy Oil spill at its refinery in St. Bernard Parish, Louisiana, despite the fact that the spill was caused by waters from Hurricane Katrina dislodging, lifting, and damaging a 250,000-barrel above ground storage tank.[72] While there are no reported cases finding the Act of God defense successful under federal environmental laws, the Environmental Protection Agency can exercise enforcement discretion in choosing not to pursue liability if it believes an Act of God defense is appropriate, so the defense may have worked despite the lack of reported cases.[73]

5.3 Contract Law Claims[74]

A design professional may also be liable for climate-related damages under contract law for breaching a term of an applicable contract. For instance, if a contract requires the design professional to build a stormwater system that can handle a certain designated storm size and the system is built below that capacity, there may be a viable breach of contract claim. It is noteworthy that a claim can be brought if a party can prove that an element of the contract was not fulfilled, even if no harm has occurred. A brief analysis of some of the standard form contractual agreements frequently used by design professionals, including engineers, illustrates some of the ways in which contractual liability for failing to adapt to climate change can arise.[75]

5.3.1 **Standard of Care**

Just as in a negligence claim, a design professional can be found contractually liable for violating the pertinent standard of care applicable to the project. In the case of a contract claim, however, the standard of care comes from the contract itself, as opposed to being determined by a court upon review of the factors articulated earlier. While many of the same considerations discussed above may also apply, they must have a "hook" in the contract.

For example, the Engineers Joint Contract Documents Committee (EJCDC) document E-500 provides that the standard will be "the care and skill ordinarily used by members of the subject profession practicing under similar circumstances at the same time and in the same locality."[76] Thus, the design professional's standard of care could include considering the effects of climate change in a project site's locality, especially as the understanding of those effects becomes more sophisticated and they manifest more frequently.

This standard of care language highlighting the locality also means that extra care must be taken for projects in some parts of the country, but not others. For example, design professionals working in a coastal city must consider rising sea levels or increased flooding as part of their standard of care, while professionals in the Midwest might need to consider increased snowfall or more frequent tornadoes. While climate hazards obviously vary by location, these standard of care provisions also include a time element. They suggest that not monitoring these conditions over time or modeling future changes might create contractual liability for failing to follow the standard of care set out in the contract.

5.3.2 **Scope of Work**

Design professionals also are obligated to conform to other standards that may be set out in the owner's requirements. The standard form documents give owners the ability to control and expand the design professional's scope of work to consider climate change impacts on the design. The EJCDC E-500 Agreement, for example, covers the design professional's services in a separate exhibit (Exhibit A), which allows greater flexibility on the owner's part to include the responsibility to account for climate change in the design engineer's work. Exhibit A also specifically requires the engineer to visit the site (thus creating a heightened awareness of local conditions), advise the owner of additional information and testing needed, propose solutions to achieve the owner's objectives, and work with governmental authorities on environmental assessment issues.[77]

The owner's ability to purchase additional services from the design professional beyond the basic requirements in the standard form contracts creates additional exposure for liability. For example, according to the EJCDC E-500 Agreement, an owner can hire the project engineer to perform the additional services of preparing or reviewing environment assessments and impact statements and obtaining required government approvals for the anticipated impact.[78] Similarly, the ConsensusDocs 240 Agreement notes that the owner can purchase from the design professional surveys, site evaluations, and soils and environmental studies, each of which creates an additional point at which climate change effects might be considered, and liability may attach for failure to do so.[79]

5.3.3 **Indemnification Clauses**

A design professional also has exposure to liability through indemnification clauses. The ConsensusDocs 240 Agreement provides that the design professional will indemnify the owner from any claims or liabilities for personal injury or property damage that arise from the performance of or failure to perform the services under the contract. The AIA B103–2017 Agreement also requires the design professional to indemnify the owner, though its scope is limited to third-party claims.[80] The EJCDC E-500 Agreement also has an indemnification provision, although it does not cover destruction of tangible property, that is, the project itself.[81]

To the extent that adapting the design to address the negative effects of climate change was contemplated by the owner's program, local statute, or ordinance, or was otherwise generally foreseeable, the design professional could be on the hook under these types of contractual indemnification clauses.

At the time of writing, we could not find any "failure to adapt" claims arising out of contract. This could be in part because these cases are settling out of court. However, we do expect these types of claims to become more frequent, especially in the form of indemnity claims, as adaptation claims start to stick to the owners.

5.3.4 **Contract Claims and the Act of God Defense**

The Act of God defense primarily appears in contract law in the form of "force majeure" clauses.[82] These clauses are contractual provisions allocating the risk of loss if performance becomes impossible or impracticable, especially as a result of an event or effect that the parties could not have anticipated or controlled.[83] Courts typically enforce these clauses as written under the principle of freedom of contract.[84] However, even if a contract does not expressly contain a force majeure clause, some courts may still relieve a party of the duty to perform under the common law doctrines of contractual impracticability and frustration of purpose.[85]

As in tort law, the Act of God defense is likely to be undermined by climate change due to the increasing foreseeability of previously "unforeseeable" events. As extreme weather becomes the new normal, courts may limit the scope of the doctrines of impracticability and frustration of purpose since contracting entities should anticipate the likelihood of extreme weather events. On the other hand, courts have traditionally based foreseeability on past experience rather than future previsions, so courts may be slow to limit the scope of the doctrine. Also, as in tort law, judicial scrutiny of the Act of God defense commonly involves a fact-specific analysis to determine whether a particular natural event qualifies as an Act of God.

For example, in cases involving delays in construction projects due to heavy rain, courts will examine short- and long-term precipitation data to determine whether the period of heavy rainfall constitutes an Act of God that makes performance of the contract impracticable.

As climate change has increasingly exposed the flaws in the Act of God defense, at least one commentator has called for the judiciary to limit the application of the doctrines of impracticability and frustration of purpose to events that can truly be classified as extraordinary, unforeseen, and contractually unassumed given today's readily available knowledge of weather patterns and climate change.[86] This policy shift might lead to a system of more accurate risk prediction and climate change cost internalization by contracting parties.[87] For example, the more narrow application of the impracticability and frustration of purpose doctrines might encourage parties to take steps to prevent the negative effects of climate change through adaptation and insurance rather than resorting to after-the-fact litigation.[88] More careful risk evaluation and acceptance of the true costs of climate change by the business sector might also increase awareness of the financial consequences of climate change at both the government and private levels.[89] Thus, there are substantial public policy grounds for limiting the application of the doctrines of impracticability and frustration of purpose in response to climate change.[90]

5.4 Conclusion

While climate adaptation liability remains a developing area of law, it is clear that these types of cases will be an important tool as the effects of climate change continue to increase in size and scope. A compounding factor for design professionals is that many codes and standards, including building codes, fail to explicitly account for climate risks. As a result, design professionals will be less able to rely on them for meeting their standard of care and preparing for foreseeable harm. Updates to existing regulatory and statutory frameworks that explicitly account for climate risks could help decrease uncertainty for the design professional community, which may otherwise be vulnerable to liability claims for failing to adequately adapt. Some jurisdictions are proactively pursuing these updates, but the inconsistency in these efforts underscores a continuing need for design professionals to incorporate climate risks into their own analyses and project designs.

Although this chapter focused primarily on design professional liability, designers are not the only relevant actors nor are they the ultimate decisionmakers in most projects. Other private actors including contractors, property owners, and others, may be subject to liability as well. In some cases, design professionals may ultimately be accountable to other actors, often their clients, who are not interested in incorporating climate risks into their projects. Reviewing the roles and responsibilities of these other actors in the climate adaptation liability context could be the subject of further research and analysis. It may also provide design professionals with a platform to advocate for more forward-looking codes, standards, and practices.

Notes

1 *See e.g., American Electric Power (AEP) v. CT,* 131 S. Ct. 2527 (2011); *Native Village of Kivalina v. ExxonMobil Corp.,* 2012 WL 4215921 (2012); *Comer v. Murphy Oil USA, Inc.,* 607 F.3d 1049 (2010).

2 *See, e.g., Uhley v. Tapio Const. Co.,* 573 So. 2d 390, 391 (Fla. Dist. Ct. App. 1991) (contractor who relied on survey prepared by third party with specifications that were marked as "assumed" and "assigned" had an independent duty (i.e., standard of care) to ensure the correctness of that information, and therefore was liable for flood damages from negligent design).

3 The rule of negligence *per se* states that if a defendant's action violated a law or regulation, then the court will consider the action to be negligent without asking whether or not a reasonable person would have done the same thing. That is, the court presumes a breach of duty occurred, and moves on to assess what (if any) damages occurred as a result of that breach. For example, in *Blake v. HI-Lu Corp.,* 781 So.2d 1122 (Fla. Dist. Ct. App. 2001), a Florida state appellate court held that a defendant negligently designed and constructed a home that collapsed during Hurricane Andrew because the underlying building code required the home to be built to withstand wind speeds of up to 120 miles per hour, which was a higher wind speed than occurred during the early stages of the storm when the home collapsed.

4 Chris Arsenault, "Canada's Building Code is Getting a Climate Change Rewrite," CBC News, Apr. 11, 2019, https://www.cbc.ca/news/canada/canada-building-code-climate-change-resilience-1.5092732.

5 Australian Building Codes Board. (2014). "Resilience of Buildings to Extreme Weather Events." Retrieved from: https://www.abcb.gov.au/ Resources/Publications/Consultation/Resilience-of-Buildings-to-Extreme-Weather-Events.

6 Council of Standards Australia and Council of Standards New Zealand. (2011). "Structural Design Actions Part 2: Wind Actions." Retrieved from: https://shop.standards.govt.nz/catalog/1170.2:2011(AS%7CNZS)/scope.

7 International Code Council. (2015). International Building Code.

8 Climate Resiliency Design Guidelines (March 2019), NYC Mayor's Office of Recovery and Resilience, https://www1.nyc.gov/assets/orr/pdf/NYC_Climate_Resiliency_Design_Guidelines_v3-0.pdf.

9 Climate Resilient Design Standards and Guidelines for Protection of Public Rights-of-Way (Oct. 2018), Boston Public Works Department, https://www.boston.gov/sites/default/files/imce-uploads/2018-10/climate_resilient_design_standards_and_guidelines_for_protection_of_public_rights-of-way_no_appendices.pdf.

10 *Ibid.,* p. 1.

11 *The T.J. Hooper,* 60 F.2d 737, 737 (2d Cir. 1932).

12 *Ibid.,* p. 739.

13 *Ibid.,* p. 740.

14 In a later case, *United States v. Carol Towing,* 159 F.2d 169 (2nd Cir. 1947), Judge Learned Hand crafted an algebraic formula for sorting whether an industry custom

met the standard of reasonableness, ($B = PL$). Negligence liability turns on the relation between investment in precaution (B) and the product of the probability (P) and magnitude (L) of harm resulting from the accident. If the product of P and L exceeds B, then the defendant should be liable. If B equals or exceeds the product of P and L, then the defendant should not be held liable.

15 228 N.W.2d (Iowa 1975).

16 *Ibid.*

17 *Laukkanen v. Jewel Tea Co.*, 222 N.E.2d 584, 587 (Ill. App. Ct. 1966).

18 *Ibid.*, p. 588. But note that the court said the analysis may have changed if there had been evidence of a tornado or other catastrophe in the neighborhood. However, given the increasing frequency and predictability of tornadoes and catastrophic weather events today, it is difficult to imagine a court making the same statement in a modern negligence decision. This very question is at issue in the *Lijo v. Costello, Inc.* case discussed below.

19 *See, e.g., L. H. Bell & Assocs., Inc. v. Granger*, 543 P.2d 428, 433 (Ariz. 1975) (although an engineer's bridge design met contractual standards, the engineer negligently failed to protect against foreseeable flooding of neighboring properties).

20 Plaintiffs' Original Petition and Request for Disclosure, available at https://www.scribd.com/document/375653781/Plaintiffs-Original-Petition-and-Request-for-Disclosure; Plaintiffs' Third Amended Petition, filed on Oct. 2, 2020, contains the same claims as the original 2018 Petition, on file with author.

21 Defendant's Original Answer Subject to Its Motion to Transfer Venue, May 18, 2018, on file with author.

22 Plaintiff's Third Amended Petition and Request for Disclosure, Oct. 2, 2020, on file with author.

23 *Texas v. Arkema Inc; Texas v. Richard P. Rowe, Texas v. Leslie Comardelle and Texas v. Michael P. Keough*, case number 1600310, 1627625, 1600311, 1600312 and 1627626, in the 339th District Court of Harris County, Texas; *see also* Rebecca Hersher, *Case Alleges Chemical Companies Should Prepare for Unprecedented Storms*, NPR, (Aug. 30, 2018), https://www.npr.org/2018/08/30/638936124/lawsuit-alleges-chemical-companies-should-prepare-for-unprecedented-storms.

24 Arkema Inc. Incident News, "Previous Update Statements," April 10, 2019, https://www.arkema-americas.com/en/social-responsibility/incident-page-2/previous-update-statements/.

25 Rebecca Hersher, *Texas Criminal Trial Highlights Liability for Factories in Floodplains*, NPR (Mar. 2, 2020) https://www.npr.org/2020/03/02/723217659/texas-criminal-trial-highlights-climate-liability-for-factories-in-floodplains.

26 Michelle Casady, *Arkema Fully Cleared in Criminal Emissions Trial*, Law360 (Oct. 1, 2020), https://www.law360.com/articles/1315816/arkema-fully-cleared-in-criminal-emissions-trial-.

27 Katie Watkins, *No Convictions in Arkema Trial After Judge Drops Remaining Charges*, Houston Public Media (Oct. 1, 2020), https://www.houstonpublicmedia.org/articles/news/energy-environment/2020/10/01/383091/no-convictions-in-arkema-trial-after-judge-drops-remaining-charges/.

28 *Ibid.*

29 Kenneth T. Kristl, "Diminishing the Divine: Climate Change and the Act of God Defense," 15 *Widener L. Rev.* 325, 325 (2010).

30 Denis Binder, "Act of God? Or Act of Man?: A Reappraisal of the Act of God Defense in Tort Law," 15 *Rev. Litig.* 1, 13 (1996).

31 *See, e.g.* Jill M. Fraley, "Re-Examining Acts of God," 27 *Pace Envtl. L. Rev.* 669, 687 (2010); Kristl, *supra* note 29, at 327.

32 Jill M. Fraley, "Re-Examining Acts of God," 27 *Pace Envtl. L. Rev.* 669, 674 (2010).

33 Kristl, *supra* note 29, at 329.

34 *Ibid.*, p. 330.

35 *Ibid.*, p. 331.

36 *Ibid.*, p. 333.

37 *Ibid.*

38 *McFarland v. Entergy Mississippi, Inc.*, 918 So. 2d 697, 698 (Miss. Ct. App. 2004) ("An Act of God, such as an unprecedented storm, will excuse liability, provided the storm is the 'sole cause of the injury'; however, 'if the injury is caused by an Act of God, in connection with which the negligence of the defendant is a concurring cause, and the injury would not have occurred except for such negligence,' the defendant is liable.") (quoting *Public Service Co. v. Sonagerra*, 208 Okla. 95, 98, 253 P.2d 169, 171 (1953)).

39 Kristl, *supra* note 29, at 356.

40 *Ibid.*

41 Binder, *supra* note 30, at 77.

42 *See* Amended Complaint at 56-60, *Conservation Law Foundation, Inc. v. ExxonMobil Corp. et al.*, No. 1:16-cv-11950 (D. Mass. filed Oct. 20, 2017), ECF No. 34.

43 *Ibid.*

44 *See* Order at 2, *Conservation Law Foundation Inc.*, ECF No. 29 (D. Mass filed Sept. 13, 2017).

45 *Ibid.*

46 *See* Transcript of Oral Argument at 135:23-24, *Conservation Law Foundation, Inc*, ECF No. 73 (D. Mass filed Mar. 19, 2019).

47 *Ibid.*, p. 136:13-16.

48 For most recent case updates, *see* http://climatecasechart.com/case/conservation-law-foundation-v-exxonmobil-corp/?cn-reloaded=1.

49 *CLF v. Shell Oil Products*, 1:17-cv-00396, Complaint, D.R.I. (2017), complaint available for download at http://climatecasechart.com/case/5619/.

50 *Ibid.*

51 *Ibid.*

52 *Ibid.*

53 *CLF v. Shell Oil Products*, 1:17-cv-00396, Memorandum and Order, D.R.I. (2020), on file with author.

54 Robert Boyd, "Virginia Beach Unanimously Votes Against Rezoning Holland Swamp," WVEC News, https://www.13newsnow.com/article/news/local/virginia-beach-tells-developer-no-swamp-for-you/291-541622871.

55 *Argos Properties II, LLC v. City Council for the City of Virginia Beach et al.*, Petition for Review and Complaint at 6, available at http://climatecasechart.com/case/argos-properties-ii-llc-v-city-council-for-virginia-beach/.

56 *Ibid.*

57 *Ibid.*

58 Order available at http://climatecasechart.com/case/argos-properties-ii-llc-v-city-council-for-virginia-beach/; *see also* Peter Coutu, "Judge Rules Virginia Beach Council Can Factor in Sea Level Rise When Deciding on New Developments," *The Virginia Pilot,* Apr. 24, 2019, available at https://pilotonline.com/news/government/local/article_f05e1e32-66e1-11e9-b07a-ff3214644d86.html.

59 33 U.S.C. § 1321(a)(12); 42 U.S.C. § 9601(1); and 33 U.S. Code § 2701(1).

60 Clifford J. Villa, "Is the 'Act of God' Dead?," 7 *Wash. J. Envtl. L. & Pol'y* 320, 326 (2017); H.R. Rep. No. 99-253, pt. 1, at 22-23 (1977).

61 Kristl, *supra* note 29, at 345.

62 *Id*; H.R. Rep. No. 91-940 (1970), at 37.

63 *See US v. Barrier Indus. Inc.*, 991 F. Supp. 678, 679 (S.D.N.Y. 1998) (involving spill of hazardous substances caused by bursting of pipes following unprecedented cold spell).

64 *See Sabine Towing & Transp. Co. v. US*, 666 F.2d 561, 562 (Ct. Cl. 1981) (involving oil spill caused by vessel striking unknown underwater object in the Hudson River).

65 *See US v. M/V Santa Clara I,* 887 F. Supp. 825, 843 (D.S.C. 1995) (rejecting an Act of God defense under CERLCA for loss of containers of arsenic trioxide overboard because of storm); *Kyoei Kaiun Kaisha, Ltd. v. M/V Bering Trader,* 795 F. Supp. 1054, 1056 n.2 (W.D. Wash. 1991) (rejecting attempt to assert Act of God defense for liability under Clean Water Act arising from oil releases after grounding of vessel: "Defendants, however, fail to put forth any evidence that the weather on the night of the grounding could not have been foreseen or guarded against, as required by this defense"); *US v. Stringfellow,* 661 F. Supp. 1053, 1061 (C.D. Cal. 1987) (rejecting the claim that heavy rainfall causing releases from hazardous waste site was act of God under CERCLA).

66 *See US v. Alcan Aluminum Corp.*, 892 F. Supp. 648, 658 (M.D. Pa. 1995) (rejecting a claim that Hurricane Gloria caused used oil emulsion discharge to Susquehanna River).

67 Kristl, *supra* note 29, at 343-344.

68 *In re Sept. 11 Litig.*, 931 F. Supp. 2d 496, 504 (S.D.N.Y. 2013), *aff'd,* 751 F.3d 86 (2d Cir. 2014).

69 *Stringfellow*, 661 F. Supp. at 1061.

70 Kristl, *supra* note 29, at 344-345.

71 Villa, *supra* note 61, at 328.

72 *Ibid.*

73 *Ibid.*, p. 331.

74 The authors would like to acknowledge the contributions of attorneys Bill Franzek and Andrew Selman to this section on contract liability.

75 The standard form contractual agreements reviewed include the Engineers Joint Contract Documents Committee (EJCDC) document E-500, Agreement Between Owner and Engineer for Professional Services (2014); the American Institute of Architects

(AIA) Document B103—2017, Standard Form of Agreement Between Owner and Architect for a Complex Project (2017); and ConsensusDocs 240—Standard Agreement Between Owner and Design Professional (2017). The EJCDC is sponsored by the National Society of Professional Engineers, the American Council of Engineering Companies, and the American Society of Civil Engineers. The ConsensusDocs contracts are developed by a coalition of national industry associations in the design and construction industry.

76 EJCDC E-500 § 6.01(A).

77 EJCDC E-500 § A1.01(A).

78 EJCDC E-500 § A2.01(A)(1).

79 ConsensusDocs 240 § 3.3.

80 AIA B103—2017 § 8.1.3.

81 EJCDC E-500 § 6.11(A).

82 Myanna Dellinger, "An 'Act of God'? Rethinking Contractual Impracticability in an Era of Anthropogenic Climate Change," 67 *Hastings L.J.* 1551, 1561 (2016).

83 *Ibid.*, p. 1572.

84 *Ibid.*, p. 1599.

85 *Ibid.*, p. 1572.

86 *Ibid.*, p. 1559.

87 *Ibid.*, p. 1560.

88 *Ibid.*, p. 1608.

89 *Ibid.*, p. 1607.

90 *Ibid.*, p. 1614.

6

The Role of Government and Regulation in Climate Change Adaptation

Stephanie Kruel

6.0 Introduction

Using the City of Boston, Massachusetts, as an example, this chapter examines how we can improve our ability to adapt the built environment to the impacts of climate change, namely increased coastal flooding, through improved governance.

The laws, regulations, and policies that underpin our design and planning efforts have been developed with the assumption that our climate would remain in relative stasis. While some variation was expected, it was assumed that past conditions basically represent present-day and future conditions. However, climate conditions have in fact been changing since the introduction of many of the regulations and policies that engineers, design professionals, and planners adhere to today, and they are on a trajectory to change more rapidly than they have since the beginning of human history. Our current governance framework likely needs to be either modified by incremental means or transformed swiftly (or perhaps a bit of both) to respond and adapt to rapidly changing climate conditions.

Due to its coastal location, Boston is vulnerable to a variety of changing climate conditions, particularly flooding exacerbated by sea level rise. This chapter will focus on a recent collaborative effort between the City and the University of Massachusetts, Boston[1] to address how existing regulatory standards can be modified to adapt our built environment to climate change.

This chapter will:

1. Describe the role of both government and governance in adapting to climate change;
2. Identify key governance capabilities needed to support adaptation;
3. Explore how both incremental changes to existing regulatory standards and transformational governance can bring about significant change with respect to adaptation; and
4. Discuss the role engineers, design professionals, and planners play not only in implementing revised standards, but in helping to formulate new standards for climate adaptation.

6.1 Government and Governance

The role of a government is to "govern" or oversee the rules, regulations, and interactions of the people to whom they apply. In the U.S. and other parts of the world, governments are composed of elected and appointed officials, as well as civil servants, who run the day-to-day operations within a political boundary and help plan for its future. "Governance" is a broader term than "government." It encompasses the processes that enable people and institutions to interact and societies to plan, make decisions, and implement activities within and across political boundaries. Governance includes all of us—public agencies, businesses, civic and academic

Figure 1. **Boston's Coastal Location Makes it Vulnerable to Flooding Due to Sea Level Rise**

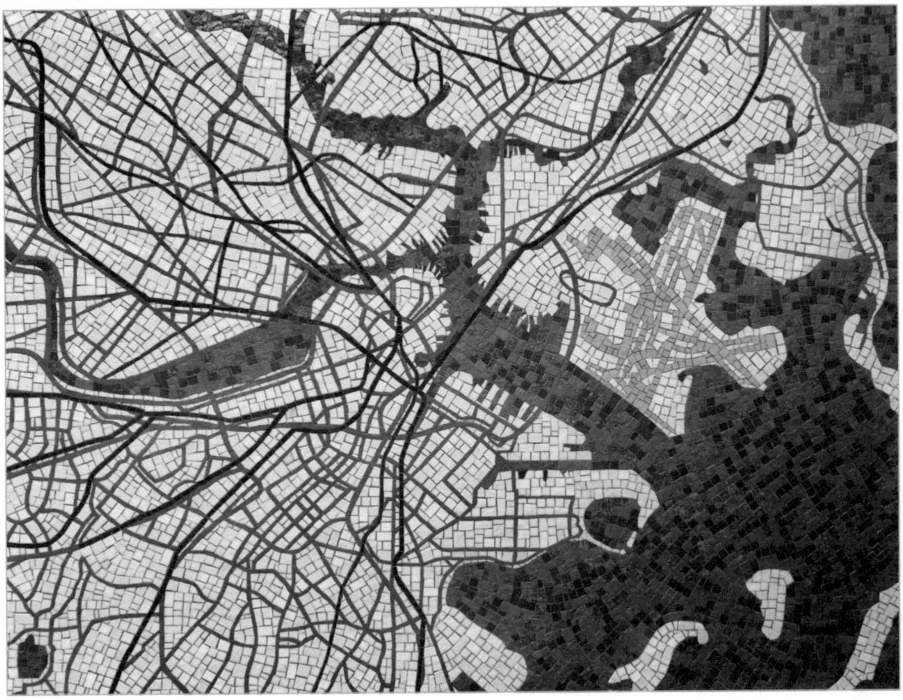

Credit: Stephanie Kruel.

institutions, and individuals—and the rules and norms that shape who we are and what we do. Societies always have multiple centers of power—in government and across the public sector; in individual firms and across the private sector; in a broad array of professional organizations and community institutions and across civil society. The choices we make in developing and organizing markets and communities are influenced not only by legal and social institutions, but also by individual mindsets, habits, and expectations.

In short, governance refers to *how* things are done, rather than *what* is done.

Governance can be thought of as a continuum, from using less-formal concepts within the realm of civil society, to more formal ones in the private and public sectors (Figure 2, next page). Effective governance requires the coordinated actions of several types of actors: public agencies, private for-profit and non-profit enterprises, communities, and individuals. Each plays a different role, and each is necessary.

In order to be successful, all components of governance must be oriented toward addressing the impacts of climate change (or any complex issue). The following section explores just one of the many components of governance for a changing climate: public agencies (i.e., the government).

Figure 2. **Components of Governance**

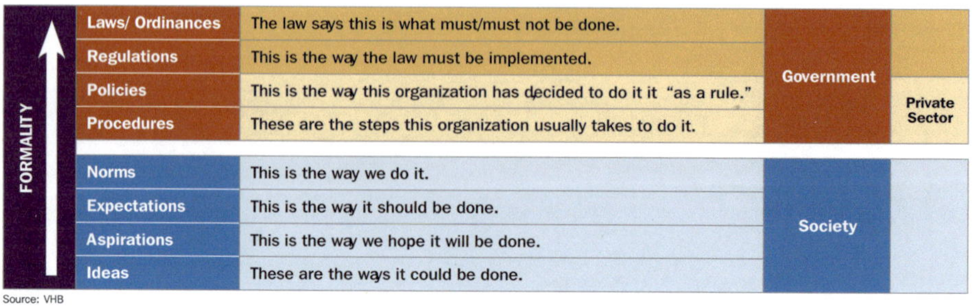

Source: Governance for a Changing Climate: Adapting Boston's Built Environment for Increased Flooding. UMass Boston Sustainable Solutions Lab. September 2018.

6.1.1 **The Government's Role**

The government's role is to develop, enact, and enforce laws, ordinances, regulations, policies, and plans (referred to herein collectively as "regulatory standards") at the local, county, regional, state, tribal, or federal levels. The regulatory standards can be understood as different tools with specific purposes, as follows:

- Laws are written statutes, passed by either the U.S. Congress or state legislatures, while ordinances are laws or statutes enacted by municipalities.
- Regulations are standards and rules adopted by administrative agencies that define how laws will be interpreted, implemented, and enforced. Regulations often have the same force as laws, since, without them, regulatory agencies wouldn't be able to enforce laws.
- Policies are statements of intent to guide decisions and achieve rational outcomes, and they are implemented as procedures or protocols. While they do not have the force of law, policies do provide more flexibility because they are easier to change.
- Plans are the product of public processes in which a location's land use, economic, environmental, and social trends are analyzed, and from which optimal regulations, infrastructure plans, and public realm visions may be established. Plans can be adopted as binding rules or regulations, but most often they serve as guidance documents that for adopting ordinances and regulations, implementing projects, and conducting additional studies.

The regulatory standards most pertinent to engineers, designers and planners are those that guide or control a variety of real estate and infrastructure-related domains, such as land use, building design and construction, water use and wastewater generation, stormwater management, air quality, accessibility, natural resource protection, historic and archaeological resource protection, health and public safety, transportation, site contamination and hazardous materials, geotechnical investigations and design, and structural design. Many of these regulatory standards establish minimum requirements to safeguard public health, safety and general welfare, while others go beyond that to promote additional goals.

Regulatory standards evolve over time. Individual laws, ordinances, regulations, and policies are often created in response to disasters and are designed to mitigate or prevent their future occurrence. Seismic and fire codes are excellent examples of this. When harm is not adequately prevented by existing regulatory standards, and as public demand for change mounts, they are revised, replaced, or repealed. Typically, past and current conditions are assessed (or reassessed), and protective regulatory solutions are devised and adopted. What has often been missing from this process is consideration of projected future conditions.

6.1.2 Changing Climate, Changing Governance

While there has always been some variability in weather patterns, the overall climatic average has remained, until recently, within a predictable range. Around the middle of the twentieth century, we began to document the dramatic impacts on the Earth's landscape and living systems caused by human activities, which had been driving significant atmospheric and hydrospheric changes. Climate variations are becoming more extreme such that we are now seeing a shift in average climate indicators, and historical conditions can no longer reliably predict future conditions.

Meanwhile, we have continued to pass laws, adopt regulations, and plan for our cities and towns under the assumption of a fixed climate based on past trends. For example, our stormwater systems were designed based on precipitation volumes and intensities that were calculated at a particular moment in time and expected to remain more or less the same. In many instances we have not been accounting for the 70% increase in heavy precipitation events that occurred in the Northeast between 1958 and 2010.[2] Our stormwater management regulations do not currently acknowledge that if greenhouse gas emissions continue to increase at the current rate, the frequency of heavy downpours will also continue to increase. Similarly, buildings erected in areas with a 1% annual chance of flooding, as modeled by the Federal Emergency Management Agency (FEMA) and based on historical flooding patterns, are required to meet certain floodproofing standards outlined in local and state building codes. But updated models that use current and projected climate conditions show that the areas subject to flooding are often more extensive, and subject to deeper flooding, than indicated on official flood maps, leaving both existing and new buildings underprepared for inundation.

We now need to adjust the way we govern to acknowledge the reality of a continuously changing climate. Rather than focusing exclusively on a system of prescriptive regulations and standards that provide consistency across time and varied geographies, we must move toward systems of governance capable of confronting landscape-scale problems in a manner that is flexible and responsive enough to incorporate the complex feedback we receive from our environment.

6.2 Necessary Governance Capabilities for Adaptation

Using the City of Boston as a case study, this section examines seven governance capabilities that support development of a built environment adapted to the impacts of climate change. These include the ability to:

1. Generate, communicate, and integrate complicated, rapidly evolving information;
2. Conduct outreach and develop plans that engage a variety of stakeholders;
3. Develop and apply transparent, objective, and equitable criteria for project prioritization;
4. Create and implement laws, regulations, and policies that are equitable, provide a balance of stability and flexibility, and promote a resilient built environment;
5. Design, finance, construct, and maintain a system of shore-based, district-scale flood protection measures;
6. Design, finance, construct, and maintain infrastructure that will continue to function in a changing climate; and
7. Institutionalize flexibility through monitoring and evaluating outcomes.

6.2.1 Generate, communicate, and integrate complicated, rapidly evolving information

Understanding of the regional and local effects of climate change has evolved quickly. The Boston region has conducted some of the most sophisticated assessments of likely impacts and made them accessible to decision makers.[3] Some gaps still remain in the timing, intensity, and specific scale of the threats, primarily because of continuing uncertainty about when, whether, and how much global emissions will be reduced, but also due to emerging understanding of processes that have not yet been included in climate models, such as certain glaciological processes, abrupt permafrost thaw, and wildfire soot emissions, among others. Nevertheless, the types of risks posed by climate change are well known.

However, even when accurate information is available and has been communicated to its intended audience, a path to action may still be elusive. This type of challenge is related to the problem of linking science and policymaking, which are historically practiced in silos. The emerging field of sustainability science argues that "too much potentially valuable knowledge produced by committed researchers languishes in libraries, unused by society; and too many of society's greatest needs for new knowledge remain relatively unexplored by researchers."[4] Climate change is inherently a multi-sector, interdisciplinary problem. To respond effectively to the challenges of climate change, we need a comprehensive view of what drives change and the responses to it, and their relationship to each other in real time. Very few, if any, academic, public, or private institutions currently have that capacity. Addressing climate change adaptation requires a new culture of research that accounts for multiple interests and seeks to create actionable findings.

While many sectors are actively using climate change research and integrating climate risks into their endeavors, some important ones, including the insurance industry, lag behind. As was noted in the Sustainable Solutions Lab's *Financing Climate Resilience* report, "market signals in insurance and property markets are not yet fully reflecting climate risks,"[5] and, as of this writing, government regulation does not yet require better climate risk information in the insurance market. Finally, in a world of rapid change, we need to figure out how to continuously integrate new information into our governance practices and to update regulatory standards as new knowledge emerges.

6.2.2 Conduct outreach and develop plans that engage a variety of stakeholders

Over the past decade, Boston has engaged in wide-ranging planning, bringing climate change to the forefront of an already crowded agenda of citizens' concerns. Recent endeavors include the work of the Green Ribbon Commission, a group of business, institutional, and civic leaders in Boston working to develop shared strategies for fighting climate change; the suite of Climate Ready Boston Reports,[6] the City's initiative to prepare for the long-term impacts of climate change; and Imagine Boston 2030,[7] which defines a vision for Boston leading up to its 400th birthday (2030) and beyond.

While multiple additional planning exercises on the state and regional levels and in the transportation and utility sectors take climate risks into account, such efforts are only beginning to be coordinated and integrated. Initial research has identified shore-based flood mitigation measures[8] and community-based education and preparation efforts as integral to protecting Boston from climate impacts while simultaneously providing communities with opportunities to improve quality of life.[9] Comprehensive planning to launch adaptation investments now, and continuing to implement them decades into the future, will be critical. In addition, long-term and deep community engagement is necessary beyond specific plans or projects to promote community resilience.

6.2.3 Develop and apply transparent, objective, and equitable criteria for project prioritization

In order to gain support for solutions that will involve multiple stakeholders, time, money, and significant tradeoffs, it is vital that projects be prioritized in a fair and transparent manner. As the Climate Ready Boston Report points out, it is critical to consistently quantify the social, environmental, and economic benefits (and detriments) of each alternative intervention—with particular attention to social equity and the needs of socially vulnerable populations—so that they can be weighed both against the costs of a project and against each other. The Climate Ready Boston Report establishes a framework to evaluate flood risk reduction benefits, public co-benefits (open space, shoreline access, public amenities, etc.), environmental impacts, and financial costs (of land acquisition, design, maintenance, construction, land ownership, permitting and regulations, and intergovernmental coordination). Other specific community goals may need to be incorporated into such a framework.

6.2.4 Create and implement laws, regulations, and policies that are equitable, provide a balance of stability and flexibility, and promote a resilient built environment

Legal systems purposely foster stability and predictability.[10] Nevertheless, building codes, zoning laws, and state and federal guidelines and regulations have evolved in response to changes in technology, demands for increased energy efficiency, and greater emphasis on safety. However, thus far they have been developed with an assumption of a relatively static climate. To cope with a climate that is already changing, and with the uncertain magnitude and timing of that change, a new regulatory framework is needed to accommodate anticipated (and unanticipated) changes in sea level, coastal and inland flooding, extreme storms, extreme temperatures, high winds, and other weather variables. Such a framework needs to acknowledge impacts on people and their communities in addition to the built environment.

For example, under the business-as-usual emissions scenario, the likely range of sea level rise by 2070 in Boston Harbor is currently estimated to be between 1.5 and 3.1 feet.[11] As we approach 2070, actual sea level rise will be measured, and more accurate projections of additional sea level rise will become available. To protect property, businesses, homes, and communities in such a dynamic world will require an equally adaptable regulatory framework that can be responsive to new information, while still providing enough predictability to encourage investment.

6.2.5 Develop the capacity to design, finance, construct, and maintain a system of shore-based district-scale flood protection measures

The idea of a harbor-wide storm barrier has gained traction since it was first contemplated in the late 1980s. However, a recent analysis of a harbor-wide barrier system to protect against the impacts of sea level rise advises against pursuing such a strategy in the coming decades.[12] The analysis found a harbor-wide barrier system to be technically impractical and less effective at mitigating the impacts of sea level rise, dollar for dollar, than investment in shore-based coastal protection solutions such as those described in the Climate Ready Boston Report. The study's lead author noted that staying focused on neighborhood, shore-based measures, moving quickly and working closely with communities and landowners are the most effective strategies that the City could pursue. Challenges associated with implementing

Figure 3. **Potential Alignments for Shore-Based Flood Protection Measures**

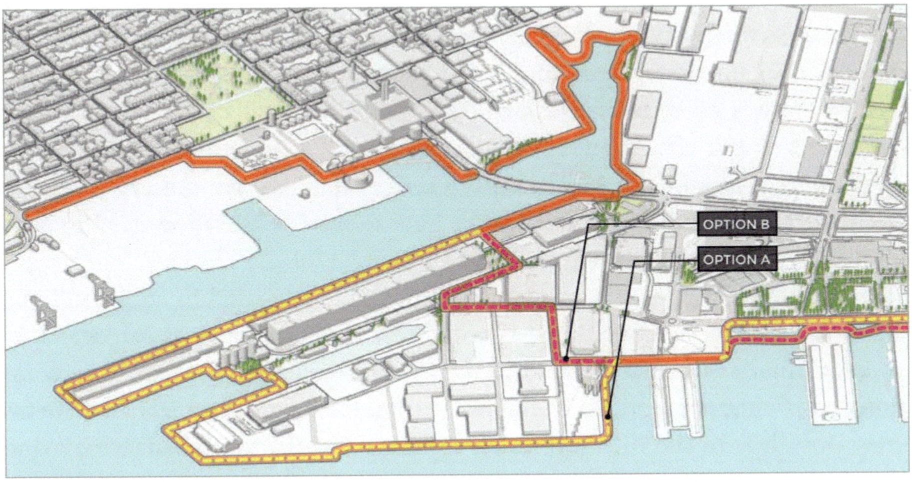

Climate Ready South Boston presents several potential alignments for shore-based flood protection measures.

Source: Coastal Resilience Solutions for South Boston Final Report. City of Boston, MA, 2018, p. 131. Available online at https://www.boston.gov/departments/environment/climate-ready-south-boston.

shore-based measures include studying, selecting, and designing the appropriate flood protection measures, as well as securing land, permits, and financing. Boston and the region will need to renovate existing governance institutions and regulatory standards, devise new ones, or combine different approaches to implement shore-based flood protection measures.

6.2.6 Develop the capacity to design, finance, construct, and maintain infrastructure that will continue to function in a changing climate

Boston's infrastructure, including water and sewer, transportation, energy, and telecommunications systems, is vulnerable to changing climate conditions.[13] These systems are not directly under municipal control, and while there is currently coordination among key private and public infrastructure owners and operators in the Boston metro area, it tends to vary in its quality and outcomes. For some issues, long-standing and well-structured coordination is well established, while for other issues, coordination is weak or absent. While many owners and operators have taken steps to understand their systems' vulnerabilities and make plans for improvements, not all have, and there are no universally shared climate design standards and no way to track investments in climate-adapted infrastructure. Bringing all infrastructure players together in a coordinated, perhaps more formal, fashion is yet another challenge, although one that is not unique to the field of climate change adaptation.

6.2.7 Institutionalize flexibility through monitoring and evaluating outcomes

For us collectively to learn, reflect on, and evolve our strategies over time we must prioritize, integrate, and fund robust monitoring and evaluation programs. Currently, there is no large-scale effort at systematic and purposeful observation or measurement of the effectiveness of climate-adaptation programs and activities in Massachusetts. Through active monitoring and evaluation, we can create a culture of learning that can provide a continuous flow of new information used to coordinate resource management across multiple systems.

The concept of "adaptive management," which is often used in conservation and international development settings, is applicable here. Adaptive management is an intentional approach to making decisions and adjustments in response to new information and changes in context. It emphasizes flexibility and requires that we avoid limiting ourselves in the future through rigid agreements today. Adaptive management can be especially useful for decision-making in the face of uncertainty, as it allows participants to change the path to achieving goals in response to changes in the environment. Who might perform this monitoring and evaluation function, at what scale, and with what tools needs to be determined.

6.3 Approaches to Adaptation-supportive Governance

Having identified the necessary capabilities of a climate-adapted governance system, the next step is to actually create such a system. One way is to take an incremental approach in which small, yet meaningful, steps are taken toward improving our existing institutions and regulatory standards. An incremental approach is useful when we want to do what we already do, but better. However, due to the rate of climate change and the extent of its impacts, it is likely that taking the incremental approach will be inadequate. More in-depth, larger scale, and/or faster changes may become necessary (although changes that meet those three criteria simultaneously are likely impossible).[14] Therefore, a transformational approach to governance should also be considered, one that involves innovative governing strategies, changes in power structures, and the introduction of new institutional arrangements and regulatory frameworks. This type of approach is useful when we want to completely change the way we do things. These two approaches are described in more detail below.

6.3.1 Regulatory Adjustments: An Incremental Approach

The small, cumulative wins that are characteristic of an incremental approach are an important start to producing the type and scale of adaptation needed and, given the often slow and complex nature of changing institutions, it is essential that we begin this incremental change now.[15] One way this can be done is by identifying ways our existing regulatory standards can be modified to produce successful outcomes (see below). When attempting to improve climate-adapted outcomes for any regulatory standard, both flexibility and predictability can often be better achieved through modifying performance standards, voluntary measures, incentives, and other forms of regulatory experimentation than through less malleable standards such as laws, ordinances, and regulations.

The following sections examine select regulatory standards at the local, regional, and state level that are related to flood protection measures on private land in Boston.[16] Each regulatory standard includes a description of its purpose; its applicability as it relates to climate change; its limitations in a changing climate, either of its applicability or the ways in which it actually limits that which it regulates; and suggestions for ways the regulatory standard can be altered to improve outcomes.

Local Government

Local governments, such as the City of Boston, have at their disposal a variety of regulatory standards that can be useful in climate change flood adaptation, including articles of the zoning code, policies, strategic planning, and land use planning, several of which are described below.

Flood Resilient Zoning Overlay District

- **Purpose:** Zoning is allowed as a proper function of Boston's police power, which is defined as the power to regulate for the advancement and protection of the health, morals, safety, and general welfare of the community as a whole (Chapter 665 of the Acts of 1956, Section 2). The Zoning Code allows the Zoning Commission to adopt regulations that control the use of land and buildings, as well as their dimensions, including first floor elevation (the elevation of the top of the first habitable floor of a structure); the maximum ratio of floor area and lot coverage; minimum lot sizes; minimum lot area for each dwelling unit or equivalent; minimum lot width; minimum dimensions of front, side, and rear yards; and maximum height of structures. It also provides requirements for open space, parking and loading, and accessory structures. As part of the City of Boston's multi-layered approach to climate adaptation, a Flood Resilient Zoning Overlay District (FRZOD) is currently under development. It is anticipated to establish a design flood elevation[17] based on the area modeled to be subject to flooding during the 1% annual chance of flood in the year 2070.
- **Applicability:** When any structure is erected, reconstructed, or substantially altered, it must be done in conformity with the zoning district in which it is located as per the city's Zoning Map. While the Zoning Code regulates massing, building placement, and use, the Massachusetts Building Code (see below) regulates the details of the structures themselves, such as load requirements and building methods and materials (for example, floodproofing materials).

- **Limitations in a changing climate:** Existing buildings that are not being altered are not subject to the Zoning Code, making it ill-suited to impact the vast majority of Boston's built environment. In addition, state and federal projects are not subject to local zoning ordinances. Finally, by law (M.G.L. 40A §3), the Zoning Code cannot be construed to regulate aspects of structures that are the domain of the Massachusetts State Building Code at 780 CMR, including floodproofing materials or methods of construction.
- **Improving climate-adapted outcomes:** Existing zoning district regulations may be amended to promote climate-adapted results. However, new districts such as an FRZOD are more likely to produce the desired outcomes, as they provide a fresh opportunity to create regulations specifically geared toward climate adaptation, rather than having to retrofit regulations that were originally created for other purposes.

Article 37 Climate Change Preparedness and Resiliency Policy

- **Purpose:** The Board of the Boston Planning and Development Agency (BPDA) approved in 2011, and updated in 2017, a Climate Change Preparedness and Resiliency Policy (the "Resiliency Policy"), which requires that certain projects subject to review under the Zoning Code consider present and future climate conditions in assessing project environmental impacts. These include carbon emissions, extreme precipitation, extreme heat, and sea level rise. The policy includes specific resilience targets for assessing project vulnerabilities and adverse impacts.

  Under this policy, each project must complete a Climate Resiliency Checklist that identifies initial strategies for reducing vulnerabilities and adverse impacts, as well as future adaptation strategies for meeting or exceeding resilience targets. The City's guidance document requires projects located within either the currently effective floodplain (FEMA's 1% annual chance floodplain) or the projected future floodplain (the area projected to be subject to flooding during the 1% annual chance flood in 2070 with 40 inches of sea level rise) to establish first floor elevations one to two feet above the future floodplain.
- **Applicability:** The Resiliency Policy must be addressed during the project review process established under the local zoning code. The Climate Resiliency Checklist must be submitted with the Project Notification Form, which is the first document submitted as part of a multi-step process.
- **Limitations in a changing climate:** While completing the checklist is mandatory, implementation of measures to increase climate adaptation is not. To effect adaptive outcomes, the City currently relies on negotiations during project review, when a variety of competing interests are under consideration. In addition, the Resiliency Policy is primarily applicable to new construction over 50,000 square feet, allowing smaller projects to avoid scrutiny. For these reasons, the beneficial impact of this policy is limited.

Figure 4. **Priority Flood Protection Areas**

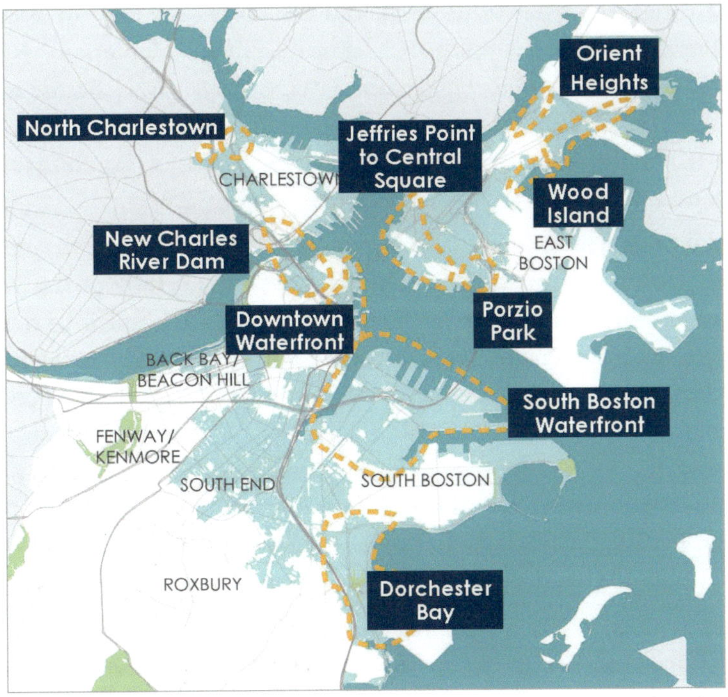

The Climate Ready Boston report prioritizes neighborhoods that need protection from future flooding.

Source: City of Boston, 2016. *Climate Ready Boston Final Report*. Boston, MA. p. 111. https://www.boston.gov/departments/environment/preparing-climate-change.

- **Improving climate-adapted outcomes:** With the adoption of the revised Resiliency Policy and Climate Resilience Checklist, the City has made progress in educating development proponents about future climate conditions and requiring those conditions to at least be considered during project design. Codifying this policy by incorporating it into the Flood Resilience Zoning Overlay District would help promote climate-adapted outcomes.

Climate Ready Boston

- **Purpose:** The 2016 Climate Ready Boston Report set the foundation for the City's ongoing climate preparedness activities through a strategic planning effort. The report included:
  - Updated projections of climate change in Boston;
  - A detailed vulnerability assessment of the city and specific focus areas; and
  - Principles, strategies, and initiatives to achieve the City's climate preparedness goals.

  Climate Ready Boston's vulnerability assessment identified East Boston, Charlestown, and South Boston as three of the neighborhoods most vulnerable to sea level rise and coastal flooding. Climate Ready Boston recommended that the City "prioritize and study the feasibility of district-scale flood protection" for these and five other focus areas, and "develop local climate resilience plans in vulnerable areas to support district-scale climate adaptation." Related planning efforts include *Coastal Resilience Solutions for East Boston and Charlestown* and *Coastal Resilience Solutions for South Boston*, both completed in October 2018.[18] Additional neighborhood-specific plans for Downtown/North End and Dorchester are anticipated to be completed in 2020.

- **Applicability:** The Climate Ready Boston Report addresses the entire city of Boston. Eight focus areas were selected where the results of the Vulnerability Assessment and the climate resilience initiatives are applied in more detail to illustrate the risks Boston faces and how Boston can address them. As a planning document, the Climate Ready Boston Report itself has no enforceability. However, it recommends multiple regulatory changes that would be enforceable, and the subsequent neighborhood-specific plans identify project costs and funding sources. The Climate Ready Boston Report identifies multiple parties responsible for implementation of various recommendations.
- **Limitations in a changing climate:** The report does not identify financial resources for many of its recommendations, which implies that those recommendations may take longer to pursue. This can be problematic where time is of the essence.
- **Improving climate-adapted outcomes:** Implementing the initiatives in the Climate Ready Boston Report is necessary but not sufficient to prepare Boston for the flooding anticipated to result from climate change. Actions at the federal, state, and regional levels, additional municipal actions, and private-sector actions are needed as well.

Imagine Boston 2030

- **Purpose:** Imagine Boston 2030 is the City's first comprehensive development plan in 50 years. It serves as both a strategic and land-use plan for Boston, and articulates the following approach to climate change adaptation:

> Boston's approach to climate adaptation will protect our existing housing and job centers and catalyze future growth that is climate-ready . . . . Boston will employ urban climate-adaptation solutions that produce multiple benefits—such as protective systems that also function as parks and active public realm. This approach will ensure that new value is captured to help fund improvements as we simultaneously leverage public funding. We will create layers of protection—from the district plan to community preparedness. We will take advantage of cycles of building and infrastructure rehabilitation and replacement to make investments, and we will design flexible, adaptable solutions that can evolve as climate conditions continue to change.

> Just as Boston has been a leader in reducing greenhouse gas emissions, we will become a global leader in protection and adaptation of waterfront cities, and in doing so, create new jobs and transferable expertise and unlock opportunities for growth.[19]

Applicability: The plan covers the geography of the entire city, and outlines programs, policies, and investments related to housing, health and safety, education, the economy, energy and the environment, open space, transportation, technology, arts and culture, land use and planning. As a planning document, Imagine Boston 2030 has no enforceability. However, an implementation strategy is woven through the plan.

- **Limitations in a changing climate:** Imagine Boston 2030 "doubles down" on waterfront growth. It declares that "the waterfront must continue to play a critical role in meeting the needs of future generations by providing spaces where new jobs can locate and where housing growth can alleviate pressure in existing neighborhoods." Four of the five growth areas identified in this plan are shown to be vulnerable to future coastal flooding. By supporting and encouraging growth in these areas, Boston is essentially committing to protecting them, which makes the development of successful adaptation strategies all the more important. Imagine Boston 2030 commits to partnering with federal, state, and private entities to invest in nature-based and hard-engineered flood protection to protect Boston's neighborhoods and strengthen its shoreline. The City plans to develop new infrastructure, prioritizing shoreline investments that integrate co-benefits such as open spaces and improved stormwater infrastructure, as part of public and private capital projects.

  The upside of this approach is that promoting these areas, which are already developed to varying degrees, as growth centers helps generate the investments that will be needed to protect them. However, this approach may end up putting more citizens in harm's way, thereby requiring further investment in emergency services. It may also end up directing resources to areas that investors, businesses, and residents ultimately reject in favor of inland neighborhoods less prone to coastal flooding. In addition, allowing new development in flood-prone areas could have a serious impact on the City's budget in the future, when its responsibility to new residents and property owners requires it to make increasingly large expenditures to either maintain infrastructure or incentivize property owners to relocate. Finally, like the Climate Ready Boston Report, this plan does not identify financial resources for many of its recommendations.

- **Improving climate-adapted outcomes:** Not all growth described in Imagine Boston 2030 is located along the waterfront. Neighborhood enhancement and expansion, as well as mixed-use core areas, are identified in inland locations. Actions in these areas could be prioritized to increase climate adaptation.

Regional Governance

Regional agencies, authorities, and organizations serve an important cross-jurisdictional role. Here we discuss the flood resilience-related policies of several regional associations, authorities, and commissions.

Watershed associations

- **Purpose:** The City of Boston lies within two major drainage basins: the Charles River Basin and the Boston Harbor Basin. The latter consists of the Mystic River, Neponset River, and Weir River Watersheds (which is outside of Boston). Independent nonprofit watershed associations have established themselves for various purposes related to each of these drainage basins, or watersheds. One of the most climate adaptation-focused is the Mystic River Watershed Association (MyRWA). In 2018, MyRWA established the Resilient Mystic Collaborative, which is currently comprised of 20 communities that are committed to working together on projects of regional significance to decrease their collective risk of harmful flooding, drought, extreme temperatures, storms, sea level rise and other climate-intensified risks. As of March 2021 the organization had secured $2.5M in regional climate resilience funding.
- **Applicability:** The work performed by each of the watershed associations is applicable to an entire watershed, which has several advantages. Flooding ignores political boundaries, and analyzing impacts and developing solutions at the watershed scale holds the promise of more holistic solutions. In addition, watershed associations can leverage resilience projects at key locations to protect larger areas that may lie in separate municipalities. For example, if upgrades to a dam were coordinated at the watershed level, upstream communities would have more input regarding the timing and extent of those upgrades, which could then be incorporated into their own flood protection programs.
- **Limitations in a changing climate:** Watershed associations are private nonprofit organizations. While they have local expertise and strong networks, they generally do not have the authority or capacity to implement large-scale adaptation efforts.
- **Improving climate-adapted outcomes:** Watershed associations could play a larger role in adaptation planning. Recognizing their local expertise and commitment and drawing them into formal processes could help grow local understanding of natural processes and related support for watershed-based adaptation planning.

Massachusetts Port Authority Floodproofing Design Guide

- **Purpose:** The Massachusetts Port Authority (Massport) is an independent, quasi-public authority enabled by the Massachusetts Legislature in 1956. New construction on Massport's properties is subject to the Massport Floodproofing

Design Guide, which aims to make buildings and operations more resilient to anticipated flooding threats. It specifies an elevation based on projected future flooding to which structures and critical facilities must be either elevated or floodproofed.

- **Applicability:** The design guide is intended to be used by Massport staff, tenants, third-party developers, design professionals, and contractors during planning, design, and construction at Logan International Airport, the Paul W. Conley Container Terminal, and other Massport properties in South Boston. Projects subject to the standards include new structures and additions, substantial repairs or improvements, and retrofits with the explicit objective to make them resilient to flooding. Critical infrastructure is subject to specific floodproofing performance standards.
- **Limitations in a changing climate:** These standards are only applicable to Massport's properties, and therefore their beneficial impact is limited. In addition, they differ in many respects from standards that are applicable to adjacent properties that are not owned by Massport, which poses potential challenges for infrastructure continuity and urban design. There are also challenges with tenant-leased properties, owned by Massport, but for which the leasing terms may pre-date the more recent climate initiatives.
- **Improving climate-adapted outcomes:** In an effort to keep the Floodproofing Design Guide relevant and current, Massport actively seeks best available information and, when appropriate, changes their standards. Massport could coordinate to align their standards with those of other state and local government entities. They could also promote their Floodproofing Design Guide as a model to those entities that do not currently have their own standards.

State Government

Massachusetts has several laws, regulations, policies, and plans that are important for adaptation to climate change-induced flooding. It is not uncommon for a single development project to be subject to many or even all of the laws and regulations listed below, which sometimes complement, and other times contradict, each other. These regulatory standards are in various stages of being updated to address climate change. Interestingly, one of the reasons why it is difficult to keep laws and regulations current is because references to studies and data must be specific at the time of their adoption: They cannot reference future anticipated updates to publications because all aspects of a law or regulations must be available for review during the statutory public comment period.

Chapter 91, The Massachusetts Public Waterfront Act

- **Purpose:** The Massachusetts Public Waterfront Act is a state law (M.G.L. c. 91) with implementing legislation (310 CMR 9), known as the Waterways Regula-

Figure 5. **Guaranteed Public Access**

Chapter 91 of the Massachusetts Public Waterfront Act guarantees public access to tidelands and waterways for the purposes of navigation, fishing, fowling, and passing over or through the water and intertidal area. *Credit:* Stephanie Kruel

tion Program or collectively as "Chapter 91." It is administered by the Massachusetts Department of Environmental Protection (MassDEP). In Massachusetts, private property ownership is recognized to the low tide line. Chapter 91 guarantees public access to tidelands and waterways for the purposes of navigation, fishing, fowling, and passing over or through the water and intertidal area. Enforcement of Chapter 91 has resulted in, for example, the creation of Boston's Harborwalk, a continuous 40-plus-mile public walkway along the shoreline.

- **Applicability:** Chapter 91 regulates activities on both coastal and inland waterways, including construction, dredging and filling in tidelands, great ponds and certain rivers and streams. Chapter 91 jurisdiction applies to flowed tidelands to the mean high water elevation and to filled tidelands (tidelands filled to create dry land) to the farther of the first public roadway open to vehicles or 250 feet from mean high water elevation along the coast, great ponds, and non-tidal rivers and streams. Activities subject to Chapter 91 include construction, placement, excavation, addition, improvement, maintenance, repair, replacement, reconstruction, demolition or removal of any fill or structures not previously licensed, or a change in use or structural alteration of fill or structures authorized under a prior license. When the mean high-water elevation reaches further inland, or if additional fill were permitted to be placed within flowed tidelands, Chapter 91 would be applicable to additional areas of land.
- **Limitations in a changing climate:** Sea level rise has long-term impacts on the protection of the public trust. While Chapter 91 does require that sea level rise during the design life of the buildings (which is undefined in Chapter 91) be considered, it currently allows this rise to be based on historical rates. However, even this provision is not consistently applied during the review

process. Also, due to sea level rise, the shoreline will change more rapidly than it has in the past, complicating the obligation of private landowners to provide long-term access to ever-narrowing intertidal areas and more frequently flooded waterfront walkways. Rising sea levels will also put property lines into question since state ownership typically moves with tide lines. Currently, resilience to climate change impacts is not recognized as a public benefit under Chapter 91 regulations, which often results in a direct conflict between implementation of climate adaptation measures and Chapter 91 requirements. For example, non–water-dependent uses, as listed in the statute, are required to reserve ground-floor spaces in facilities of public accommodation, provide ground-level usable open space, and preserve access or sight lines to the water. These requirements often directly conflict with implementing climate-adapted measures such as raising first occupiable floors and seawalls. While Chapter 91 does not categorically restrict the use of fill for flood protection purposes, it does not expressly allow it for that purpose either. Under the existing regulations, in-water coastal flood protection measures could potentially be licensed as a water-dependent use project, as a water-dependent infrastructure project, or through the variance process.

- **Improving climate-adapted outcomes:** MassDEP is currently seeking to determine what actions are appropriate to accommodate predicted sea level rise. An advisory group has crafted several possible approaches to increasing resilience, such as using projected rates of sea level rise in the engineering standards; creating a regulatory definition of sea level rise; adjusting licensing requirements to account for sea level rise; and allowing building height exemptions/modifications for non–water-dependent buildings. Other ideas for increasing resilience include allowing climate-adapted structures and measures to be considered public benefits, and to be counted toward, rather than against, open space requirements, and to specifically categorize flood protection structures as either water-dependent infrastructure or a water-dependent use. Once climate adaptation is better incorporated into the Waterways regulations, there will be more provisions that can be substituted or amplified with a direct effect on climate adaptation through Municipal Harbor Plans, thereby strengthening that regulatory standards as well.

Wetlands Protection Act

- **Purpose:** The Wetlands Protection Act (WPA) (M.G.L. c 131 §40) is a state law, is regulated under 310 CMR 10, and is administered in Boston by the Boston Conservation Commission. The WPA protects eight important public values, or "interests," including protection of public and private water supply, protection of groundwater supply, flood control, storm damage prevention, prevention of pollution, protection of land containing shellfish, protection of fisheries, and protection of wildlife habitat.

Figure 6. **Flooding in South Boston During Winter Storm Riley**

Land subject to coastal storm flooding in South Boston during Winter Storm Riley, March 2, 2018. *Credit:* Stephanie Kruel

The WPA applies to inland and coastal resource areas, such as land under the ocean, designated port areas, coastal banks, and land subject to coastal storm flowage (LSCSF). The boundary of LSCSF is coterminous with the 1% annual chance flood elevation identified on the FEMA Flood Insurance Rate Maps (rather than the precise boundaries depicted on those maps) or flooding from the storm of record, whichever is higher. Unlike virtually all other resource areas defined in the WPA, LSCSF has no dedicated preambles, definitions, or performance standards in the WPA. There is also no buffer zone to LSCSF.

- **Applicability:** The WPA is applicable to projects that constitute work within jurisdictional resource areas. Project proponents must seek and obtain an Order of Conditions, Determination of Applicability, or Order of Resource Area Delineation prior to the start of work. The Order of Conditions describes the environmental protections that must be in place before, during, and after the proposed work is completed.
- **Limitations in a changing climate:** The main limitation to the WPA in a changing climate is that it assumes resource areas are static. However, today's resource areas may not have the same boundaries as tomorrow's resource areas. For example, LSCSF will encroach landward as time passes, and buildings constructed today outside of LSCSF may soon be within that resource area. Another limitation is that the Act does not distinguish between urban, developed LSCSF, which often is no longer available to perform most natural floodplain functions, and less-developed or undeveloped LSCSF, which often continues

to perform natural floodplain functions: These resource areas will be impacted differently by climate change. Any conditions placed on work in LSCSF are required to regulate the impact of development on the wetlands resource interests protected by the WPA, and not the impact of the environment on development, or the development's ability to function under different climate conditions. This makes it challenging to address projected flooding impacts to development using this regulation. Another limitation is the regulations' silence on the categorization and treatment of flood control projects. Finally, while fill is not prohibited in many of the resource areas that would be impacted by shore-based flood protection measures, Conservation Commissions would be unfamiliar with reviewing such projects and could benefit from guidance on how to apply the WPA's performance standards in such instances.

- **Improving climate-adapted outcomes:** It is generally accepted that LSCSF is significant to the interests of storm damage prevention and flood control and that building appropriately in LSCSF can be accomplished by following the State Building Code at 780 CMR 3107. If the State Building Code were to require that buildings in the future floodplain be constructed to reduce flood damage, LSCSF would be positively impacted. In April 2014 MassDEP convened an advisory group to discuss revisions to the WPA related to developing performance standards for LSCSF. The group was disbanded prior to issuance of such standards, reconvened in 2018 for further work, and is anticipated to produce recommended revisions at some point in the future. Regarding flood control, in concert with categorizing flood control structures as water-dependent uses in the Chapter 91 regulations, guidance should be developed for how impacts to jurisdictional coastal resources, including Land Under Ocean, Coastal Banks, Rocky Intertidal Shores, Land Containing Shellfish and Fish Runs, should be evaluated related to projects involving the creation of fill within those resource areas for the purpose of adapting to climate change impacts. Finally, the WPA allows municipalities to adopt a local wetlands bylaw or ordinance to further regulate wetlands. The City of Boston could draft its own ordinance to help increase protection of wetland resource areas from the impacts of climate change.

The Massachusetts Building Code

- **Purpose:** The Massachusetts Building Code is a state regulation (780 CMR: Massachusetts Amendments to the International Building Code 2009) adopted by the Board of Building Regulations and Standards. It is locally administered by the City of Boston's Inspectional Services Department. The stated purpose of the building code is "to establish the minimum requirements to safeguard the public health, safety and general welfare through structural strength, means of egress facilities, stability, sanitation, adequate light and ventilation, energy conservation, and safety to life and property from fire and other hazards attributed to the built environment and to provide safety to fire fighters and emergency responders during emergency operations." The building code provides minimum requirements for flood-resistant design and construction of applicable structures, as well as energy efficiency.

- **Applicability:** The flood-related portions of the building code are applicable to structures within the Special Flood Hazard Area delineated on the currently effective FEMA Flood Insurance Rate Map (FIRM). Compliance is required for construction that includes: (a) new structures, including subsequent work to such structures; (b) work classified as substantial repair or substantial improvement of an existing structure that is not an historic structure; and (c) substantial repair of foundations. New and replacement mechanical and electrical systems must comply with the elevation and floodproofing requirements, even if not part of a substantial repair or improvement.
- **Limitations in a changing climate:** Since the building code only applies to new construction or substantial repairs or improvements of existing structures, it cannot be used to address the resilience of existing buildings that do not need substantial improvements. Also, while the code emphasizes that the standards are only minimums, there is generally little incentive to build to a higher standard, particularly if doing so would increase the cost of the project. Another limitation is that requiring code compliance only for buildings within Special Flood Hazard Areas indicated on the FIRMS, which are modeled on past flooding events, does not promote adaptation to future climate conditions. Finally, with the exception of fire codes and the energy "stretch codes" (a code with higher energy standards that municipalities may voluntarily adopt), Massachusetts communities cannot develop local building codes that are stricter than the State's. For example, municipalities cannot require buildings to use specific floodproofing materials or mechanisms or require building design to be adapted to future flood conditions.
- **Improving climate-adapted outcomes:** In order to address climate change–induced flooding, one of the following three actions must occur:
  - The State Building Code would need to be revised;
  - A new Stretch Code for Flood Resistant Construction would need to be adopted under the State Building Code; or
  - More stringent requirements for Flood Resistant Construction in the current and future 100-year floodplain in Boston would have to be adopted in accordance with the Massachusetts Building Code enabling legislation (M.G.L. c.143 §98).
- Code amendments could include basing first floor and floodproofing elevations on the flood conditions that are projected to be present during the lifetime of the structure rather than the currently effective FEMA base flood elevation. Another amendment could be to require, over time, flood-resilient retrofitting of existing structures that are currently, or will become within their lifetime, vulnerable to flooding. (San Francisco's Mandatory Soft Story Retrofit Program could serve as a model for this type of regulation.)

The Massachusetts Environmental Policy Act—Draft Climate Change Adaptation and Resiliency Policy

- **Purpose:** The Massachusetts Environmental Policy Act (MEPA) (M.G.L. c. 30 §61) and its implementing regulations at 301 CMR 11 are intended to provide meaningful opportunities for public review of the potential environmental impacts of projects for which an agency action is required, and to assist each agency in using all feasible means to avoid, minimize, and mitigate damage to the environment. MEPA is administered by the Massachusetts Office of Energy and Environmental Affairs MEPA Office. MEPA staff review environmental notification forms (ENF) and environmental impact reports (EIR); interpret the MEPA regulations; and publish the *Environmental Monitor*, an on-line publication that provides information on projects under review by MEPA.

  In 2014, MEPA released a Draft Climate Change Adaptation and Resiliency Policy for public review. (It has not been finalized or adopted yet.) The goal of the policy is to meet MEPA's obligations pursuant to the Massachusetts Global Warming Solutions Act of 2008, including incorporating the evaluation of "reasonably foreseeable climate change ... effects, such as predicted sea level rise" into projects within its purview. The MEPA Office intends to revise and update the Draft Policy on a periodic basis, with input from stakeholders, to reflect new data and resources as they become available and to incorporate benchmarks and standards for review and mitigation, including specific design criteria, where appropriate.

- **Applicability:** MEPA is applicable to projects that meet or exceed review thresholds related to impacts on: land; state listed species; wetlands, waterways, and tidelands; water use; wastewater generation; transportation; energy; air quality; solid and hazardous waste; historical and architectural resources; areas of critical environmental concern; and regulations. It also applies to projects that meet the following jurisdictional criteria:

  - Projects that are undertaken by a state agency (department/office, board/commission of the Commonwealth including municipal redevelopment authorities);
  - Projects that are subject to state or federal review;
  - Projects that include financial assistance by a state agency; or
  - Projects that involve a land transfer by an agency.

  The Draft Climate Change Adaptation and Resiliency Policy is applicable to those projects that are required to submit an EIR. For projects in Boston, MEPA review is often conducted in parallel with project review under the local zoning code.

- **Limitations in a changing climate:** While the Draft Climate Change Adaptation and Resiliency Policy has not been adopted, MEPA typically applies it through its scoping determinations. The draft policy stops short of requiring specific conditions or actions to adapt to changing climate conditions. Finally, MEPA review is applicable to only a limited array of projects, leaving many projects without the benefit of this type of review.

- **Improving climate-adapted outcomes:** The policy could be revised to require evaluation of the potential to adapt to specific sea level rise scenarios, and to apply to all projects requiring an ENF, rather than only to those subject to an EIR. In addition, specific types of analyses and procedures for undertaking evaluations could be established and communicated to applicants so that they have a better understanding of what level of effort is expected during the environmental review process. Information, data, and technical assistance should also be made readily available to applicants to encourage and support adequate consideration to future conditions.

Commonwealth of Massachusetts Integrated State Hazard Mitigation and Climate Adaptation Plan

- **Purpose:** This plan serves as an update to the Commonwealth's existing State Hazard Mitigation Plan and is also driven by Executive Order 569 (E.O. 569). It addresses the Commonwealth's federally mandated hazard mitigation plan requirements and establishes a statewide climate adaptation plan.
- **Applicability:** In accordance with E.O. 569, this plan includes strategies for incorporating the best available climate science; guidance and strategies for state agencies and municipalities to address climate impacts through adaptation and resilience measures; clear goals, expected outcomes, and a path for implementation; approaches for increasing the resiliency of government operations; policies and strategies for ensuring that adaptation and resiliency efforts complement efforts to reduce greenhouse gas emissions; and strategies that conserve and sustainably employ the natural resources of the Commonwealth to enhance climate adaptation, build resilience, and mitigate climate change. The plan is intended to serve as a roadmap for state agencies, as well as cities and towns, that adopt climate adaptation and resilience strategies.
- **Limitations in a changing climate:** While the plan provides information and data for understanding and prioritizing statewide climate adaptation efforts, it does not have the force of a law or regulation. Integrating the statewide climate adaptation plan with the state's hazard mitigation planning requirement for the Federal Emergency Management Agency does not impose any legally enforceable rights or obligations. Some of the priorities or strategies identified in the plan may require buy-in or action from outside state government, which could present challenges because participation in implementation efforts would be voluntary.
- **Improving climate-adapted outcomes:** The plan provides a necessary foundation for local climate adaptation and hazard mitigation planning efforts; provides a consistent, statewide climate projection dataset for use by state agen-

cies, cities and towns, and private sector entities; and analyzes and makes recommendations for modernizing state regulations and policies to better address climate risks. Finally, the plan will guide the selection and prioritization of activities and funding for climate adaptation activities in accordance with Chapter 209 of the Acts of 2018, including infrastructure projects, natural resource restoration, and conservation projects.

6.3.2 Watershed Governance: A Transformational Approach

The incremental changes described above are necessary but may not be sufficient to adapt our built environment to the changing climate. More in-depth, larger scale, and faster changes may be needed (although changes that meet those three criteria simultaneously are likely impossible).[20] The scope, pace, and significance of climate change may warrant unique solutions that can only be realized through transformational changes that involve innovative governing strategies, changes in power structures, and the introduction of new institutional arrangements and regulatory frameworks. One transformational approach that could significantly improve our ability to thrive in a changing climate is to establish watersheds as the geographic and conceptual basis for governing, enabling us to align land-use decision-making with management of water resources.[21] Governing in Massachusetts at the watershed scale, particularly to implement climate change adaptation measures, may be advantageous for several reasons as described below.

Ideal scale for Integrated Governance

Research indicates that watersheds provide an ideal context for integrated governance that addresses many factors such as environmental, socioeconomic, and health priorities. The United Nations promotes the concept of Integrated Water Resources Management (IWRM) as the way forward for efficient, equitable, and sustainable development and management of water resources. The widely accepted definition of IWRM as developed by the Global Water Partnership is "a process which promotes the coordinated development and management of water, land and related resources, in order to maximize the resultant economic and social welfare in an equitable manner without compromising the sustainability of vital ecosystems."[22] As a discipline, watershed management is similar to IWRM because it also focuses on coordinated management of water, land, and related resources. Watershed models provide the capability to simulate the responses of natural systems to natural forces or human activities, promoting a social-ecological perspective. Watershed-level governance can be part of transformational governance, providing a structure that could eventually be used to address a broader range of climate change adaptation, and other issues as appropriate.

History of Watershed-level Planning and Policy

There is already a long history in the Commonwealth of planning and policy at the watershed level. In 2004, the Executive Office of Environmental Affairs (EOEA) issued the *Massachusetts Water Policy*, consisting of 10 recommendations and 34 actions intended to:

- Keep water local and seek to have municipalities live within their water budgets by addressing issues from a watershed perspective;
- Protect clean water and restore impaired waters;
- Protect and restore fish and wildlife habitats; and
- Promote development strategies consistent with sustainable water resource management.

MassDEP's companion *Guide to Water Resource Management Planning* encouraged municipalities to develop their own IWRM plan, comprehensive water resource management plan, or engineering report, while also recognizing that many municipalities had limited resources to do so. Creating a governance structure at the watershed-level could help municipalities truly address water-related issues, including stormwater management and flooding, from a watershed perspective.

Filling a Regulatory Void

Many water-related issues are currently coordinated at the watershed-level in Massachusetts. For example, drinking water supply and water quality are overseen at the watershed level by agencies and programs such as the Water Resources Commission, the Department of Conservation and Recreation's Office of Watershed Management, and MassDEP's Watershed Planning Program, with laws to support them, such as the Watershed Protection Act (Chapter 36 of the Acts of 1992). In addition, dozens of private watershed organizations monitor trends in water quality, conduct river cleanups, restore habitats, launch information and advocacy campaigns, operate recreational alert systems, and even act as conveners around climate change issues for their individual watersheds. However, there is currently no governing body with regulatory powers, staffing capacity, or financing capabilities dedicated to stormwater and flood control at the watershed level, where these issues can be most meaningfully addressed.

Matches Current Climate Data Collection Scale

Climate data are currently being reported at the watershed level in the Commonwealth. In March 2018, the Executive Office of Energy and Environmental Affairs (EOEEA) published the *Massachusetts Climate Change Projections–Statewide and for Major Drainage Basins* report. As indicated in the title, the drainage basin is one of the levels at which the authors, researchers from the Northeast Climate Science

Figure 7. **Massachusetts Includes 27 Major Drainage Basins**

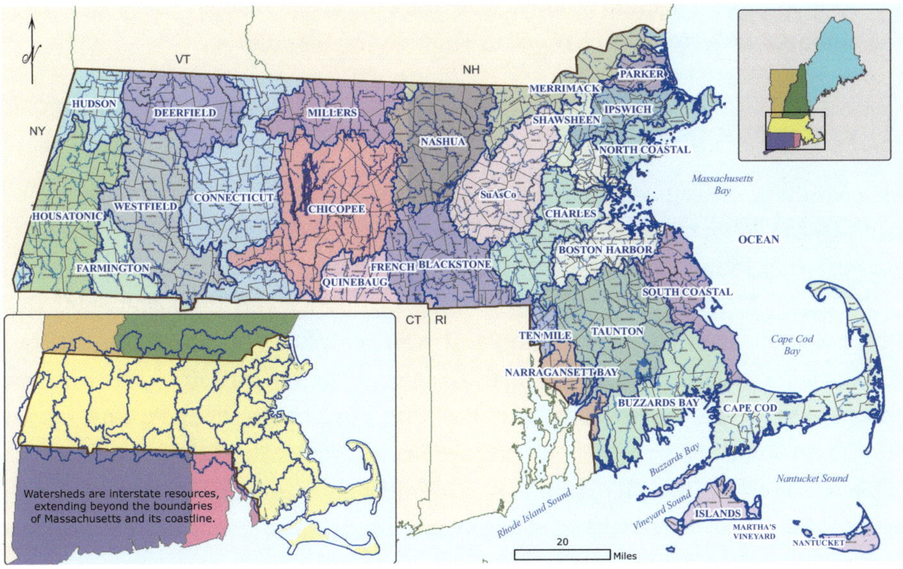

Source: MassGIS (Bureau of Geographic Information)

Center at the University of Massachusetts Amherst, developed procedures to take information from a larger scope and apply it to local scales for changes in temperature, precipitation, and sea level rise. The availability of data for each of the state's 27 watersheds helps support locally appropriate science-based planning at a smaller geographic scale.

Ability to Garner State-wide Support

Because each municipality in the Commonwealth falls into at least one watershed, an entity that governs at a watershed scale can be organized to serve the entire state. It would make it easier to garner support for legislation by spreading the voter base over the entire state rather than concentrating resources in specific locations. Establishing an entity at this scale would allow contextual governance, where watersheds could focus on the issues of importance to them, from droughts, to intense precipitation, to coastal flooding and beyond.

The Role of Engineering in Climate Adaptation

Whether we take an incremental approach or a transformational one, engineers, design professionals, and planners will play important roles in improving governance for climate change adaptation. They will be called upon to help guide the rules and norms that will shape how we create a climate-adapted built environment. They will be instrumental in taking steps such as developing and promoting rating systems that go above and beyond minimum code requirements and developing and influencing new and revised regulatory standards to implement sustainable and resilient solutions. This section summarizes three popular rating systems and discusses the potential trajectory of changes to engineering standards.

6.4.1 Rating Systems

Sustainability rating systems can help provide design professionals with forward-thinking standards and bolster the legitimacy of designs that exceed minimum regulatory requirements. One of the best-known sustainability rating systems is the U.S. Green Building Council's Leadership in Energy and Environmental Design (LEED) program. First launched in 1994, it now includes rating systems for the design, construction, operation, and maintenance of green buildings, homes, neighborhoods, and cities that aim to help building developers, owners, and operators be environmentally responsible and use resources efficiently.[23] It also offers Climate Resilience Screening Tool that helps users identify climate sensitivities and prioritize opportunities to promote enhanced resilience through the green building outcomes rewarded in LEED credits.

Another concept focused primarily on buildings is Passive House (also known as Passivehaus). Organizations such as the Passive House Institute and the Passive House Institute US have formalized a set of passive building design principles so they can be used to attain a quantifiable and rigorous level of energy efficiency within a specific quantifiable comfort level.[24] Compliance with Passive House principles results in buildings that can both mitigate and adapt to changing climate conditions.

A third rating system is the Institute of Sustainable Infrastructure's Envision program, which is primarily focused on infrastructure. It provides an objective framework of criteria designed to identify sustainable ways to plan, design, construct and operate infrastructure projects. Envision is a decision-making guide rather than a set of prescriptive measures that provides industry-wide sustainability metrics for all types and sizes of infrastructure. It uses 64 sustainability and resilience indicators organized around five categories: Quality of Life, Leadership, Resource Allocation, Natural World, and Climate and Resilience. These indicators help users assess and measure the extent to which their project contributes to conditions of sustainability and resiliency across a range of social, economic, and environmental indicators.

These and other rating systems support engineers, design professionals, and planners as they produce sustainable and resilient projects that mitigate and are adapted to climate change.

6.4.2 Regulations and Policies

As decision makers become more aware of the risks associated with climate change, engineering standards, as well the standard of care for engineers, will begin to reflect and mitigate those risks. We can expect engineering standards to:

- Be in the spotlight more than they previously have been;
- Be informed by performance monitoring and evaluation;
- Be revised more frequently to reflect changing conditions and projections; and
- Become more tailored to geographically based climate risks than to political boundaries.

Similarly, local and state governments are responding to this growing awareness by adjusting regulations and policies. For example:

- Local governments are updating their zoning, development review, and environmental protection requirements;
- The Commonwealth's environmental review policies and permitting requirements are beginning to align around anticipated future conditions; and
- Changes to the State Building Code that raise the bar for risk reduction are also being contemplated in many circles.

As creators of the built environment, engineers, architects, design professionals and planners must be partners in revising existing and establishing new regulations and policies to be able to successfully plan for, design, and construct climate-adapted buildings and infrastructure. These well-trained, highly regarded professionals also have the power to affect the ideas, aspirations, expectations, and norms that shape design processes. They can help create a culture that prioritizes adaptation to changing climate conditions, and can advocate for addressing climate mitigation, adaptation, and resiliency in every project from its inception.

6.5 Conclusion

Climate change presents many challenges to effective governance. We can no longer base our regulations and policies on the past because it is no longer a reliable indicator of future conditions. Humans live in a complex adaptive system where we affect and are affected by the natural environment. We therefore need to modify one of our most fundamental social systems—governance—so that our built environment can be made more resilient to changes in our natural environment. We will need governance structures that allow us to gather adequate information about ecological resources and social values, obtain feedback through monitoring, and use this data to inform policies, programs, and projects.

Given the slow and complex nature of changing institutions, cultivating incremental change in existing legal institutions will be necessary while more transformational changes are developed. While the governments of the City of Boston and the Commonwealth of Massachusetts can, and should, begin to address adapting to our changing climate by improving our existing regulatory standards, new approaches to governance capable of confronting landscape-scale problems will soon become necessary. We will need governance that supports the conditions under which ecosystems can continue to function as climate changes. One promising approach is governing at the watershed scale. Instituting changes in power structures and introducing new institutional arrangements and regulatory frameworks is always challenging, but the extreme challenge of climate change adaptation may warrant such actions.

The regulatory framework within which engineers, design professionals and planners operate has guided the formation of today's built environment. Going forward, these professionals will need to design projects that exceed the existing standards to create a climate-adapted future. Utilizing sustainable rating systems and participating in the creation of regulatory standards that respond to a changing environment are two ways they can begin the process. Most importantly, they can help create a culture that prioritizes and advocates for addressing climate change mitigation, adaptation, and resiliency throughout our built environment.

Works Cited

CH2M Hill Engineers, Inc. 2015. Wastewater and Storm Drainage System Facilities Plan. Section 7 Climate Change. Prepared for Boston Water and Sewer Commission.

Chaffin, B. C., H. Gosnell, and B. A. Cosens. 2014. A decade of adaptive governance scholarship: synthesis and future directions. *Ecology and Society 19*(3): 56. Available online at http://dx.doi.org/10.5751/ES-06824-190356.

City of Boston. 2016. *Climate Ready Boston Final Report*. Boston, MA.

City of Boston. 2017. Imagine Boston 2030: A Plan for the Future of Boston. Boston, MA.

City of Cambridge, 2015. *Climate Change Vulnerability Assessment*. Cambridge, MA.

Cosens, B.A., R.K., Craig, S.L. Hirsch, et al. 2017. "The Role of Law in Adaptive Governance," *Ecology and Society, 22*(1): art. 30. https://doi:10.5751/ES-08731-220130.

Douglas, E., P. Kirshen, V. Li, C. Watson, and J. Wormser, *Preparing for the Rising Tide*. Boston: The Boston Harbor Association, 2013. Available online at https://www.bostonharbornow.org/wp-content/uploads/2017/02/Preparing-for-the-Rising-Tide-FINAL.pdf.

Douglas, E., P. Kirshen, et al. 2016. *Climate Change and Sea Level Rise Projections for Boston*. City of Boston, MA. Available online at https://www.boston.gov/sites/default/files/document-file-12-2016/brag_report_-_final.pdf

Hirokawa, K.H. "Driving Local Governments to Watershed Governance" *Environmental Law,* Vol. 42 Issue 1 (Winter 2012), pp. 157–200.

Kirshen, P. 2018. *Feasibility of Harbor-wide Barrier Systems: Preliminary Analysis for Boston Harbor*. Boston: Sustainable Solutions Lab, UMass Boston. Available online at https://www.greenribboncommission.org/wp-content/uploads/2018/05/Feasibility-of-Harbor-wide-Barriers-Report.pdf.

Kruel, S. 2016. "The Impacts of Sea-Level Rise on Tidal Flooding in Boston, Massachusetts." *Journal of Coastal Research, 32*(6). Available online at https://www.

researchgate.net/publication/295081247_The_Impacts_of_Sea-Level_Rise_on_Tidal_Flooding_in_Boston_Massachusetts.

Levy, D. I. 2018.*Financing Climate Resilience: Mobilizing Resources and Incentives to Protect Boston from Climate Risks*. Boston: Sustainable Solutions Lab, UMass Boston.

Massachusetts Office of Coastal Zone Management. 2013. *Sea Level Rise: Understanding and Applying Trends and Future Scenarios for Analysis and Planning.* Boston, MA.

MassDOT-FHWA. 2015. *Pilot Project Report: Climate Change and Extreme Weather Vulnerability Assessments and Adaptation Options for the Central Artery*. Boston, MA.

Termeer, K., Dewulf, A., and Biesbroek. G.R. 2017. "Transformational Change: Governance Interventions for Climate Change Adaptation from a Continuous Change Perspective." *Journal of Environmental Planning and Management, 60*:4, 558–576.

Notes

1 Kruel, Stephanie, David Cash, and Rebecca Herst, *Governance for a Changing Climate: Adapting Boston's Built Environment for Increased Flooding*. Boston: Sustainable Solutions Lab, UMass Boston, 2018. Available online at https://www.greenribboncommission.org/document/governance-for-a-changing-climate-adapting-bostons-built-environment-for-increased-flooding/.

2 Melillo, Jerry M., Terese (T.C.) Richmond, and Gary W. Yohe, Eds., 2014: *Climate Change Impacts in the United States: The Third National Climate Assessment*. U.S. Global Change Research Program, 841 pp. doi:10.7930/ J0Z31WJ2.

3 Massachusetts Office of Coastal Zone Management, *Sea Level Rise: Understanding and Applying Trends and Future Scenarios for Analysis and Planning.* Boston, MA, 2013; Douglas, E., P. Kirshen, V. Li, C. Watson, and J. Wormser, *Preparing for the Rising Tide*. Boston: The Boston Harbor Association, 2013. Available online at https://www.bostonharbornow.org/wp-content/uploads/2017/02/Preparing-for-the-Rising-Tide-FINAL.pdf. Boston Water and Sewer Commission (BWSC), 2015. *Wastewater and Storm Drainage System Facilities Plan, Final Report*. Boston, MA; City of Cambridge, 2015. *Climate Change Vulnerability Assessment*. Cambridge, MA; Douglas, E., P. Kirshen et al., *Climate Change and Sea Level Rise Projections for Boston*. City of Boston, MA, 2016. Available online at https://www.boston.gov/sites/default/files/document-file-12-2016/brag_report_-_final.pdf. City of Boston, 2016. *Climate Ready Boston Final Report*. Boston, MA; Kruel, S., 2016. "The Impacts of Sea-Level Rise on Tidal Flooding in Boston, Massachusetts." *Journal of Coastal Research, 32*(6). Available online at https://www.researchgate.net/publication/295081247_The_Impacts_of_Sea-Level_Rise_on_Tidal_Flooding_in_Boston_Massachusetts. Levy, D. I., *Financing Climate Resilience: Mobilizing Resources and Incentives to Protect Boston from Climate Risks*. Boston: Sustainable Solutions Lab, UMass Boston, 2018; Kirshen, P., *Feasibility of Harbor-wide Barrier Sys-*

tems: Preliminary Analysis for Boston Harbor. Boston: Sustainable Solutions Lab, UMass Boston, 2018. Available online at https://www.greenribboncommission.org/wp-content/uploads/2018/05/Feasibility-of-Harbor-wide-Barriers-Report.pdf.

4 Clark, William C., Lorrae van Kerkhoff, Louis Lebel, and Gilberto C. Gallopin, April 26, 2016. "Crafting Usable Knowledge for Sustainable Development." *PNAS 113* (17) 4570-4578. Available online at https://doi.org/10.1073/pnas.1601266113.

5 Levy, David L., and Rebecca Herst. April, 2018. *Financing Climate Resilience: Mobilizing Resources and Incentives to Protect Boston from Climate Risks*, Boston: Sustainable Solutions Lab, U Mass Boston, 2018, p. 3. Available online at https://www.greenribboncommission.org/wp-content/uploads/2018/04/Financing-Climate-Resilience-April-2018.pdf.

6 Climate Ready Boston reports are available at https://www.boston.gov/departments/environment/preparing-climate-change.

7 City of Boston, 2017. *Imagine Boston 2030: A Plan for the Future of Boston*. Boston, MA.

8 Levy, D. I., Financing Climate Resilience: Mobilizing Resources and Incentives to Protect Boston from Climate Risks. Boston: Sustainable Solutions Lab, UMass Boston, 2018.

9 City of Boston, 2016. *Climate Ready Boston Final Report*. Boston, MA.

10 Cosens B.A., Craig R.K., Hirsch S.L., et al., 2017. "The role of law in adaptive governance," *Ecology and Society, 22*(1): art. 30. https://doi:10.5751/ES-08731-220130.

11 Douglas, E., P. Kirshen et al., *Climate Change and Sea Level Rise Projections for Boston*. City of Boston, MA, 2016. Available online at https://www.boston.gov/sites/default/files/document-file-12-2016/brag_report_-_final.pdf.

12 Kirshen, P., *Feasibility of Harbor-wide Barrier Systems: Preliminary Analysis for Boston Harbor*. Boston: Sustainable Solutions Lab, UMass Boston, 2018. Available online at https://www.greenribboncommission.org/wp-content/uploads/2018/05/Feasibility-of-Harbor-wide-Barriers-Report.pdf.

13 City of Boston, 2016. Climate Ready Boston Final Report. Boston, MA.

14 Termeer, K., Dewulf, A., and Biesbroek. G. R. "Transformational Change: Governance Interventions for Climate Change Adaptation from a Continuous Change Perspective." *Journal of Environmental Planning and Management, 60*:4 (2017), 558–576.

15 Chaffin, B. C., H. Gosnell, and B. A. Cosens. 2014. A decade of adaptive governance scholarship: synthesis and future directions. *Ecology and Society 19*(3): 56. Available online at http://dx.doi.org/10.5751/ES-06824-190356.

16 Public land and public entities are not subject to many of the regulatory standards described in this section. Flood adaptation measures on public land are planned and implemented through a different set of processes.

17 Defined by FEMA as the height of the lowest occupiable floor (when wet floodproofing), or the height of the lowest structural member of an inhabitable floor (when elevating a building).

18 City of Boston, *Coastal Resilience Solutions for East Boston and Charlestown: Final Report*. October 2017, and City of Boston, *Coastal Resilience Solutions for South Boston: Final Report*. October 2018. Available online as PDF files via links at https://www.boston.gov/departments/environment/preparing-climate-change.

19 City of Boston, *Imagine Boston 2030: A Plan for the Future of Boston*. Summer 2017. p.128. Available online at https://www.boston.gov/departments/mayors-office/imagine-boston-2030.

20 Termeer, K., Dewulf, A., and Biesbroek. G. R. "Transformational Change: Governance Interventions for Climate Change Adaptation from a Continuous Change Perspective." *Journal of Environmental Planning and Management*, *60*:4 (2017), 558–576.

21 Hirokawa, Keith H. "Driving Local Governments to Watershed Governance." Environmental Law 42, no. 1 (2012): 157–200. Accessed September 27, 2020. http://www.jstor.org/stable/43267515.

22 Global Water Partnership, "What Is IWRM?" Available online at https://www.gwp.org/en/GWP-CEE/about/why/what-is-iwrm/.

23 *See* https://www.usgbc.org/leed

24 *See* https://passivehouse.com and https://www.phius.org.

7

Contracting Practices and Risk Management

Patricia B. Gary

PART 3

7.0 Introduction

As concerns over climate change grow, the construction sector is being called upon to design and construct infrastructure that is sustainable, adaptive, and resilient. This requires design professionals to pursue a dual path: reducing the causes of climate change through mitigation of GHG emissions and simultaneously implementing preparedness and adaptation measures. In the face of this challenge, it is important to understand how contract terms can increase or decrease risk to the project participants. Risks are shared through an allocation of risk in the contracts. A party's risk tolerance on a project will be affected by its ability to transfer the risk, and parties must be prepared to negotiate their contract. An improper allocation of risk in a contract term may jeopardize insurance coverage and expose the design professional to an uninsured loss, often without the parties realizing it. There are numerous contract terms and conditions that should be considered by each stakeholder.

This chapter proceeds in three parts. Part 1 discusses project delivery methods, and addresses how achievement of sustainability goals may be optimized by the owner's selection of a delivery method that allows for early involvement of the constructor. In larger projects, the ability to leverage the constructor's specialized knowledge early in design maximizes the potential for meeting or exceeding sustainable objectives and can result in meaningfully greener structures and systems. Part 2 discusses some of the important terms and conditions that design professionals should consider when drafting and negotiating a professional services agreement. Though not an exhaustive list of important contract provisions, Part 2 highlights key clauses that may create, or mitigate, risk. Part 3 is a brief overview of insurance considerations.

PART 1

7.1 The Backdrop of Green Building and Sustainable Design

Broadly referred to as "green" or "sustainable" construction, green building reduces the carbon footprint of buildings, increases resiliency, and safeguards the environment, public health, safety, and general welfare. The green building movement began with voluntary programs such as Leadership in Energy and Environmental Design (LEED).[1] However, LEED applies only to vertical buildings and does not address horizontal systems for roads, bridges, rail, transportation, energy, wastewater, stormwater, drinking water, or other civil infrastructure.

The mandatory standards that govern how infrastructure is built are generally developed at the international and national levels through standard-setting organizations. For vertical projects, the International Code Council (ICC) has developed a family of 13 model building codes, including the International Building Code (IBC), the International Energy Conservation Code (IECC), and ICC codes for fire, electrical, structural, mechanical, and plumbing systems. Most U.S. jurisdictions have adopted some form of the I-codes, with amendments that address local concerns. The ICC updates its model codes every three years, yet most states have not adopted new building codes on a three-year cycle that mirrors the ICC's code development process. For example, Massachusetts is typically 3–6 years behind the ICC's development cycle, but the state's Board of Building Regulation and Standards (BBRS) voted in 2019 to adopt the 2021 model codes in the proposed 10th edition of the Massachusetts Building Code, which will be effective in 2022.[2]

There are professional organizations for individual sectors, such as the American Society of Civil Engineers (ASCE), the American Society of Heating, Refrigerating and Air-Conditioning Engineers (ASHRAE), and the American Association of State Highway and Transportation Officials (AASHTO). These organizations develop minimum standards that state or local jurisdictions adopt and codify into design guidelines, manuals, or codes.[3]

Civil infrastructure must be designed to withstand all structural load types that it will potentially face during its lifecycle. The nationally adopted consensus standard, ASCE/SEI 7, "Minimum Design Loads and Associated Criteria for Buildings and Other Structures," provides load requirements—dead, live, soil type, flood, wind, snow, rain, ice, and earthquake loads and their combinations—that are included in building codes and other documents. Climatic loads are traditionally defined under the assumption of a stationary climate, and although the assumption of stationarity worked in the past, it is no longer valid due to anthropogenic climate change. According to the Fourth National Climate Assessment (NCA4), "the assumption that current and future climate conditions will resemble the recent past is no longer valid."[4]

ASCE/SEI 24-14, "Flood Resistant Design and Construction," provides the minimum requirements for the design and construction of buildings and structures in flood hazard areas that are subject to building code requirements. However, the identification of flood-prone structures is based on flood hazard maps, and if flood maps are inaccurate, minimum building codes will apply only to an area that is smaller than the true 100-year floodplain, leaving infrastructure vulnerable. ASCE's Committee on Adaptation to a Changing Climate provides the following guidance to engineers regarding climate change impacts on design: "For each performance requirement, engineers need to consider changes in demands due to climate change (e.g., increased storm frequencies and intensities, environmental loads and fire hazards, etc.) and changes in strength or capacity (e.g., increased corrosion rates)."[5] Standards and regulations usually set only minimum design criteria; thus ASCE advises that engineers should "inform owners . . . of the likelihood of greater than specified extremes, and recommend (when warranted) more conservative designs" using a low-regret approach.[6]

In civil engineering design, ASCE recognizes that "sustainability is too important to merely be implied or suggested," and that the engineering community needs a benchmark for the threshold of sustainability it is seeking to accomplish.[7] ASCE is developing a new technical sustainability standard for the civil engineering community, ASCE/COS 73-XX *Standard Requirements for Sustainable Infrastructure*, that it states "will be applicable across all infrastructure sectors, providing coherent and consistent performance-based objectives that can be included in procurement documents by owners, regulators, stakeholders and policymakers."[8]

Many states and local jurisdictions are taking proactive steps to minimize the risks of climate change by enacting new codes and other standards to promote safe and sustainable construction. A few states, such as California, have enacted mandatory codes that set a precedent by going beyond the minimums of national standard-setting organizations.[9] Governments are requiring construction to meet higher standards, including energy efficiency and net zero targets.[10]

Climate change is transforming the way infrastructure is built. *Is the professional standard of care changing? Will climate change increase the risk of liability, and if it does, will it be insurable under an insurance policy*? There is no doubt that the standard of care is changing because design professionals are solving climate change issues today in ways that are different from past practice. New codes are beginning to address the impacts of climate change, but a dilemma exists when outdated minimum standards conflict with an evolving professional standard of care that requires more robust design. In that scenario, ASCE advises that engineers must use their professional judgment to bridge the gap: "incorporating climate change into engineering practice will require engineering judgment to balance costs and potential consequences of failure."[11]

There is a legal distinction between a change in the required skill and learning that must be reflected in design, and elevation of the professional standard of care in a contractual provision. The professional standard of care does not require perfection in the performance of a design professional's services. Design professionals must be wary of contract terms that attempt to transfer project risks (including climate-related risks) with language that elevates the standard of care, or creates an obligation that is uninsurable. Climate change poses new risks associated with design and construction, but most of these risks can be managed through the parties' contracts with precise, well-tailored contractual language.

7.2 Project Delivery

Design and construction undertakings are complex and highly technical due to the interdependence of many diverse project participants, roles and responsibilities, and financial arrangements. The owner (whether a public or private entity) initiates the project and the manner in which it is delivered. Public owners, including federal, state, and local governments, are required to follow statutory and regulatory frameworks to procure and manage a project. There are many variations of project delivery systems, but they usually fall into four categories: design-bid-build (DBB), construction manager at risk (CMAR), design-build (DB), and innovative approaches such as public-private partnerships (P3).

The selection of the project delivery method is an important decision that should fit the goals and requirements of the owner and the project. Governments are beginning to enact building codes with carbon-neutrality policies. Boston's *Climate Action Plan, 2019 Update* details specific actions the city will take over the next five years to reach carbon neutrality by 2050, including shifting to zero net carbon (ZNC) new construction by 2030. Boston will set an example by constructing new municipal buildings to a net zero standard and adopting a net zero standard for city-funded affordable housing.[12] In March 2021, the Massachusetts Legislature overwhelmingly passed a climate policy bill that requires the state to achieve net zero GHG emissions by 2050, and Governor Baker signed the bill into law. By adopting net zero standards, the U.S. can transition from a fossil-fuel based economy to a cleaner one.

Achieving these decarbonization goals will require innovative architectural and engineering designs that unlock deep energy cuts in all types of infrastructure, and enhance the affordability of energy saving measures. The owner's decision to use traditional DBB delivery (in which a project is designed to 100% construction documents without contractor involvement) or an alternative delivery method which integrates or provides for more contemporaneous collaboration in the roles of designer and constructor (such as CMAR or DB) determines the timing of when the owner, the designer, and the contractor are engaged, as well as the contractual relationships between them, and the allocation of risk.

DBB is the traditional U.S. building delivery method and in theory should deliver a low-cost project, but the contractor is not involved in the design process, and in larger, more complex projects with sustainability goals that are dependent upon innovative high-performance systems and technologies, there is a risk of cost increase due to conflicts between the design documents and constructability in the field. The alternative delivery methods of CMAR and DB allow early involvement of the contractor in the design, and have the potential to deliver deeply green, high performance projects faster and at lower costs. In large projects, the involvement of the constructor during design can result in meaningfully greener structures and systems. Contractor expertise can provide constructability analysis of renewable energy options and technologies, provide value engineering input for specialty materials and subcontractor pricing, and capitalize on the opportunity for fast-tracking and pre-purchase of equipment.

7.2.1 **Design-Bid-Build**

In the United States, the DBB delivery method is the most widely used. State law typically requires competitive bidding for public construction projects. In DBB procurement, the criteria for contractor selection—the lowest, most responsible bid—fosters competition and makes it reliable and predictable to both the participants and the public.[13] The owner engages a design consultant to prepare the design for a project and then bids and awards the construction contract to a contractor to construct the project in accordance with the design.

The traditional DBB project delivery approach involves separate contracts between the owner and design professional for design, and between the owner and the constructor for construction of the project. In some projects, the owner may also engage consultants, such as a geotechnical consultant for investigation of the site, soil, and subsurface properties. The lead design professional is usually an architect when the project is a building, and may hire a design team of engineering specialists such as civil and structural consultants, and mechanical, electrical, and plumbing (MEP) consultants to assist with the design.[14] For civil construction projects, an engineer is usually the prime design professional.

Design-Bid-Build

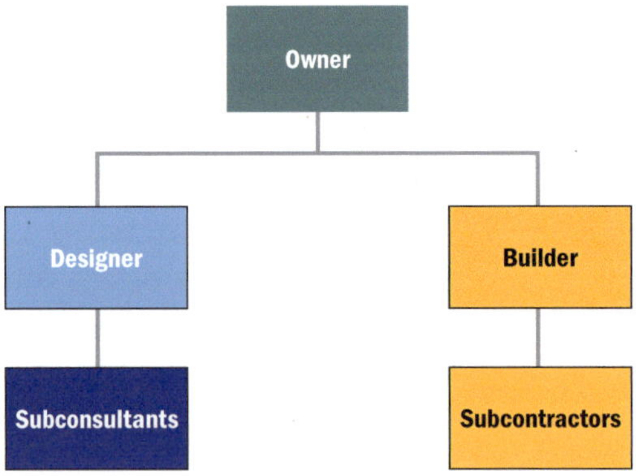

In DBB delivery, a well-established implied warranty of design adequacy, the *Spearin* Doctrine, provides that the owner impliedly warrants the adequacy of the design the owner provides to the contractor, who is bound to construct according to the plans and specifications. The *Spearin* Doctrine arose from a 1918 Supreme Court case, *United States v. Spearin*, in which the court ruled that "if the contractor is bound to build according to plans and specifications prepared by the owner, the contractor will not be responsible for the consequences of defects in the plans and specifications."[15]

7.2.2 Construction Management

The term construction management (CM) applies to several project delivery approaches. The particular tasks undertaken by a construction manager will vary depending upon the project, but can include cost estimating, project oversight, construction scheduling, value engineering, and constructability analysis. There are two primary contracting arrangements for construction management: agency and construction manager at risk (also abbreviated CMAR). In agency CM, the construction manager is retained by the owner in an advisory role to provide professional management services, but does not guarantee the price of the project or assume traditional general contractor responsibilities.

7.2.3 CMAR and CM/General Contractor

CMAR is a hybrid of agency CM and DBB. In a two-phase process, the construction manager acts as a consultant to provide preconstruction advice during the design phase and to provide design input, constructability review, and other preconstruction advice intended to optimize the design process. When the design is 60% to 90% complete, the construction manager has an opportunity to bid on the construction. If the parties agree upon a price, the construction manager retains subcontractors and is responsible for overall construction of the project at a specified maximum price, much like the general contractor in the DBB approach.[16]

Design input regarding constructability is the most-often cited advantage of CMAR project delivery, but a CM's preconstruction scope may include any services an owner or agency needs to facilitate the successful execution of a project, including offering ideas regarding sustainable materials and methods, and energy solutions.[17] Preconstruction services commonly include:

- cost estimates,
- schedule analysis,
- work sequence,
- risk identification/mitigation/pricing,
- constructability reviews,
- development of work packages for bid, and
- development of a GMP that meets owner requirements and budget restraints.[18]

The CM will often assist the project team with design development, value engineering, constructability review, and developing Critical Path Method (CPM) schedules. The CM may review the cost estimates of the designer's estimator and independent cost estimators, and evaluate information such as pricing from suppliers and subcontractors in formulating its own bid, typically for a guaranteed maximum price (GMP). If the bid is acceptable, the parties will enter a construction contract, and the CM becomes the general contractor. The GMP may include funding for mutually agreed upon contingencies.

As in the DBB project delivery method, the owner in a CMAR project retains control over design decisions and retains design risk. In *Coughlin Electrical Contractors, Inc. v. Gilbane Building Co.*[19] the Massachusetts Supreme Judicial Court (SJC) determined that the *Spearin* doctrine, which applies in DBB projects, also applies in the context of a CMAR project because "the owner, through the designer, ultimately controls the design and is the final arbiter of it." The SJC held that the owner impliedly warrants the sufficiency of the plans and specifications to the CM-constructor. However, the implied warranty "is more limited than in a design-bid-build project," because the CM-constructor is only protected "where it has acted in good faith reliance on the design and acted reasonably" in light of its own design responsibilities.[20]

CMAR and CM/GC

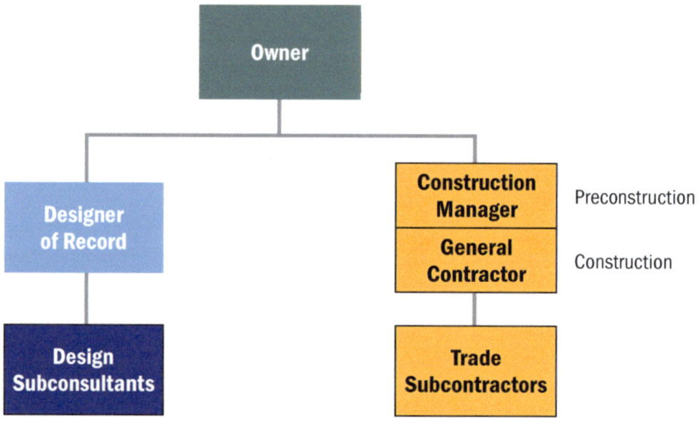

Adapted from USDOT, Federal Highway Administration, "Construction Manager/General Contractor (CM/GC)."

CMAR is sometimes also known as Construction Management/General Contractor (CM/GC).[21] Although the terms are often used synonymously by industry experts, not all states consider them to be the same. In Massachusetts, CMAR is regulated by statute in M.G.L. c. 149A, and is subject to the subcontractor bidding requirements of Chapter 149A, but CM/GC is still novel. In June 2012, the Massachusetts Bay Transportation Authority (MBTA) received legislative authority to use CM/GC as a pilot program for the $2.3 billion Greenline Extension Project (GLX Project), but ultimately, MBTA did not proceed with CM/GC for the transit project.

On January 3, 2017, the FHWA implemented a final rule authorizing CM/GC delivery. A recent paper by David J. Hatem, "Improving Risk Allocation on Design-Build Subsurface Projects," discusses the advantages of the CM/GC approach for subsurface projects, including early contractor involvement in the design development process.[22]

7.2.4 **Design-Build**

The design-build (DB) project delivery method places responsibility for both design and construction activities in one contractual entity, the design-builder.[23] This offers the owner a single point of responsibility for any defects arising out of design and/or construction. Although DB was historically prohibited under federal procurement statutes, in the 1960s, government agencies at federal, state and local levels began to experiment with DB, based upon success achieved in the private sector. Eventually, federal agencies received legal authority to utilize DB in the 1996 Federal Acquisitions Reform Act [Clinger-Cohen Act].[24] A majority of states have enacted legislation allowing use of DB to deliver public construction projects.

Design-Build

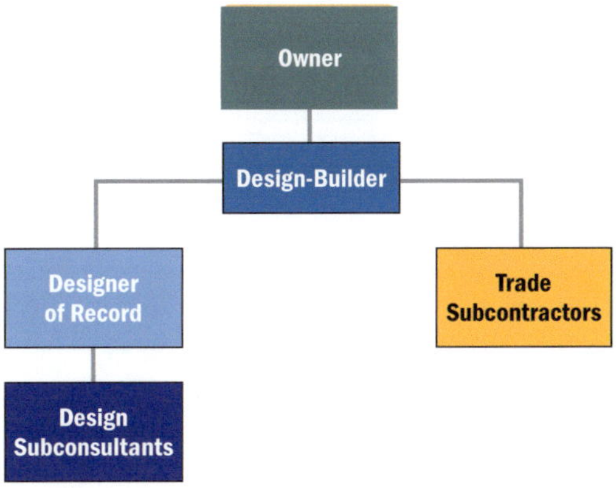

Adapted from USDOT, Federal Highway Administration, "Design-Build (D-B)."

As with CMAR, the DB delivery method provides opportunity for the design-builder to provide preconstruction services during the proposal phase and provide input to the owner and designer to help guide and inform project development. The design-builder accepts the owner's design criteria, and a final cost estimate for the project is established. The project can be completed faster because construction can begin before the design is compete and construction drawings are finalized.[25] The design-builder has responsibility for the adequacy of the final design and construction of the project.

7.2.5 **Design Risk in Design-Build**

The design-builder's single-point of contractual responsibility provides significant advantages to the owner. The design-builder has responsibility for design and construction, including quality, cost, and schedule. Fast-tracking accelerates the project because long lead time items can be procured and field work begun on early work portions of the project before design of the entire project is completed. Risk is shifted to the design-builder for the final design and construction, as the owner relinquishes design control after defining the performance and prescriptive criteria.

DB enabling legislation differs throughout the country, resulting in variations of DB formats. The traditional two-step best-value selection process (BVS) includes: a request for qualifications (RFQ) to identify a short list of the most qualified teams (step one), followed by a request for proposal (RFP) issued to the team(s) selected in the qualifications phase (step two). The owner develops the RFP documents in-house or utilizes a consultant to perform front-end tasks (e.g., surveys and geotechnical/environmental investigations).

The RFP issued to design-build teams may contain elements such as project program, site analysis, performance and/or prescriptive specifications, drawings, reports, sustainability goals or criteria (e.g., net zero, solar, or wind solutions to improve energy efficiency), and budget and schedule requirements. Design-build teams usually begin with a teaming agreement that defines the roles and responsibilities of the design professional and the contractor during the proposal preparation phase, and if the DB team is selected, will conclude with a second design agreement between the design-builder and the design professional. Responses to the RFP will generally include a conceptual design and GMP.

DB projects present unique risks to design professionals because the design-builder's estimate for a fixed price, or GMP, is usually established when the design is only partially developed (typically 30% design development). The design-builder will use the design professional's preliminary design to prepare quantity take-offs and materials costs, create construction schedules, and prepare its bid. Thus, one distinction between DBB and DB is that the design-builder calculates and submits a bid on a project that is not fully designed. The fact that the design is incomplete when the design-builder prepares an estimate and submits its bid creates a risk of cost growth as design advances during design development. The final 100% drawings and specifications issued for construction may be significantly different than the preliminary design. The design-builder may attempt to shift this risk to the design professional based on requirements of the RFP, and attempt to recover the cost overruns from the design professional.

The design professional can mitigate its exposure to this risk through contract language in the teaming agreement, or Phase One Agreement, disclaiming any liability for the preliminary design supplied to the design-builder and related to the RFP because it is not a "final" design. A design contingency fund is another important risk-management tool on a DB project. The design professional should require that the design-builder establish an adequate design contingency fund for design development. The design contingency fund is a monetary fund that is intended to reduce the risk and address the costs of design development.

7.2.6 **Progressive Design-Build**

Progressive design-build (PDB) is an alternative type of the DB method that helps to mitigate the imbalanced risk allocation of traditional DB approaches. PDB uses a qualifications-based (QBS) selection process. For instance, if the project entails sustainability challenges, the design-builder's expertise in sustainability might be paramount in the selection process. The owner selects a design-build team early in the process, before any design development, and without a cost or schedule commitment. The project then "progresses" in a collaborative process until the owner and the design-builder negotiate a price for the second phase. The core feature of PDB is use of two distinct phases: phase one, which includes preconstruction services, design development, and negotiation of the contract price (lump sum or GMP); and phase two, for final design and construction.[26] The design-builder typically issues its proposal and GMP at 50%–90% design development, resulting in a GMP that more accurately reflects the actual project cost.

Similar to CMAR, the design-builder's bid is not part of the initial qualifications-based selection process. The PDB approach facilitates early involvement and collaboration between the owner, designer, and contractor, and allows the owner to control decisions of scope, schedule, and cost in phase one. It leverages the benefits of CMAR while retaining a single point of accountability for the owner. If the parties cannot reach an agreement on the phase two services, the owner has the option to move forward with the project through another contract strategy. The article, "Improving Risk Allocation on Design-Build Subsurface Projects," notes the advantages of PDB for subsurface and tunneling projects, to better evaluate subsurface conditions prior to reaching contractual commitments.[27]

Progressive Design-Build

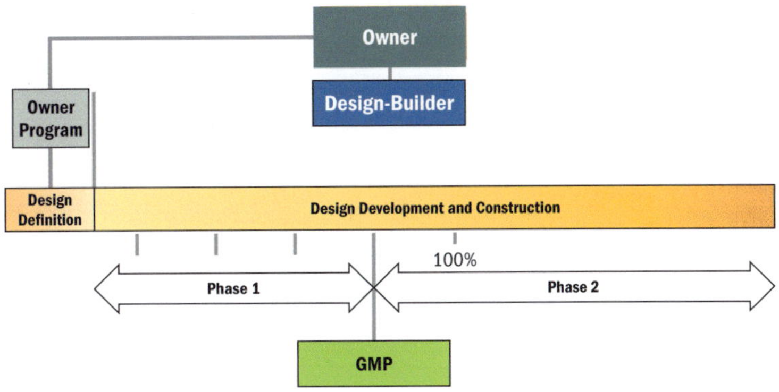

Adapted from Awezec Doller, and Hughes, "Developing Model Procurement Documents for progressive Design-Build Projects," https://www.slideshare.net.

7.2.7 Public-Private Partnerships

The term public-private partnership (P3) describes a variety of contractual arrangements whereby a private entity delivers infrastructure or services that are traditionally the responsibility of the public sector.[28] In a P3, the private partner assumes a greater role in acquiring the necessary financing than in traditional procurement methods.[29] There is no universal definition, but a P3 often entails a design-build (DB) contract with other components added, depending upon the extent of the private sector's participation in financing (F), operations (O), and maintenance (M) of the project, with a large number of potential combinations of components, such as design-build-finance (DBF), design-build-finance-operate (DBFO) and design-build-finance-operate-maintain (DBFOM).[30] Long-term lease arrangements are another form of P3 that involves the operating and maintenance of an existing facility.

P3 Availability Payment Model

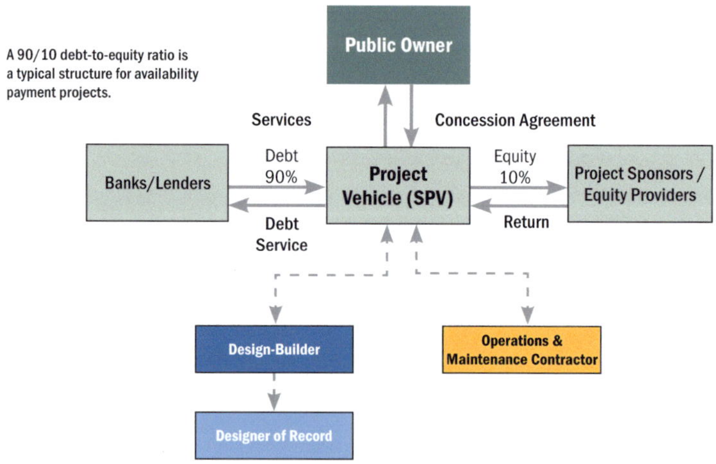

Adapted from Hawaii P3 Workshop, https://planning.hawaii.gov/wp-content/uploads/P3_Jun-6-17_02_Panel-P3-101.pdf.

While not appropriate for every public infrastructure project, P3 project delivery is an option for municipalities and states facing funding shortfalls and critical infrastructure needs. If properly procured with an RFP that includes sustainability criteria and/or energy benchmarks, P3s can promote sustainability and resilience in the built environment. In a DBFOM project, in addition to transferring the risk of construction costs, the P3 also transfers the risk for the facility's long-term operations and maintenance (O&M) to the private sector. The public owner benefits from guaranteed performance standards that are locked in during the term of the P3 agreement. In this sense, the P3 "lifecycle" is a driver that incentivizes private sector investments in climate resilience, and encourages innovative, sustainable designs and solutions.

In a P3 infrastructure project, the concessionaire has an interest in maintaining a high level of service over the duration of the contract term. Although sustainable design may increase initial costs, they are offset by reduced lifecycle costs. In an availability-payment P3, the concessionaire earns regularly scheduled "availability" payments from the public partner based on the availability of a facility at the specified performance level. Performance-based specifications can be built into the contract, including incentives and penalties tied to public-sector goals for climate resilience and sustainability. P3s can drive innovation and align the incentives of the concessionaire with public-sector resilience and sustainability goals.

7.2.8 **Community-Based Public-Private Partnerships**

A community-based public-private partnership (CBP3) is a partnership between a local government and a private entity for the delivery of stormwater infrastructure. The CBP3 model includes features of the traditional P3 model, but focuses on green infrastructure (GI) approaches to stormwater rather than "gray" infrastructure (the system of sewers, pipes, and treatment plants that collects and treats stormwater and wastewater before it is released to waterways).[31] Stormwater runoff is an especially acute concern for the 722 cities in the United States with combined sewer systems, which bring together stormwater and sewage in the same pipes, resulting in combined sewer overflows (CSOs). A growing number of U.S. cities are using green solutions to reduce CSO discharges and meet the environmental requirements of their consent decrees.[32] Green infrastructure uses vegetation and soil to capture rainwater,

Community Based Public-Private Partnership (CBP3)

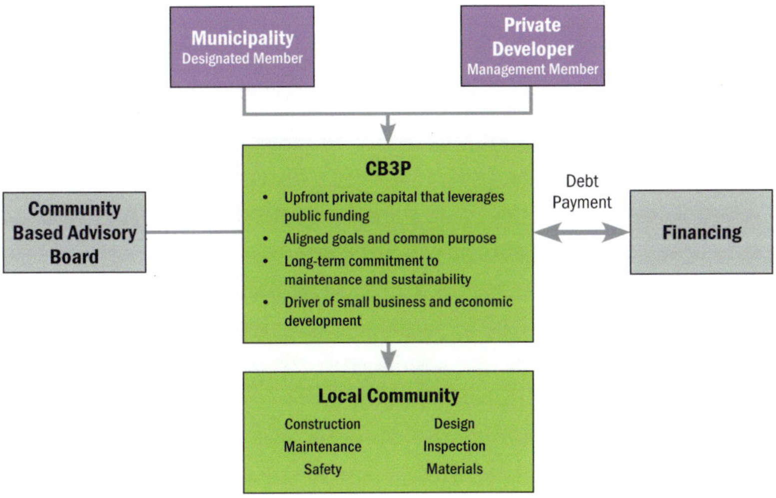

Adapted from EPA, https://www.epa.gov/G3/inancing-green-infrastructure-community-based-public-private-partnerships-cbp3-right-you.

and includes applications such as green roofs, rainwater harvesting systems, raingardens/bioretention facilities, planter boxes, bioswales, and permeable pavements.

Each CBP3 will have a customized financial approach, but the EPA encourages a DBFOM model that leverages federal funding tools such as State Revolving Fund (SRF) and Water Infrastructure Finance and Innovation Act (WIFIA) loans. Under the availability-payment structure, a fixed incentive-based fee is tied to the achievement of performance criteria. In March 2015, Prince George's County, Maryland, implemented one of the first stormwater CBP3s.

7.2.9 **Owner Commitment Is the Key to Sustainability Outcomes**

All of the project delivery methods described above can be used for constructing high-performance infrastructure with climate resilience and sustainability objectives. High owner commitment is the key to achieving sustainability targets regardless of the project delivery method chosen. When owners believe in the sustainable concept as the "right thing to do," the potential for success is maximized. The sustainable objectives must be compatible with the owner's budget within which the project must be delivered. Apart from owner commitment, the timing of constructor involvement is the next most important factor affecting performance outcomes. Early involvement of the constructor in the delivery process can optimize outcomes on projects with sustainability goals.

7.2.10 **Project Delivery Method Can Influence Sustainability Outcomes**

A study undertaken in 2009 (ten years after the introduction of LEED in 1998), titled "Sustainable, High Performance Projects and Project Delivery Methods: A State of Practice Report," sought to determine how project delivery and procurement methods influence an owner's ability to achieve LEED goals and sustainability objectives.[33]

The study, conducted by a team of researchers from the University of Colorado, University of Oklahoma, Pennsylvania State University, and Michigan State University, evaluated the use of three project delivery methods—DBB, CMAR, and DB—and also considered four primary procurement procedures: low-bid, best-value, qualifications-based, and sole-source procurement.

The researchers found that all of the project delivery methods and procurement procedures have been used to achieve every level of LEED certification, but the success rate was higher for the CMAR and DB models, because these delivery methods allow integration of the constructor early in design, and this advantage is pronounced in the areas of mechanical systems and energy modeling. The ability to leverage the constructor's specialized knowledge early in design maximizes the potential for achieving or exceeding sustainable objectives. The study also analyzed contract pricing methods and found that success is enhanced by GMP payment provisions, in which the price of the work is not fixed until the constructor has substantive input in the sustainable design process and can verify that the scope can be delivered within the budget. GMP provisions may also provide for inclusion of agreed-upon contingencies that become "a pool for incentivizing the sustainability features of the final constructed facility."[34]

A 2019 international study, "Procurement Requirements for Carbon Reduction in Infrastructure Construction Projects," similarly finds that collaborative contracting models lead to better sustainable outcomes and greater carbon reductions:

> Collaborative contracting models are a flexible option to encourage innovation and integrate knowledge of different participants. Many interviewees state the importance of breaking silo-thinking and integrating the supply chain in order to reach greater carbon reductions. Also, long-term alliances allow for continuous learning and more transformational innovation, including incentivising contractors to find ways of fulfilling client goals while building less. However, it should be emphasised that strong client leadership and commitment is essential both to legitimise collaborative contracting models and to achieve more fundamental behavioural change within collaborative projects and alliance schemes.[35]

For larger projects, early contractor involvement in the design process allows for valuable contributions to energy efficiency and other sustainability objectives, and may significantly improve project outcomes, including cost and quality. But regardless of the project delivery method selected, strong client leadership and commitment is the key to achieving sustainability goals.

PART 2

7.3 The Contracts

The second part of this chapter will cover some of the important terms and conditions that design professionals should consider when drafting and negotiating their professional services agreement. The discussion is focused on the professional services agreement of the lead design professional in a DBB project, but includes snapshots of some contractual issues faced by design professionals contracted at the highest level in a DB project, such as flow-down provisions and quantity risk.

Parties must decide whether to use a standard form or a custom-made (manuscripted) contract. Industry organizations and trade associations have created their own families of form contracts, including the American Institute of Architects (AIA), the Engineering Joint Contracts Design Committee (EJCDC), the Associated General Contractors of America (AGC), and ConsensusDocs. Industry forms are an acceptable starting point in drafting an agreement, but parties must understand how to modify them. All contracts will require negotiation and customization.

In *Sweet on Construction Industry Contracts: Major AIA Documents*, Jonathan J. Sweet offers this advice about form contracts: "It is not possible to emphasize sufficiently the need for careful modification of any AIA document. If this cannot be done, it is better to use the document unchanged or to use a custom-made contract."[36]

It is important for all parties to identify and fully understand the risks they assume through their contract. Risk should be allocated to the party in the best position to control the risk. When an owner attempts to transfer risks to a design professional or a contractor that the party cannot control, the owner should expect the design professional or contractor to increase its price, or include a contingency to cover the possibility of the risk materializing. Alternatively, if a design professional takes on a high risk, the design professional may seek to balance the risk allocation with a limitation of liability clause or another contractual mechanism, discussed below.

7.4 The Design Professional's Agreement

In private construction, AIA and EJCDC forms are often used for professional services agreements. There are several different AIA B-Series forms of the owner-architect agreement which vary by type of project. The AIA revises its documents periodically, usually every 10 years, and the 2017 editions are the most recent. The EJCDC recently released the 2020 editions of the E-500 Agreement Between Owner and Engineer for Professional Service and the E-570 Agreement Between Engineer and Subconsultant for Professional Services. In public construction, A&E contracts are awarded in a qualifications-based selection process (QBS), and government agencies often require the use of their own contract templates.

7.4.1 Standard of Care

The standard of care clause in a design professional's agreement is arguably the most important provision because it defines the standard which the design professional must meet in providing its professional services under the agreement. When a claim arises against a design professional, the first questions asked are "What is the standard of care?" and "Was it met?" The failure of a design professional to provide a perfect result, standing alone, does not amount to legal liability. Instead, the test of liability for professional negligence is whether the performance met the standard of care. The earliest case to define the architect's standard of care was *Beede v. Coombs*, 36 A. 104 (Me. 1896). The Massachusetts Supreme Judicial Court's statement of the professional standard of care is as follows:

> An architect's efficiency in preparing plans and specifications is tested by the rule of ordinary and reasonable skill usually exercised by one of that profession . . . [I]n the absence of a special arrangement he does not imply or guarantee a perfect plan or satisfactory result." [citations omitted] "Architects, doctors, engineers, attorneys, and others deal in somewhat inexact sciences and are continually called upon to exercise their skilled judgment in order to anticipate and provide for random factors which are incapable of precise measurement. The indeterminable nature of these factors makes it impossible for

> professional service people to gauge them with complete accuracy in every instance . . . Because of inescapable possibility of error which inheres in these services, the law has traditionally required, not perfect results, but rather the exercise of that skill and judgment which can be reasonably expected from similarly situated professionals."[citations omitted] ... We believe that unlike a manufacturer, an architect does not impliedly guarantee that his work is fit for its intended purpose. Rather he impliedly promises to exercise that standard of reasonable care required of members of his profession.[37]

The common law standard of care applies to any professional activity a design professional undertakes, regardless of whether the standard of care is stated in the contract for design services. Beginning in 2007, AIA added a standard of care provision to its B-Series of agreements between the owner and architect, as follows:

> The Architect shall perform its services consistent with the professional skill and care ordinarily provided by architects practicing in the same or similar locality under the same or similar circumstances. The Architect shall perform its services as expeditiously as is consistent with such professional skill and care and the orderly progress of the Project.

What is the purpose of this clause in the face of the common law rule? Prior AIA forms (B141–1997 and B151–1997) did not contain a standard of care clause, yet parties often modified their agreements by adding poorly worded standard of care language. The AIA document commentary for the B101-2007 form states that AIA revised its standard form of agreement to include the standard of care clause because of the "high occurrence of parties misstating the standard of care in their contracts."[38] A similar definition of the standard of care for engineers can be found in EJCDC E-500, Paragraph 6.01.A, Agreement between Owner and Engineer for Professional Services.

Design professionals should be aware that they can inadvertently agree, by contract, to perform a higher standard of care. They should carefully review the standard of care language in their contract, remove any inappropriate wording, and replace it with a clause that mirrors the common law negligence-based standard in the AIA or EJCDC forms. Professional liability policies provide coverage only for conduct that is judged by the common law negligence-based professional standard of care. Therefore, agreeing by contract to a heightened standard could result in the design professional's insurer denying coverage.

Is the standard of care evolving due to climate change? Climate change impacts are exposing design professionals to new risks. For example, when design professionals prepare specifications and drawings for a project that is located within, or in proximity to, an existing floodplain, their failure to anticipate and adjust to climate change in the design could potentially expose them to legal liability. If the locality's floodplain maps contain historic information that is outdated and not reflective of future climate conditions, and the design professional's peers are solving climate change issues in the same locale with more robust designs, the professional standard of care may require the design professional to provide an enhanced, resilient design. But the standard of care is only a benchmark of reasonable expectations. The design professional is not required to exercise the "highest" level of skill, or deliver a design "with no material errors."

Are there risks to design professionals in using a new or emerging technology for which a standard of care does not yet exist? Climate change is posing new risks for design professionals because they are being asked (and in some instances, required by statute) to lead the way in reducing GHG emissions with innovative projects that incorporate climate action in forward-thinking designs. The new paradigm requires design professionals to evaluate complex energy performance issues and select cost-effective options, such as new products and emerging technologies that may entail a measure of risk. If a specified product or system fails to perform as intended, the designer may be exposed to liability:

> [Sustainable design] involves the potential for specification and use of many innovative, untested and experimental products, the selection of which may significantly increase the professional liability risk exposure of design professionals. The absence of any or any reliable data regarding product experience or performance; the availability of the product; the potential discontinuance of product manufacture if experimental or marketing goals are not met—are all factors accounting for that increased professional liability risk exposure.[39]

Under these circumstances, the design professional should keep the owner informed about design options and should clearly document its risk–benefit discussions with the owner about any experimental materials or systems.

Design professionals should be wary of contract clauses that attempt to elevate the standard of care. Examples of inappropriate clauses in a contract for professional services are language calling for the "highest degree of skill and care," "best services," "first class services," "with no material errors," "free from defects," or "fit to perform its intended purpose." This type of language amounts to a warranty or guarantee, which elevates the professional standard of care and should be removed from the contract. In addition, as discussed below, owners often attempt to insert explicit contractual warranties for achieving LEED levels or other energy performance goals that expose the design professional to liability for breach of contract or warranty. Warranty provisions should either be stricken from the contract or explicitly tied to the professional standard of care to ensure that coverage is available under a professional liability policy.

The professional standard of care is a method of allocating risk between the owner and the design professional for potential deficiencies in the design professional's services. The standard of care does not require perfection in the performance of services. Under the standard of care, the design professional is only allocated the risk of deficient performance that falls below the standard of care, and the project owner retains the risk of deficient designer performance that does not amount to professional negligence.

In addition, if a design professional contractually agrees to a higher standard of care, its professional liability insurance carrier may deny coverage. A claim alleging that an architect or engineer breached the professional standard of care falls within the scope of professional liability coverage, but contract language creating a higher degree of care would be excluded under the contractual liability or warranty exclusions of the policy. Insurability is in the interest of both parties, and the owner may not realize that language creating a heightened standard of care could make it unable to collect on a claim.

7.4.2 **Flow-Down Provisions in DB**

Unlike design professionals, contractors routinely warrant their work. The prime contract typically requires the contractor to construct and deliver a project that is free from defects. The contractor views these risks as justified by the contract price and profit it will make on the project. The prime contractor then attempts to transfer as much of this risk as possible downstream, and does this by including a warranty that the work will be "free from defects" in all of its subcontracts. Most subcontracts include standard flow-down clauses that incorporate by reference the provisions of the prime contract into the subcontract and obligate the subcontractor to assume toward the contractor all obligations which the contractor assumes toward the owner. For example, the AIA401-2017 Standard Form of Agreement Between Contractor and Subcontractor contains standard flow-down provisions. In this way, "back-to-back" contracts operate to pass through all of the prime contractor's obligations to its subcontractors for the scope of their services.

On a DB project, the design-builder will often seek to "flow down" its own warranty of defect-free performance in the design-build agreement (DBA) to the design professional. For example, in a DB project with energy benchmarks or performance goals that increase resilience to weather-related phenomena, the design-builder's agreement with the owner may require it to warrant achievement of the project's performance. Contractors familiar with the *Spearin* doctrine[40] are used to owners in DBB and CMAR projects impliedly warranting the suitability of the design, and may expect a warranty from the designer. The design professional should not agree to assume the same risk of defect-free performance that the design-builder assumes in the prime agreement, and should remove flow-down provisions and references to the prime agreement. A warranty of defect-free design can make the design professional contractually liable even though it was not negligent in its performance. Flow-down of prime agreement terms in the design subcontract exposes the design professional to risks that far outweigh its reward, or profit, which is typically modest in comparison to the profit reaped by the design-builder.[41] No design is ever completely "defect free," and the professional standard of care does not require perfection or hold design professionals to unattainable results.

Professional liability policies contain exclusions which preclude coverage for express guarantees or warranties. Therefore, the design professional should carefully review its professional services agreement to ensure that its exposure to liability is limited to claims for negligence. A professional liability insurer may deny a breach of warranty claim, but it will ordinarily defend a claim for negligence and will reimburse damages that are incurred due to the design professional's negligence.

7.4.3 Projects with LEED, Green, or Sustainable Objectives

As more and more states and cities codify net-zero and carbon neutral policies into mandatory laws, environmentally focused "green" building is becoming the norm rather than a niche. Green design and construction presents a risk of unrealized or frustrated owner expectations if a project fails to meet an energy efficiency benchmark or other green goal. In a LEED project, the failure to achieve a specific certification level may result in damages that include higher operating costs, loss of a tenant or sale, loss of government incentives and tax credits, and other penalties.

Design professionals must avoid giving warranties, guarantees, or certifications of their services, and should be aware that a warranty of defect-free performance or a guarantee to achieve a particular LEED, green, or other sustainable result increases their risk exposure. Under the professional standard of care, some degree of imperfection is tolerated, but a warranty expands the design professional's potential liability, and may also be an uninsurable risk, excluded from coverage under the contractual liability exclusion of the professional liability policy.

The agreement should be consistent with the design professional's role. The designer cannot control all of the factors needed to obtain LEED credits and/or certification and should not inadvertently assume risks that are controlled by the contractor or owner. Risks controlled by the owner and contractor include owner-driven value engineering changes, unworkmanlike performance by subcontractors, on-line submission of documentation to USGBC, and the owner's use and operation of the completed project.

In negotiating LEED design services or similar sustainable design services, design professionals should include terms that tie the design services to the standard of care, and disclaim any warranties, similar to the following:

> 1.1 The Owner acknowledges that the LEED Green Building Rating System, and other similar environmental guidelines, utilize certain design and usability recommendations in order to promote an environmentally friendly and energy efficient facility. In addressing these guidelines the Architect shall perform its services with that degree of skill and care ordinarily exercised by similarly situated members of it profession involved in the design of similar projects in the same locale as the Project. The Owner acknowledges and understands that compliance with the guidelines may involve factors beyond the control of the Architect, including but not limited to the Owner's use and operation of the completed Project.

> 1.2 The Architect will use reasonable care consistent with the foregoing professional standard of care in Paragraph 1.1 to meet the intent of the LEED Green Building Rating Systems credit requirements, but does not warrant or represent that the project will achieve LEED certification. The Architect shall not be responsible for the Contractor's failure to adhere to the Contract documents and any applicable laws, codes, and regulations incorporated therein, nor for any changes to the design made by the Owner without the direct participation and written approval of the Architect. Likewise, the Architect shall not be responsible for any environmental or energy issues arising out of the Owner's use and operation of the completed Project.[42]

In short, a contract for green or sustainable design services presents the same traditional concerns, namely, that warranties, guarantees, or certifications of services expand the risk of liability, and may remove or nullify insurance coverage.

When LEED and green construction arrived 20 years ago, many within the industry predicted a new paradigm of risks for design professionals, a change in the standard of care, and a flood of litigation. In 2008, USGBC convened a working group to discuss the "conjecture, anecdote, and even rumor . . . circling the question of what legal claims may be based on the design, development and construction of sustainable buildings." In a paper, "The Legal Risk in Building Green: New Wine in Old Bottles?" it concluded that the risks and legal theories associated with building green are not a new paradigm, but instead, were "simply new wine in old bottles."[43] The working group found that the same legal theories were being asserted as in traditional construction litigation, but the failure to achieve LEED certification was a new risk. As one attorney noted,

> From an architect's perspective, or more specifically, from the perspective of legal counsel representing an architect in contract drafting, the biggest concern is not to guarantee a certain level of LEED certification. There are just too many factors that go into certification for the architect to be able to guarantee a certification level.[44]

The design professional cannot control all of the factors that are necessary to achieve a specific certification level under a rating program or an energy objective (whether voluntary, regulatory, or statutory), and for this reason, its contract should not warrant or guarantee that a sustainable goal will be met.

The number of LEED registrations has steadily increased from just a few in 1993, to 3,156 registrations in 2005, to 32,396 registrations in 2010, and to 69,066 in 2019. Despite this proliferation, green building has not resulted in much appellate case law.[45] Patrick J. O'Connor, author of the treatise, *Bruner & O'Connor on Construction Law*, noted that, "[f]or all the concern and ink spilled on the question, at the present time there have been precious few suits directly dealing with sustainable design or construction gone awry."[46] The dearth of appellate decisions arising out of green building and LEED is explained by the fact that many disputes have been resolved outside of the court system, through arbitration and mediation. One such arbitration matter is discussed below.

The case, *Animal Rescue League of Boston v. ARQ Architects and Hunter Construction Corp.*,[47] illustrates the perils of contract language that can be construed as a warranty or guarantee. The dispute, filed with the American Arbitration Association (AAA), arose out of a project for renovation of a nonprofit animal care, rescue, and adoption facility—an animal shelter. The agreement for architectural services was a standard AIA B101-2007 form which included a clear statement of the professional standard of care. However, an introductory paragraph of the form agreement included the following language that was not tied to the standard of care:

> The project is to be renovated to current shelter standards and environmental sustainability goals equivalent to meet LEED certification ... Formal LEED certification will be considered and decided at the completion of the schematic design phase of work.[48]

During construction, the project encountered HVAC commissioning problems and LEED certification was not achieved because the final on-line submissions to USGBC were not completed. Animal Rescue League (ARL) filed a demand for arbitration against the architect and contractor alleging claims for negligence, breach of contract, and violation of M.G.L. c. 93A.[49] The architect filed a motion for summary judgment seeking dismissal of the action based on the three-year statute of limitations for negligence in G.L. c. 260, §2B. In response, ARL argued that the claims should not be dismissed because the architect's agreement included two warranties requiring that the animal shelter meet "current shelter standards" and certification under LEED.

A 3-member arbitration panel issued a decision which dismissed a majority of the negligence claims, finding them time-barred by the statute of limitations for negligence. However, the panel's decision found that the above-quoted language in the architect's agreement included two express warranties:

1. The Facility will be renovated (that is, designed) to "current shelter standards."
2. The Facility will be renovated (that is, designed) to "environmental sustainability goals equivalent to meet LEED certification."[50]

The arbitration panel held that the two express warranties were not barred by the statute of limitations, and survived dismissal, because they are subject to a longer, 6-year statute of limitations for contract actions. The case was resolved through settlement, but illustrates an additional reason why warranties should be avoided, viz., the statute of limitations for breach of contract claims is sometimes longer than the limitation period for a negligence action, depending upon the jurisdiction.

When the architect or engineer assumes a contractual warranty obligation, its risk exposure potentially increases in three ways: it will not have its conduct judged by the professional standard of care; it may not be able to transfer its contractually assumed liability to its insurance carrier; and, depending on the jurisdiction, it may not receive the benefit of a short statute of limitations that could potentially bar the claim.

The AIA's latest green building document, E204-2017, Sustainable Projects Exhibit, includes a provision that expressly excludes a warranty from the architect that the project will achieve the sustainable objective.[51]

7.4.4 Warranty or Guaranty

New laws are requiring that buildings achieve net-zero or other emissions targets in the near future. Many contracts, whether custom drafted or edited versions of standard forms, require design professionals to "comply with all laws, codes, and regulations," or other standards. While the design professional will strive to be in full compliance, laws are sometimes unclear, or conflict, or may not apply to the project, and thus the word "all" should be stricken. If a contract includes warranty language regarding laws, codes, and regulations, it should be deleted and replaced with revised language that ties the obligation to the standard of care. The following language is suggested:

> The design professional shall perform its services in accordance with laws, codes, rules, and regulations applicable for the location of the Project consistent with the generally accepted professional standard of care stated in Paragraph __ [the AIA statement of the professional standard of care is recommended] and shall perform its services as expeditiously as is consistent with such professional skill and care and the orderly progress of the Project. No warranty or representation express or implied is included or intended in this Agreement, or in any report, document, or otherwise.

The best solution for problematic warranty language is to tie the obligation to the professional standard of care, and to include a clause which disclaims or excludes any warranty.

7.4.5 Indemnification

In general, indemnification clauses should be avoided in a professional services agreement with the owner. There is an endless variety of contractual indemnification language that is unacceptable because it will expose the design professional to liability beyond the scope of its professional liability policy. Consequently, it is important to be on the alert for this language and to carefully negotiate the scope of indemnification provisions. The indemnity provision is arguably the most dangerous risk-shifting term in the contract, as hidden risks lurk within them that have an enormous impact on insurance coverage and resolution of claims.

In an indemnification provision, the owner (the indemnitee) seeks to obligate the design professional (the indemnitor) to reimburse it for losses it incurs due to the claims of third parties, and the owner may also attempt to shift onto the design professional a duty to defend it against such claims. A duty to defend is broader than a duty to indemnify, and in the vast majority of jurisdictions, must be explicitly set forth in the contract. Therefore, the word "defend" should be removed if it appears in an indemnification provision, and any other language which requires the designer to "assume the defense of" a claim. Unlike the duty to indemnify, which is not triggered until there is an actual judgment, a duty to defend the owner may be triggered the moment the owner is sued by a third party (depending upon the allegations of the complaint), and could require the design professional to begin paying the owner's legal fees and defense costs.[52] The contractual obligation is not covered under professional liability insurance and exposes the design professional to costs that could equal or exceed the amount of the claim itself.

A broad form indemnity provision attempts to transfer to the design professional the risk of the owner's and the contractor's own negligence or poor performance. Language that requires the designer to indemnify the owner for "any and all losses . . . related to, or in any way connected with the Project or the design professional's services" makes the designer liable for injuries and losses that are not its own fault, but are caused by the mistakes of other parties, such as the contractor or subcontractors, over whom it has no control. Similarly, a clause that requires the design professional to indemnify the owner from losses arising out of its "acts" or "performance of the contract," or "acts or omissions of the designer, whether negligent or not," extends liability beyond negligence by including acts and omissions where there is no requirement for proof of a violation of the standard of care. If the indemnity obligation is not negligence-based, the clause is uninsurable under a professional liability policy, which does not provide coverage for contractual undertakings beyond liability imposed by law.

If an owner insists on including an indemnification clause in the professional services agreement, the design professional should endeavor to make the clause consistent with, or to use language which precisely mirrors, the language in AIA B103-2017, § 8.1.3, as follows:

> The Architect shall indemnify and hold harmless the Owner and the Owner's officers and employees from and against damages, losses and judgments arising from claims by third parties, including reasonable attorneys' fees and expenses recoverable under applicable law, but only to the extent they are caused by the negligent acts or omissions of the Architect, its employees and its consultants in the performance of professional services under this Agreement. The Architect's duty to indemnify the Owner under this provision shall be limited to the available proceeds of insurance coverage.

The above language *"but only to the extent they are caused by the negligent acts of the Architect"* is the pivotal phrase from a risk management standpoint. If this language is used, the design professional's indemnity obligation is the same as the preexisting duty the architect already has, and the design professional has the comfort of knowing that its professional liability insurer will cover the claim.[53]

The design professional may also seek to negotiate a reciprocal indemnity obligation from the owner, and make sure that its agreements with subconsultants contain comparable indemnities.

7.4.6 **Limitation of Liability**

When a design professional is engaged by an owner or design-builder to prepare a design for an infrastructure project with climate resilience or sustainability goals, the project's ultimate performance will depend upon many variables, including construction components—which are controlled by the contractors and subcontractors—and operation components, which are controlled by the owner. The project is a collaborative process involving all members of the project team. A limitation of liability provision balances the owner's or design-builder's climate resilience or sustainability goals with the design professional's concern about assuming responsibility to achieve a goal that may be unachievable as the result of causes beyond the design professional's control. The rationale for capping liability is that the small fee paid to the design professional is disproportionate to its risk exposure. The project owner or the design-builder benefits from the allocation of risk in a limitation of liability provision because it is able to obtain innovative designs at a cost-effective price.

Limitation of liability clauses are a method of apportioning the risk. The design professional should seek to limit its liability on projects where its risk exposure far outweighs its fee, or where risks assumed are beyond the coverage afforded by its professional liability insurance. On a DB project, a limitation of liability provision is a way to manage flow-down risks.

Generally, there are three types of contractual limitation of liability clauses that limit the design professional's liability to the total fee received by the design professional, the amount of insurance available, or a dollar amount which the parties agree upon. Model language for all three types of clauses may be found in AIA B503-2017 *Guide for Amendments to AIA Owner-Architect Agreements.*

Most jurisdictions enforce limitation of liability provisions, and an often-cited case is *Valhal Corp. v. Sullivan Associates, Inc.*, 48 F.3d 760 (3d Cir. 1995). However, some states find them unenforceable or void as against public policy, and other jurisdictions have not directly addressed the enforceability of limitation of liability clauses contained in professional services contracts. A limitation of liability clause is not an exculpatory clause which relieves one party from its entire liability. Design professionals should seek legal counsel to help draft a limitation of liability clause to maximize the likelihood that it will be enforced. It is also important to recognize that limitation of liability provisions only protect a design professional from claims by its client—who is a party to the agreement—and that third parties are not bound by such a contractual provision.

7.4.7 Waiver of Consequential Damages

Consequential damages, also known as "special damages," are damages that are not direct and are less traceable to the breach of duty than direct compensatory damages. Black's Law Dictionary defines consequential damages as "Such damage, loss or injury as does not flow directly and immediately from the act of the party, but from some of the consequences or results of such act . . . Those losses or injuries which are a result of an act but are not direct and immediate." The English case, *Hadley v. Baxendale*, 9 Exch. 341, 56 Eng. Rep. 145 (1854) set forth the rule for determining the scope of consequential damages: a party is liable for all losses within the reasonable contemplation of the contracting parties at the time of the contract.

When a building fails to meet an energy benchmark or the desired level of certification under LEED or another rating system, the potential damages could be enormous—statutory penalties, increased design and construction costs, loss of future rents or sales, inability to obtain tax credits, increased energy costs, or more amorphous impacts such as decreased employee productivity. Including a waiver of consequential damages provision in a contract can help to protect the design professional on a sustainability project, but it is not a guaranteed solution, because it is not always clear whether a particular damage is direct or consequential. This difficulty in identifying damages as "direct" or "consequential" is illustrated by a recent Florida case:

In Keystone Airpark Authority v. Pipeline Contractors, Inc., a government airport authority in Florida (the owner) hired a contractor to build airplane hangars and taxiways. The owner separately contracted with an engineering firm (the engineer) to provide engineering services, including material testing and observations of the contractor's work, to ensure compliance with the plans and specifications. The engineer's professional services agreement contained a standard waiver of consequential damages provision. After construction was completed the owner discovered that the hangar's concrete slabs and the concrete and asphalt runways were deteriorating because the contractor used substandard materials. The owner sued the engineer and the contractor for the costs to remove, repair, and replace the hangars, taxiways and underlying subgrade.

Florida's First District Court of Appeals affirmed a trial court's summary judgment decision that a waiver of consequential damages provision in the engineer's professional services agreement limited the owner's recovery against the engineer to the costs of its inspection services.[54] The appellate court found that all of the repair damages were caused by the contractor's deficient work, and while the damages were reasonably foreseeable and direct damages as to the contractor, they were consequential damages as to the engineer, and were therefore excluded by the waiver of consequential damages provision.

A waiver of consequential damages provision helps to control risks but as the above example illustrates, it is often difficult to predict how the waiver will be applied to a particular damage. Therefore, the design professional should endeavor to list all foreseeable damages in the provision. The following is an example of a mutual waiver of consequential damages:

> Notwithstanding any other provisions of this Agreement, each Party waives all claims against the other for any indirect, incidental, special, exemplary, punitive, or consequential damages, whatsoever incurred by such Party, however caused, whether or not the other party had or should have had knowledge of such damages, arising out of, or connected in any way to the Project or this Agreement. This mutual waiver includes, but is not limited to, those consequential damages resulting from failure of the Project to achieve the Sustainable Objective or one or more Sustainable measures, including unachieved energy savings, unintended operational expenses, lost financial or tax incentives, or unachieved gains in worker productivity, damages for loss of profits, loss of income, loss of reputation, unrealized savings, or diminution of property value, and shall apply to any cause of action including negligence, strict liability, breach of contract, and breach of warranty.

Owners will often agree to mutual waivers of consequential damages, but if an owner resists such a clause, another approach is to include a limitation of liability provision which sets the limit at the available insurance proceeds of the design professional, the amount of its fee, or another negotiated amount (discussed above). A waiver of consequential damages is not generally binding against third parties.

7.4.8 **Waiver of Subrogation**

A waiver of subrogation clause is a risk management tool that can eliminate the design professional's liability for losses paid in the first instance by the owner's builder's risk or property insurance policy. Subrogation is a concept whereby the insurer who has paid a loss or liability of its insured is entitled to step into the shoes of its insured and assert the rights of its insured against third parties in order to recoup the amounts it paid. Waiver of subrogation clauses are intended to leave the loss with the insurer who initially pays it, and to eliminate the subrogation actions that can be filed by the insurance company.

A waiver of subrogation can be illustrated by the following hypothetical example: The designer of a 28-story apartment complex located in a coastal city vulnerable to tidal flooding from climate change was required to follow the city's Climate Change Design Guidelines for sea-level rise and storm surge (SLR/SS) during the permitting process. The ground floor was raised two feet above the projected flood level, and flood barriers including vegetated berms were added between the complex and the waterfront. The project's design specifications adhered to the Climate Change Design Guidelines, which required that components of fire suppression systems, such as pumps and controls which are inherently vulnerable to flood damage, be elevated above the flood protection level.

In an incident unrelated to seawater or rainwater flooding, a sprinkler supply pipe of the fire protection system failed, causing several million dollars of property damage. The owner filed a claim with its insurance company and the insurer paid the loss. The insurance company subsequently filed a subrogation action against the contractor, seeking to recover the costs it had paid on the loss, and the contractor brought a third-party action against the architect, seeking contribution and indemnity. The agreements between the parties incorporated the AIA A201-2017 general conditions, which includes a waiver of subrogation clause, as follows:

> The Owner and Contractor waive all rights against (1) each other and any of their subcontractors, sub-subcontractors, agents, and employees, each of the other; (2) the Architect and Architect's consultants; and (3) Separate Contractors, if any, and any of their subcontractors, sub-subcontractors, agents, and employees, for damages caused by fire, or other causes of loss, to the extent those losses are covered by property insurance required by the Agreement or other property insurance applicable to the Project, except such rights as they have to proceeds of such insurance. The Owner or Contractor, as appropriate, shall require similar written waivers in favor of the individuals and entities identified above from the Architect, Architect's consultants, Separate Contractors, subcontractors, and sub- subcontractors. The policies of insurance purchased and maintained by each person or entity agreeing to waive claims pursuant to this section 11.3.1 shall not prohibit this waiver of subrogation. This waiver of subrogation shall be effective as to a person or entity (1) even though that person or entity would otherwise have a duty of indemnification, contractual or otherwise, (2) even though that person or entity did not pay the insurance premium directly or indirectly, or (3) whether or not the person or entity had an insurable interest in the damaged property.

Courts in many jurisdictions, including Massachusetts, have held that the waiver of subrogation clause in the AIA A201 general conditions is enforceable in favor of contractors, architects, and engineers. The AIA B101 professional services agreement also contains a waiver of subrogation clause, as follows:

> To the extent damages are covered by property insurance, the Owner and Architect waive all rights against each other and against the contractors, consultants, agents, and employees of the other for damages, except such rights as they may have to the proceeds of such insurance as set forth in AIA Document A201–2017, General Conditions of the Contract for Construction. The Owner or the Architect, as appropriate, shall require of the contractors, consultants, agents, and employees of any of them, similar waivers in favor of the other parties enumerated herein.

Design professionals should always consider including a mutual waiver of subrogation provision in their contracts.

7.4.9 **Force Majeure**

Construction projects are complex and often span a number of years. During that time period, the design professional may be exposed to delay claims, disruption claims, or other types of claims made by its client. The doctrine of *force majeure* excuses nonperformance or delay under certain circumstances. A project designed with the sustainable objective to withstand the impacts of hurricane wind speeds and flooding from a 500-year storm may still be vulnerable to other types of unprecedented shocks (e.g., an earthquake, war, or disease outbreak), or to extreme weather events that occur before the project is completed. *Force majeure* is a French term that means "superior force," and the types of events traditionally range from "acts of God" (hurricanes, floods, tornadoes), to acts of people (labor strikes, terrorist attacks, government actions). There is little case law addressing whether a disease outbreak such as COVID-19 constitutes a *force majeure* event under the common law. A *force majeure* clause is a contractual means of allocating risk.

Two commonly used construction forms, the AIA A201-2017 "General Conditions of the Contract for Construction" and the ConsensusDocs "Standard Agreement and General Conditions Between Owner and Constructor," both include *force majeure* events under the heading, "Delays and Extensions of Time," and provide for an equitable adjustment of the contract time. The ConsensusDocs provision explicitly includes "epidemics" and "adverse weather conditions not reasonably anticipated." However, these standard clauses apply only to contractors and not to design professionals. There are many reasons why a design professional should consider including a *force majeure* clause in its own contract, including the recent worldwide COVID-19 pandemic. The following is an example of a *force majeure* provision:

> The [Owner or Contractor] agrees that the Design Professional is not responsible for damages arising from any circumstances beyond the Design Professional's reasonable control. For purposes of this Agreement, such causes include, but are not limited to, severe weather disruptions, natural disasters, disease outbreak, epidemics or pandemics, or other declaration of public health emergency, quarantine restriction, fire, or other acts of God; riots, war, or other emergencies; failure of any government agency to act in a timely manner, adverse government actions; failure of performance by the Contractor or the Owner or the Contractor's or Owner's subcontractors or consultants; or discovery of any hazardous substances or differing and unforeseeable site conditions.

A contractual *force majeure* provision can help protect a design professional against claims that are beyond its control.

PART 3

7.5 Insurance Program

A construction project involves significant risk for all project participants, including the risk of design errors, construction defects, work-site injuries, and property damage, to name a few. If these risks materialize, they may result in substantial damages and disputes. To help control and manage risk, owners, design professionals, and contractors typically maintain insurance. Insurance is a mechanism by which risk is transferred to a third-party insurer under a contract of insurance in exchange for payment of a premium. The parties' contracts specify requirements for insurance with limits appropriate to the type and size of the project, and services involved. This may include builder's risk insurance, commercial general liability (CGL) insurance, workers' compensation insurance, automobile insurance, professional liability insurance (PL), and environmental liability insurance. On larger projects, there is an option for an owner to purchase an owner-controlled insurance program (OCIP), or for a contractor to purchase a contractor-controlled insurance program (CCIP). OCIPs and CCIPs can include professional liability insurance, but typically do not.

7.5.1 Professional Liability Insurance

Architects and engineers professional liability insurance policies provide coverage for negligent errors and omissions (E&O) in the services a design professional performs as an architect, engineer, land surveyor, construction manager, or technical consultant. Coverage is triggered by claims alleging any negligent act, error, or omission in rendering or failing to render professional services. Professional liability coverage is written on a "claims made" basis, meaning that the policy in force for the year in which the claim is made affords coverage and controls the limits of liability. After a project is completed, the design professional must continue to maintain PL coverage until the statute of repose has run, or it will be exposed to uninsured claims that may arise months or years after the project is completed.

Litigation defense costs erode the policy limits of professional liability policies. Eroding limits are a concern because the designer's policy may have other claims against it in a given year. Project-specific professional liability insurance, which has limits dedicated to one project, is recognized as a cost-effective mechanism for transferring risk and eliminating gaps and duplication in coverage. A project-specific liability policy provides professional liability insurance for the collective design team, and is issued for the duration of the project plus an extended reporting period. All claims against the design team must be defended jointly, and there is an "insured v. insured" exclusion, which prohibits one insured design professional from making a claim against another. Project-specific policies have a deductible or self-insured retention (SIR), which is allocated among the design team.

7.5.2 CGL Insurance

Commercial general liability (CGL) insurance is the primary coverage for damages due to bodily injury or property damage. Both designers and contractors carry this coverage. Unlike professional liability insurance, CGL coverage operates on an "occurrence basis," meaning that the policy will respond to an event that occurred when the policy was in force, even if the claim is not made for months or years after the incident and after the policy has expired. CGL policies rarely afford coverage for professional liability claims due to a professional services exclusion.

7.5.3 Additional Insured Coverage

Owners often require the general contractor to name them and the design professional as additional insureds on a CGL policy. In turn, contractors typically request the same from their subcontractors. The Insurance Services Office (ISO) provides standard additional insured endorsement forms, which vary significantly in language and scope. Contractors may work with multiple clients in any given year, making it impractical to keep scheduling additional insureds on a policy. ISO offers various "blanket" or automatic additional insured endorsements, but they have changed frequently over the years to limit the scope of coverage. For example, the ISO has developed separate forms to provide coverage for ongoing operations. Since ongoing operations do not constitute completed operations, another additional insured form is needed for completed operations. Upstream parties should draft their contracts to take these different forms and their language distinctions into account.

A&E professional liability policies are written specifically for architects and engineers, and will not list owners or contractors as additional insureds. Therefore, a design professional should not agree to any contract terms that require it to name the owner, contractor, or other entities as additional insureds on a professional liability policy. On large projects, the design team may be insured under a project-specific professional liability policy. A design-builder can be added to a project-specific professional liability policy by means of an endorsement that specifically names the design-builder as an "indemnified party." This type of design-builder's indemnification endorsement will provide coverage for damages incurred as a result of third-party claims, to the extent caused by the design professional's negligent professional services.

7.6 Surety Bonds

Surety bonds are a means of reducing risks for the owner. The most common types of bonds are bid bonds, performance bonds, payment bonds, and lien bonds. Statutes governing public construction projects of appreciable size require payment and performance bonds to provide security for the contractor's performance. By issuance of a bond, the surety agrees to guarantee completion of the work and payment to subcontractors, laborers, and material suppliers who are not paid by the contractor. Although surety bonds are issued by insurance companies, they are not insurance. Instead, a suretyship is a tripartite relationship in which the surety guarantees to the obligee (owner) that if the principal (contractor or subcontractor) fails to construct the project or meet its payment obligations, the surety will do so.

7.7 **Conclusion**

As our climate continues to change, there are new uncertainties and risks in design and construction. Effective risk management entails preparing for these changes in contracting practices. A comprehensive analysis of all the important terms in a professional services agreement is beyond the scope of this chapter, but the above has identified some of the key contractual issues and solutions. Careful negotiation of contract terms can mitigate the risks that architects and engineers face in providing professional services. Risk management mechanisms, such as insurance and well-drafted contractual provisions, will minimize the design professional's risk exposure and the incidence of claims and disputes.

Notes

1 Patricia B. Gary, "The Future of Megaprojects," chapter 13 in *Megaprojects: Challenges and Recommended Practices*, D. Hatem and D. Corkum, eds. Washington, DC: American Council of Engineering Companies 2010, pp. 275–277.

2 ACEC Matters, "Board of Building Regulations and Standards Votes to Change 10th Edition of MA Building Code Schedule," American Council of Engineering Companies, June 12, 2019. Available online at https://www.acecma.org/about/news/board-of-building-regulations-and-standards-votes-to-change-10th-edition-of-ma-building-code-schedule-2535.

3 J. Rolf Olsen, ed., *Adapting Infrastructure and Civil Engineering Practice to a Changing Climate*, Reston, VA: American Society of Civil Engineers, 2015, p. 23. Available online at https://ascelibrary.org/doi/pdf/10.1061/9780784479193.

4 *Fourth National Climate Assessment*, Chapter 1: Overview, p. 4, November 2019, https://nca2018.globachange.gov/chapter/1/.

5 Olsen, *Adapting Infrastructure and Civil Engineering Practice to a Changing Climate*, p. 30.

6 *Ibid.*

7 Ben Walpole, "ASCE Developing New Sustainability Standard," *ASCE News*, April 23, 2020, https://news.asce.org/asce-developing-new-industry-wide-sustainability-standard/.

8 *Ibid.*

9 Climate-Safe Infrastructure Working Group to the California State Legislature and the Strategic Growth Council, "Governance: Changing the Rules to Enable Climate-Safe Infrastructure," chapter 7 in *Paying It Forward: The Path to Climate-Safe Infrastructure*

in California, 2018, p. 84. Available online at https://files.resources.ca.gov/climate/climate-safe-infrastructure-working-group/.

10 UCLA Luskin Center for Innovation, "Progress Toward 100% Clean Energy in Cities and States Across the U.S.," p. 6, Nov. 2019.

11 Olsen, *Adapting Infrastructure and Civil Engineering Practice to a Changing Climate*, p. 17.

12 City of Boston, "2019 Climate Action Plan Update," 2019, p. 36.

13 Patricia B. Gary, "Public Interest Considerations in P3 and Design-Build Projects, a Historical Perspective," chapter 5 in *Public-Private Partnerships and Design-Build, Opportunities and Risks for Consulting Engineers*, 3d ed., D. Hatem and P. Gary, eds. Washington, DC: American Council of Engineering Companies, 2020, p. 102.

14 Gary, "The Future of Megaprojects," pp. 261–262.

15 *United States v. Spearin*, 248 U.S. 132 (1918).

16 Ross J. Altman, "Project Delivery Systems," chapter 4 in *Construction Law*, 2d ed., Carol J Patterson, Ross J. Altman, Stephen A. Hess, and Allen L. Overcash, eds., Chicago: American Bar Association, 2009, pp. 71–75.

17 National Cooperative Highway Research Program, "NCHRP Synthesis 402, Construction Manager-at-Risk Project Delivery for Highway Programs, A Synthesis of Highway Practice," Transportation Research Board of the National Academies (2010), p. 51.

18 *Ibid.*

19 *Coughlin Electrical Contractors, Inv. v. Gilbane Building Co., 472 Mass. 549, 558 (2015).*

20 *Ibid.*

21 Federal Highway Administration, "Construction Program Guide, Construction Manager/General Contractor Project Delivery," available online at https://www.fhwa.dot.gov/construction/cqit/cm.cfm.

22 David J. Hatem, "Improving Risk Allocation on Design-Build Subsurface Projects," *Tunnel Business Magazine*, June 17, 2020. Available online at https://tunnelingonline.com/improving-risk-allocation-on-design-build-subsurface-projects/.

23 *Ibid.*, p. 265.

24 Gary, "Public Interest Considerations in P3 and Design-Build Projects," pp. 103–104.

25 Gary, "The Future of Megaprojects," pp. 265–266.

26 DBIA, "Progressive Design-Build: Design-Build Procured with a Progressive Design & Price," Design-Build Done Right Primer, October 2017, https://dbia.org/wp-content/uploads/2018/05/Primer-Progressive-Design-Build.pdf.

27 Hatem, "Improving Risk Allocation on Design-Build Subsurface Projects," *Tunnel Business Magazine*.

28 Gary, "Public Interest Considerations in P3 and Design-Build Projects, a Historical Perspective," pp. 97–98.

29 *Ibid.*

30 *Ibid.*

31 Gary, "Public Interest Considerations in P3 and Design-Build Projects, a Historical Perspective," pp. 159–163.

32 *Ibid.*

33 Keith Molenaar, Nathaniel Sobin, Douglas Gransberg, Tamera McCuen, and Michael Horman, "Sustainable, High Performance Projects and Project Delivery Methods: A State-of-Practice Report," the Charles Pankow Foundation and the Design-Build Institute of America September 1, 2009, p. 1. Available online at https://www.researchgate.net/publication/327976573_Sustainable_High_Performance_Projects_and_Project_Delivery_Methods_A_State-of-Practice_Report.

34 *Ibid.*, p. 10.

35 Anna Kadefors, Stefan Uppenberg, Johanna Alkan Olsson, Daniel Balian, and Sofia Lingegård, "Procurement Requirements for Carbon Reduction in Infrastructure Construction Projects, An International Case Study," Project Report, June 2019, p. 100.

36 Jonathan J. Sweet, *Sweet on Construction Industry Documents: Major AIA Documents*, 5th ed., New York: Wolters Kluwer, 2009, p. 17.

37 *Klein v. Catalano*, 386 Mass 701, 718, 437 N.E.2d 514 (1982).

38 AIA Commentary, B101-2007 Standard Form of Agreement, available online at https://www.aiacontracts.org/contract-documents/25141-owner-architect-agreement.

39 David J. Hatem, "Green and Sustainable Design, Part I: Professional Liability Risk and Insurability Issues for Design Professionals," *Design and Construction Management Professional Liability Reporter*, June 2010, p. 10.

40 The *Spearin* doctrine originated from a 1918 United States Supreme Court decision, *United States v. Spearin*, 248 U.S. 132 (1918), which held that a contractor will not be liable to an owner for loss or damage that results from defects in the plan, design, or specifications provided to the contractor. Under the *Spearin* doctrine, the owner impliedly warrants that the plans and specifications, if followed, will result in a functioning system, and the contractor will not be responsible for the consequences of defects in the plan.

41 ACEC Risk Management Committee, "Design-Build Projects—Achieving Fair Allocation of Risk between the Contractor and Design Professional," American Council of Engineering Companies, 2015.

42 Robyn L. Miller and James K. Lowe, Jr., "Key Contracting Issues from The Designer's Perspective," chapter 8 in *Building Better Construction Contracts 2012*, Patrick J. O'Connor and Timothy R. Twomey, eds. New York: Practising Law Institute, 2012, p. 298.

43 U.S. Green Building Council, "The Legal Risk in 'Building Green': New Wine in Old Bottles?" (panel discussion transcript) Building Green (November 2008) https://sallan.org/pdf-docs/legal_risk_in_building_green.pdf.

44 *Ibid.*

45 T. Wang, "U.S. Green Building: Number of LEED-certified Projects 2000–2019," Statista, October 25, 2009, https://www.statista.com/statistics/323383/leed-registered-projects-in-the-united-states/.

46 Patrick J. O'Connor, Jr., "Legal Considerations in Sustainable Design and Construction," Arizona State Bar Convention, June 11, 2010, p. 35.

47 Demand for Arbitration, Animal Rescue League of Boston v. ARQ Architects, American Arbitration Association Case No. 01-15-0004-9011.

48 American Arbitration Association Case No. 01-15-0004-9011, Order Regarding ARQ

and Hutter Motions for Partial Summary Judgment and Hutter Motion to Dismiss Certain Counts of ARL's Second Amended Complaint for Arbitration, dated August 29, 2016, p. 4.

49 Demand for Arbitration, *supra*.

50 Order on Motions for Partial Summary Judgment, p. 6.

51 AIA E204-2017, Article 6.1.

52 *Ferreira v. Beacon Skanska Constr. Co.*, 296 F.Supp.2d 28, 33 (D. Mass. 2003) (whether a duty to defend is triggered under an indemnification agreement is determined by reviewing the allegations of the complaint); *Herson v. New Boston Garden Corp.*, 40 Mass. App. Ct. 779, 787 (1996) ("the duty to defend in the subject clause is a duty independent of and broader than the duty to indemnify"); *Urban Inv. & Dev. Co. v. Turner Constr. Co.*, 35 Mass. App. Ct. 100 107 (1993).

53 Theodore D. Levin and Mehrdad Farivar, "Design Professional Responsibilities—Pre-Design, Design, and Pre-Construction Phases," chapter 3 in *Design Professional and Construction Manager Law*, Stephen A. Hess, Jerome V. Bales, P. Douglas Folk, and Tyrone Holt, eds. Chicago: American Bar Association, 2007, p. 76.

54 *Keystone Airpark Authority v. Pipeline Contractors, Inc.*, 266 So.3d 1219, 1222 (Fla. 1st DCA, January 25, 2019) review denied, 2019 WL 1371949, Fla., Mar. 27, 2019.

PART III

Professional Practice and Experience

8

Approaches to Resilient Infrastructure Planning and Design

Kelly C. Maiorana and Nicole Boothman-Shepard

8.0 Introduction

Colorado has experienced our fair share of disasters in the past decade. We've seen blizzards, tornadoes, wildfires, floods, and even an earthquake. Each time, Colorado—our communities and our people—have shown a great resolve not simply to bounce back from disaster, but to make ourselves better, stronger, more resilient.

— Governor John Hickenlooper, 2015

When you stretch a rubber band and let it go, it tends to bounce back to its original form due to its elasticity. However, its capability to recover is finite. If you stretch the rubber band to its yield point, it will not recover to its original form. If you stretch the rubber band to its breaking point, it will snap. An infrastructure asset functions much like a rubber band. It has been designed and built with a certain elastic capacity, but that also means it has yield and breaking points.

We are more aware of climate conditions than ever before, but how we accommodate the increasing frequency and severity of natural hazards by developing and managing our infrastructure assets is lagging. That leaves more assets at risk of reaching their breaking points. If our infrastructure is designed and built to function as a rubber band does—with the capability to accommodate increased natural risks—it creates opportunities to mitigate both damage to the assets and the cascading impacts and consequences communities face during and after disaster. The key to designing the most elastic "rubber band" infrastructure is resilience which comes from the Latin word "resilire," meaning "to rebound." If designing, constructing, operating, and maintaining infrastructure had no fiscal limits, every structure could (in theory) be "futureproofed" to withstand any natural hazard. In the absence of unlimited funding, resilient infrastructure design and construction necessarily center around finding the appropriate levels of investment where the benefits of resilience outweigh the costs.

This chapter focuses on the definition, approach, quantification, prioritization, and decision-making involved in resilient infrastructure design. Pragmatically, it also recognizes that most infrastructure owners operate in fiscally constrained environments. Policies that mandate resilient design standards and specifications for asset performance are important. However, an infrastructure owner can only be successful with an understanding of the costs and benefits of building or retrofitting—as well as operating and maintaining—the asset in order to ensure that a project (or system) meets the intent of any resilience policy. To do this, resilience must be defined, and the question answered, "How does one measure infrastructure resilience?"

Colorado Department of Transportation's (CDOT's) shift to resilience began in the aftermath of devastating floods in 2013. The 2013 disaster and CDOT's growth as a resilience-centric organization is used as the fulcrum through which Chapter 8discusses the definition, approach, prioritization, and decision-making for resilient infrastructure design. After the 2013 flood, the State of Colorado adopted the definition of resilience as:

> The ability of communities to rebound, positively adapt to, or thrive amidst changing conditions or challenges–including disasters and climate change–and maintain quality of life, healthy growth, durable systems, and conservation of resources for present and future generations.

This chapter will not substantially discuss specific climate adaptation approaches, but it begins by acknowledging that the data is clear–frequency and severity of extreme weather disasters are increasing and are exacerbated by climate stresses. For many asset owners, this increased probability and severity of stresses and shocks has resulted in commensurate increases in embedded risk that are compounded across a portfolio of infrastructure assets–sometimes creating millions or billions of dollars in exposure. Therefore, threat and vulnerability analyses from natural hazards, including climate threats, need to be included in infrastructure planning, design, construction, operations, and maintenance as part of an owner's larger asset and risk management approach. To do this, current and projected risk conditions and vulnerabilities need to be identified and measured through a uniform process to help drive decision-making on capital expenditure through rigorous and consistently applied methodologies. This chapter will take the reader through the evolution of how the Colorado Department of Transportation (CDOT) developed and implemented its resilience policy, "Building Resiliency into Transportation Infrastructure and Operations."

Among its responsibilities, CDOT serves as an innovation lab for formulating resilient design approaches for state-owned assets. CDOT's resilience mindset began as a reaction to the widespread September 2013 flood that damaged 486 lane miles of roadways and over 200 bridges and structures and caused over $4 billion in multi-sector damages statewide. The Federal Highway Administration's (FHWA's) $724 million investment of Emergency Relief (FHWA ER) funding in Colorado's transportation assets and enabled CDOT to transform its approach not only to build back stronger following the flood, but to incorporate proactive, risk-based planning throughout its asset management approach. This chapter shares CDOT's quantifiable methodologies to analyze benefits and costs of resilient infrastructure investments and a systematic approach to prioritizing limited transportation funds aligned with the State of Colorado's definition of resilience, above.

The FHWA Emergency Relief program was established to financially assist states when a state demonstrates that damages exceed its capacity to support response and recovery on its own. FHWA ER program funding comes with tight eligibility limitations and must only be utilized to recover from direct disaster damage. In many cases, FHWA ER funding supports in-kind repair or replacement of damaged infrastructure. However, FHWA has certain policy provisions that allow resilient solutions to be considered on severely damaged on-system assets, as happened in the case of CDOT.

8.1 Background

On September 9, 2013, a slow-moving cold front stalled over Colorado, clashing with warm humid air from the south, bringing a week of severe rain and flooding that did not subside until September 30, 2013. The heavy rains brought on a snow melt, which produced catastrophic flooding along Colorado's Front Range of the Southern Rocky Mountains and extended from Colorado Springs (south of Denver) northward to Fort Collins, near the border with Wyoming. The flooding progressed from the Front Range to the lower-lying Eastern Plains along the South Platte River to the Nebraska border, leaving an unprecedented disaster zone that covered an area almost 200 miles (north-south) by approximately 50 miles (east-west). The 2013 Colorado flood was estimated to have peaked between a 100-year and 500-year event (return rate) at numerous locations along affected rivers and streams.

Four-hundred and eighty-six miles of roadway were affected, including damage to over 200 bridges and large culverts. The rainfall and snow melt caused rockslides, landslides, and mudslides that blocked or damaged roadways. It washed out segments of roadway corridors, bridge structures, slopes and embankments, and right-of-way fences. Culverts plugged by debris as well as destroyed bridges and abutments contributed to water overflow and roadway overtopping. The destruction of bridges and dam overflows, such as at Lake Estes Dam, added an additional 6,000 cubic feet per second (cfs), increasing flood flows to as much as 19,600 cfs into the Big Thompson River. Flooding was so severe in some locations that it rerouted sections of riverine waterways.

Colorado Governor John Hickenlooper issued an executive order on September 12, 2013, declaring a Disaster Emergency; this was immediately followed by a (Presidential) Disaster Declaration two days later. In total, 24 counties were approved for federal disaster assistance. The disaster declarations set the groundwork for CDOT's request for FHWA ER to establish a program of projects to support emergency and permanent disaster repairs and costs. Due to the scale of damages, Congress quickly provided relief. FHWA initiated a rapid transfer of funds which was followed by substantial tranches of funding that totaled $724 million by the end of CDOT's flood recovery program. Of this, $63 million was included to fund local agency projects.

CDOT commenced emergency repairs immediately after floodwaters receded to restore essential traffic to impassable and damaged roadways. Chief among these were repairs to US Highway 34 (see "Case Study 2," page 293, below) and US Highway 36 to re-establish transportation to communities cut off from access and supply lines. The emergency repair program met the Governor's challenge to make surface transportation passable within 60 days of when the flood ended on September 30, 2013.

With emergency repairs complete, CDOT's attention shifted to the flood recovery's permanent repair (PR) program structured around the governor's challenge to "build back better." Accomplishing this demanded decisive action, including new ways of doing business. Governor Hickenlooper convened the Colorado Resiliency Working Group (CRWG), with over 130 participating organizations, including CDOT, which led the CRWG infrastructure committee. Together, the CRWG formulated the *Colorado Resiliency* (*Framework*), which developed cross-sector and sector-specific goals and objectives. This made Colorado's accelerated process for incorporating risk identification and resilience planning in key government sectors across the state possible.

Due to severe flood impacts, Northern Colorado urgently needed to implement resilient design alternatives and use adaptive capacity to absorb future acute shocks and chronic stresses that accompany a changing environment. A major concern addressed in the *Framework* was Colorado's exposure to natural hazards–including climatic threats–and increased risk in the future. Climate models project that in Denver, the number of days with a daily temperature high above 90°F will increase from 5 per year to over 30 per year in the future (*Framework*, pp. 3-9). These temperature changes are anticipated to result in strain on seasonal water availability and in the intensification of drought cycles, leading to increased wildfire threats. Not only did Colorado's permanent repairs (PR) need to be resilient to the traditional suite of natural hazards, but the resilience of critical assets needed to withstand more frequent and severe future stresses and shocks! A small glimpse at what the future might hold following the ravaging 2020 wildfire season can be seen through an example of 2013 flood effects in an area with burn scars from a pre-flood wildfire. With no healthy vegetation to attenuate erosion, floodwaters estimated at a roughly 10-year flood (return rate) created flows that washed out roadway segments. CDOT transportation professionals and consultants anecdotally concluded that the burn scar area impacts were comparable to typical damages to a roadway exposed to a 100-year flood return rates for slopes with healthy natural features.

The 2013 flood response and recovery efforts broke down silos in state government. A new network of relationships—sustained and cultivated by the CRWG—became the Framework's hallmark for planning holistic, cross-sector resilience for housing, infrastructure, economic development, communities and social health, watersheds and natural resources, and health. The purpose of the CTWG was to identify chronic stresses, examine impacts from acute shocks, and develop a resilient vision for each sector to include in the Framework and then bring it to life. Signed in May 2015, the Framework's two main purposes are to (1) provide guidance and tools for community stakeholders while calling for a commitment to partnership and resilience action, and (2) identify and implement strategies to increase resiliency in Colorado. One example of a robust resilience result coming from collaboration between CRWG infrastructure and watershed committee members is a *Unified Design Review Process on Watershed and Infrastructure Projects*. The *Unified Review* supports (1) preservation and access to scenic and recreational resources, (2) environmentally appropriate expansion of rivers and stream beds to "make room for the water" by allowing higher flow rates of water to pass within a river's banks, (3) building new infrastructure assets outside of the floodplain where reasonable and feasible, and (4) promoting the restoration of riparian habitats along waterways..

The state's topography of steep mountains with rivers adjacent to highways amplifies the transportation infrastructure's exposure to natural hazards. The catastrophic failure of roadways during the 2013 floods made CDOT realize that in the future the planning, design, and construction of roadways must take external factors, including climate change into account to avoid systemic disasters. The permanent repair phase of the 2013 flood recovery gave CDOT the opportunity to answer the governor's challenge to "build back better, stronger, and more resilient."

8.2 Resiliency: Approach, Prioritization, and Decision-Making

Resiliency planning demands attention to both current and future threats–including climate threats–when evaluating the vulnerabilities of infrastructure assets and resilience alternatives. This includes considering diverse mitigation approaches such as optimizing an asset's adaptive capacity, that is, its capacity to be modified in the future to meet changing threat levels during its useful life. CDOT collaborated with the FHWA on two resilience pilots—one to evaluate risk and resilience alternatives to advance the permanent repair phase following the 2013 flood disaster, and a subsequent pilot to evaluate risk and resilience during a "steady state" in anticipation of future shocks and climate stresses. These pilots have helped CDOT promote resilience-centric changes to its business processes.

Building on the work of the pilots, CDOT also improved its operational readiness to respond to future disasters and is incorporating resilient approaches in other aspects of its business, including environmental management and asset management. For example, in its Planning and Environmental Linkages (PEL) study, CDOT committed to proactively analyzing threats, vulnerabilities, and mitigation alternatives by focusing on resiliency throughout a project's lifecycle. PEL studies, discussed further below, link transportation issues with environmental concerns in order to prioritize future projects. In doing so, they help CDOT shape resilient solutions in terms of acute threats like wildfires and floods within the broader resiliency context that considers stresses like land use changes, increased traffic demand, and the new development that is driven by population growth.

PEL studies help CDOT meet its commitment to meet the state's resilience vision by:

- Identifying threats to assets;
- Documenting vulnerabilities;
- Quantifying risk exposure by forecasting potential impacts and consequences;
- Prioritizing assets by criticality and other key factors;
- Developing and evaluating resilient design alternatives and lifecycle costs;
- Considering opportunities to provide "triple bottom-line" benefits of environmental sustainability, social benefit for people and communities, and support for economic stability and growth; and
- Providing recommendations for future projects.

CDOT evaluates the capacity of an asset to be modified in the future to accommodate ongoing climate change. In addition to considering stresses, CDOT also considers various opportunities to achieve triple bottom-line benefits for environmental sustainability, for people and communities, and for economic stability and growth as part of this process. By necessity, that includes prioritizing resilience projects in light of current and anticipated financial constraints on available capital expenditures.

The following describes CDOT's resilience pilots and discusses the resilience benefits integrated in PEL studies to shed light on CDOT's approach, prioritization, and decision-making.

8.2.1 CDOT 2013 Flood Risk and Resiliency Pilot Study

In CDOT's first FHWA-supported resilience pilot study, Risk and Resiliency (RnR), corridor segments or structures with severe damage from the 2013 flood were selected to be considered for resilient design and construction investments in lieu of simply restoring them to pre-disaster condition. CDOT used this tool to identify the projects with the greatest benefit-cost ratio to be derived from resilient design. RnR enabled CDOT to develop and apply rigorous scientific analyses to projects that

Figure 1. **Benefit-Cost Analysis**

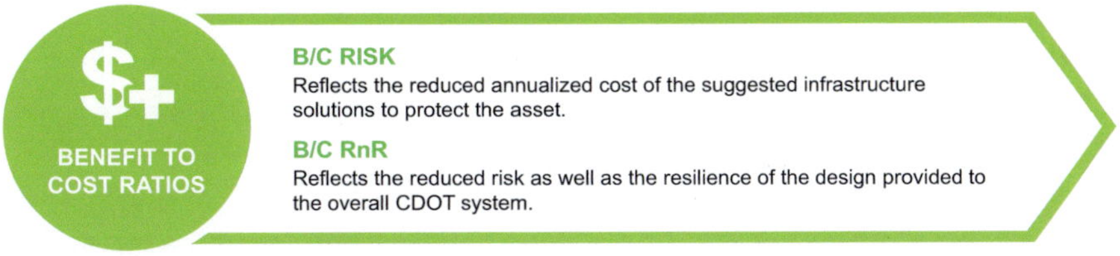

included (among other factors) asset criticality and projected physical and financial impacts due to future hazard impacts; this methodology is discussed in detail below.

Prioritizing severely damaged assets was a natural starting point for RnR because it focused on disaster damage that had already weakened their integrity (e.g., corridor segments, bridge approaches) and were consequently in need of urgent solutions. Pragmatically, it also made financial sense for CDOT, because resilient solutions for transportation assets with severe damage are potentially eligible for FHWA ER support for resilience improvements. RnR shifted planning away from only *traditional solutions*, such as increasing bridge and structure capacities to effectively convey the flow of water and debris, to *new ways of doing business*, such as resilient recovery solutions that change the alignment of a roadway outside of the floodplain (horizontally) onto bedrock or move a roadway away from river bends, where significant erosive force can be anticipated in future flood disasters.

Working closely with FHWA to understand the limitations of eligibility as outlined in the *FHWA Emergency Relief Manual* (May 31, 2013), CDOT developed a benefit-cost analysis (BCA) (Figure 1) approach to evaluate and substantiate innovative resilience improvements. The BCA evaluated risks to each federally supported "on-system" asset (Risk) by analyzing the probability of the event occurring again (Threat), the damage or losses if an event were to occur (Vulnerability), and the resulting costs to recover from the event (Consequences).

By rigorously quantifying the potential value of resilient improvements, CDOT was able to substantiate the return on FHWA's investment of approximately $200 million in FHWA ER funds for resilient infrastructure solutions. The specific details of the risk equation are described further in section 8.3, "Quantification, Prioritization, and Decision-Making" (page 218).

After quantifying a BCA ratio, CDOT compiled a pool of eligible *candidate* assets for resilience improvements. These each had a positive (greater than 1.0) BCA ratio calculation. On average, three resilient design solutions were developed for each project. Design consultants worked in cooperation with CDOT at the schematic-design level (roughly 25% design), and also prepared rough order of magnitude engineer's estimates for costs. Having a positive BCA ratio did not guarantee that a resilient design solution for a candidate asset would be implemented. Fiscal constraints limited the feasibility of implementing every resilient solution that shows a net benefit to CDOT's infrastructure. In short, just because a particular resilient solution met the eligibility criteria for FHWA ER funding did not mean that there was funding available from FHWA's overall CDOT 2013 flood appropriation from Congress to support it.

The PR phase of the flood repair was filled with tough decisions about how and where to prioritize repairs to maximize the resiliency of Colorado's infrastructure while staying within the available budget. When candidate projects met the BCA calculation and schematic design alternatives provided promising resilience solutions, roundtable-style discussions began with CDOT's Region Transportation Director/Flood Recovery Manager, CDOT engineers, engineering consultants, and FHWA to provide the necessary professional engineering judgment to stay within budget.

A key example of a project with a significant, positive BCA ratio and interesting and complex resilient design challenges involved US 34 in Big Thompson Canyon. Severe flooding in this area resulted in the second largest aerial evacuation effort in the U.S. history (after Hurricane Katrina), with 2,256 aerial evacuations and 707 ground evacuations. A total of 23 miles of US 34 was damaged and 8.5 miles of roadway lost within some of the steepest canyon walls in the U.S. highway system. For two and a half months, canyon access was severely limited or totally cut off while temporary emergency repairs were made, at a time when winter conditions were fast approaching.

The priorities for resilient recovery plans after a future flood were clear: *evacuation*—provide emergency egress to get people out of the Canyon—and *ingress*— get emergency vehicles in and restore lifelines and essential services in the Canyon—in order to save lives. Prioritizing emergency access required relocating segments of the roadway onto bedrock to protect it and make room for the river to flow at high cfs rates, rather than rebuild the roadway on the pre-disaster roadway location which is

SH 60 and SH 257 Case Study

SH 60 and SH 257 is an example of a flood recovery project whose damage cost and risk comparison did not justify FHWA ER funding 100% of the improvements to expand the bridges capacity on both highways as the BCA was only 0.87 for that design alternative. However, CDOT decided to fund the difference (27%) of the project to bring the BCA to 1.0, as documented in a concurrence memo with FHWA as this was the best resilient solution for the project. This decision is a prime example of maximizing the power of the RnR model and by adding CDOT dollars to fully fund this project to sustain function during a significant flood event demonstrates sound engineering judgement which is a key component of the RnR model. CDOT also partnered with the Little Thompson Watershed Coalition to restore the riparian habitat. The conclusion of this project is a bridge which can withstand a significant increase in river flow and a watershed capable of responding to extreme weather events placing less pressure on the built infrastructure.

SH 60 and SH 257 area after 2013 flood damage and the same area after permanent repairs.

made up of alluvial soil, clay, and aggregate rock and likely to fail as it did in 1976 and 2013 floods. Counterintuitively, rather than protecting the entire roadway from future flooding, some sections were reinforced and designed to flood with water or "overtop" the roadway where water would recede quickly, and any damages could be quickly and easily repaired. It also meant protecting the two-lane portions of the roadway in narrower areas of the Canyon where protecting the entire four-lane roadway segment was cost-prohibitive. This is how CDOT navigated resiliency within the project's budget. "Case Study 2: US 34–Big Thompson Canyon Case Study" (see pp. 293–304, below) outlines how resilient design solutions were considered and describes additional innovations that were incorporated into Colorado's largest flood repair project in its history.

8.2.2 **I-70 Risk and Resiliency Pilot Study**

The 2013 flood permanent repair effort—bolstered by the RnR Pilot Study — pushed CDOT to become more forward-thinking and to start applying resiliency criteria on a statewide level. Instead of only reacting to disaster damage with resilient design BCA methods and approaches, CDOT began to consider how it could proactively attenuate risks to the state's infrastructure, including climate-related stresses. This undertaking met with obvious challenges. Defining the aspects of risk, including threats, vulnerabilities, anticipated direct impacts, and consequences, is a straightforward exercise after a major or catastrophic disaster. However, anticipating the types and impacts of future threats to Colorado's infrastructure (what hazard? what severity? what location? when?) is complex and difficult despite good modeling on probabilistic risks. Constructing improvements to protect all of Colorado's infrastructure from every potential threat is neither possible, nor a prudent use of taxpayer dollars. When trying to define its approach for resilient design, construction, operations, and maintenance of the State's most critical and threatened transportation assets, CDOT turned again to its partner in resilience innovation, FHWA.

In August 2016, CDOT and the Colorado Division of FHWA developed another Risk and Resiliency Pilot Study on the I-70 corridor (RnR: I-70). This approach proactively identified, and is in the process of addressing, vulnerabilities and probable disruptions to the system that could result from natural and human-caused threats, such as rockfall, flooding, and landslides, and from other stresses such as utility failures. This proactive approach was tested on the I-70 corridor, stretching 450 miles from the Utah to the Kansas border. The I-70 corridor is widely known for (importantly) conveying tourists—as well as Coloradoans—from the Denver International Airport and smaller regional airports to Eagle County's ski areas like Vail and Breckenridge, which made it the ideal corridor to study for many reasons, including:

1. The I-70 corridor is one of the most utilized to move people, goods, and services in Colorado. I-70 is a critical part for the state's economy and quality of life; four segments on I-70 were studied.

a. Site 1 of the pilot study, located at milepost (MP) 112.9, experiences an average daily traffic (ADT) of 24,000 vehicles with 12.8% truck traffic.
b. Site 2, located at MP 123.7, experiences an ADT of 26,000 vehicles, with 13.5% truck traffic.
c. Site 3, located at MP 244.5, ADT is approximate 40,000 vehicles with 6% truck traffic
d. Site 4/5 located at MP 356, ADT is 11,000 vehicles with 25% truck traffic.

2. The I-70 corridor is subject to a range of the state's typical natural threats, such as rockfalls and landslides, debris flows, and avalanches, as well as floods, fires, severe winter storms, and tornadoes. It is also subject to potential human-caused threats such as high vehicle bridge strikes, major accidents, utility failures, and terrorist attacks.
3. The I-70 corridor crosses a wide array of environments, traveling through both mountain and flat terrain.
4. The I-70 corridor runs through communities with widely varying population densities, including the Denver Metropolitan Area and numerous suburban and rural communities, and includes diverse socioeconomic mixes.

The RnR: I-70 Pilot Study began in August 2016 to evaluate multiple significant threats and the capacity of I-70 to continue to function in the event of a hazard. The results of the study inform day-to-day design, construction, operations and maintenance, and asset management decisions. The segments along I-70 were ranked in priority based on the workshops conducted within CDOT, which determined the areas at greatest risk for future damage. The threats included avalanches, fires, floods, landslides, rockslides, high winds, tornadoes, and bridge strikes from vehicles. The maximum credible impacts and consequences for each threat were considered using a case scenario for each event. The vulnerability of a road segment was calculated as the conditional probability that a consequence would be realized. Data that CDOT evaluated found that rockslide events posed greatest risk specifically in the Glenwood Canyon section of I-70. The results from the workshops were placed into a map of the CDOT system, along with potential mitigation strategies. A total of five sites on I-70 were identified; mitigation options were analyzed from a design and an economic ROI perspective. In addressing the risk of flood, CDOT focused on improving minor culverts along I-70 to improve the flow of water under the roadway. The resilient recovery priority for I-70 is clear: Minimize risks to transportation reliability and maximize financial return on investment (ROI) using resilience solutions on I-70. That means implementing resilience solutions that resist damage impacts and can be brought back into emergency service and permanent service quickly and safely when damage does occur.

It was critical to approach the corridor's resiliency in holistic terms, given Colorado's growth projections and economic forecasts and increasing threats from climate change. For example, CDOT has integrated transportation reliability strategies for the I-70 corridor to address peak ski season ADT and winter weather stresses, along with other resilience strategies. The state is integrating triple bottom-line benefits: environmental sustainability, improved quality of life for people and communities, as well as economic growth and stability into its resilience approach.

To date, CDOT has added Bustang interregional rapid daily bus services to meet the needs of both skiers and service workers who need to commute to communities like Vail and Brackenridge that largely lack affordable housing. Increased train and transit services along the corridor are planned, such as the round-trip service running from Denver Bus Center to Vail Transportation Center every day of the week. CDOT has worked with communities about increased traffic along I-70 to prepare economic development strategies that include community residents in job creation, such as hiring qualified drivers for the bus service. The RnR: I-70 study is helping CDOT reimagine all aspects of its infrastructure. In order to do so, it must also understand its risks statewide.

Figure 2. **Building Resilience into Transportation Infrastructure and Operations**

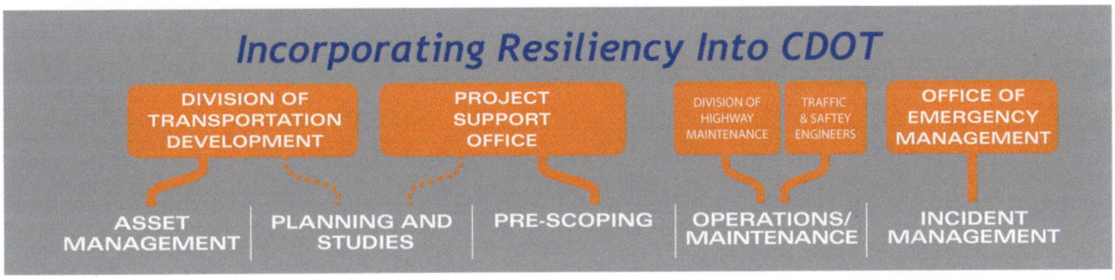

8.2.3 **Incorporating Resiliency Across CDOT's Enterprise**

CDOT understood early on that resiliency is neither limited to the practice of building the strongest and most conservative design, nor limited to capital projects. On November 15, 2018, CDOT adopted Policy Directive 1905.0, "Building Resilience into Transportation Infrastructure and Operations" (Figure 2). This Policy Directive formalized the focus on resilience activities at CDOT. Its purpose is to enable CDOT to "proactively manage risks, minimize disruptions and adapt to changing conditions in order to provide continuous transportation service in Colorado." Transportation infrastructure is critical to the state. Without secure and resilient infrastructure, key aspects of Colorado's economy, including the movement of goods, people, and information, would be jeopardized. Some activities CDOT supports as part of resilience mindset include:

1. Developing a Statewide Criticality Map (discussed below);
2. Incorporating resiliency into Planning and Environmental Linkages (PEL) studies (discussed below)l
3. Improving operational readiness for rapid response and resilient recovery, such as developing its debris management planning;
4. Utilizing procurement and contracting efficiencies to aid in disaster recovery;

5. Developing organizational structures for incident command systems that are appropriate for CDOT's mission and for the phase, scale, and severity of the disaster; and
6. Refining policy and compliance procedures to dramatically decrease audit risks.

The following describes CDOT's Statewide Criticality Map and CDOT's efforts at incorporating a holistic resilience approach into its PEL as representative examples of ways CDOT has increased the depth and breadth of its resiliency mindset.

Statewide Criticality Map

As an outrowth of the RnR: I-70 Pilot Study and in support of Policy Directive 1905.0, CDOT recognized the need to comprehensively map threats and vulnerabilities in relation to the criticality of the surface transportation asset, which is crucial for effective prioritization is. CDOT developed the Statewide Criticality Map (Figure 3), which assigned a high, medium, or low criticality level for each lane mile of CDOT roadway across Colorado. The criticality level is determined through a quantitative assessment of each corridor's social and economic vitality to the state and its communities. This information is instrumental for CDOT to understand potential cascading impacts and consequences associated with a natural or human-caused disaster in a specific location.

When combined with other available data, the Statewide Criticality Map better

Figure 3. **Statewide Criticality Map**

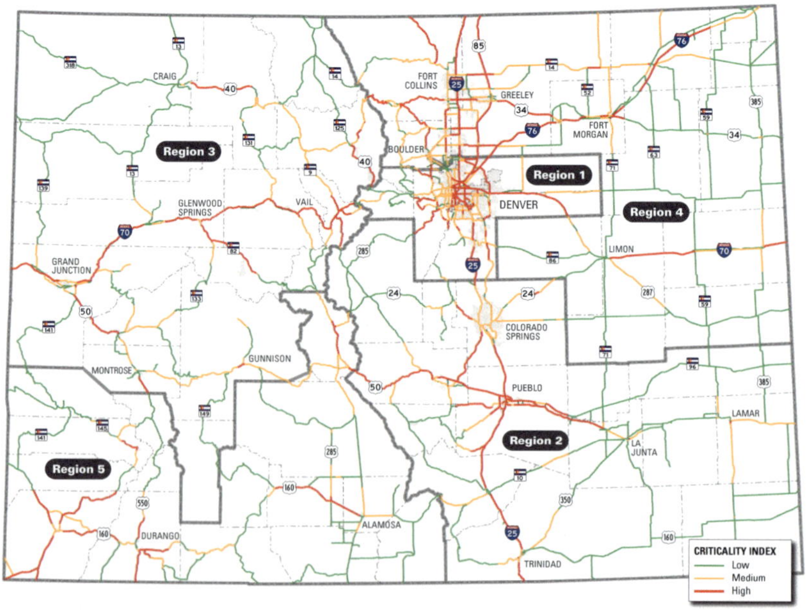

Figure 4. **Integrating Resiliency in PEL Study**

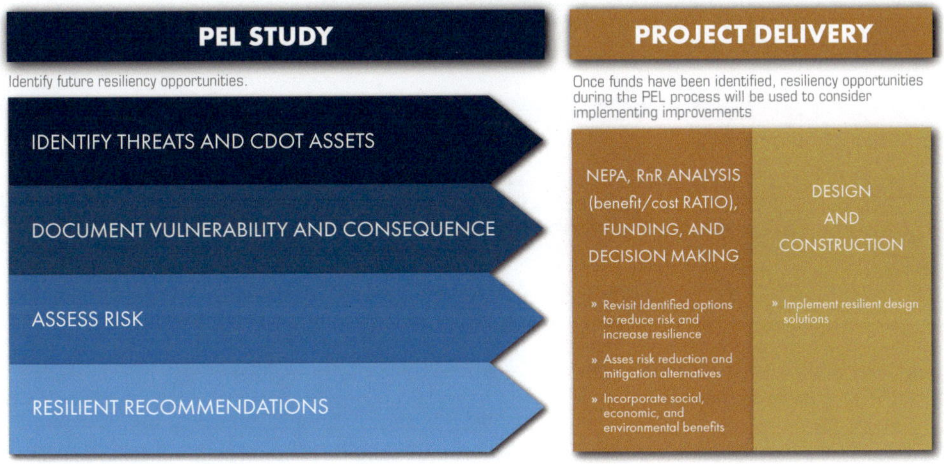

equips CDOT to make decisions and find cost-effective ways to implement resiliency where the system needs it most. CDOT collects targeted data of its assets, and as the speed and quality of information increases, data compilation tools like geographic information systems (GIS) will be extremely useful for targeted decision-making at a statewide level.

Including Resiliency in Planning

CDOT has been expanding its focus to incorporate resiliency in all phases of project delivery, from planning and design through construction. Figure 4 shows that by integrating resiliency into planning efforts, natural, human-caused, and operational threats to a corridor can be pro-actively identified, creating an opportunity for CDOT to reduce its risk by including scope within its standard planned projects to address those threats. CDOT uses the Planning and Environmental Linkages (PEL) approach on highway corridors for transportation decisions that consider environmental, community, and economic goals early in the project planning stage and carries them through project development, design, and construction. During the early stages of planning, nat

CDOT uses the PEL approach on highway corridors for transportation decision that considers environmental, community, and economic goals early in the project planning stage and carries them through project development, design, and construction. During the early stages of planning, natural and human-caused threats to transportation assets are identified along the corridor. This allows CDOT to include potential resilience measures in projects as they progress into National Environmental Policy Act (NEPA) reviews and design. To be consistent with previous projects, CDOT follows the RnR: I-70 data analysis methods, but also layers in the data and

Figure 5. **Identifying Future Resiliency Opportunities in the PEL Process**

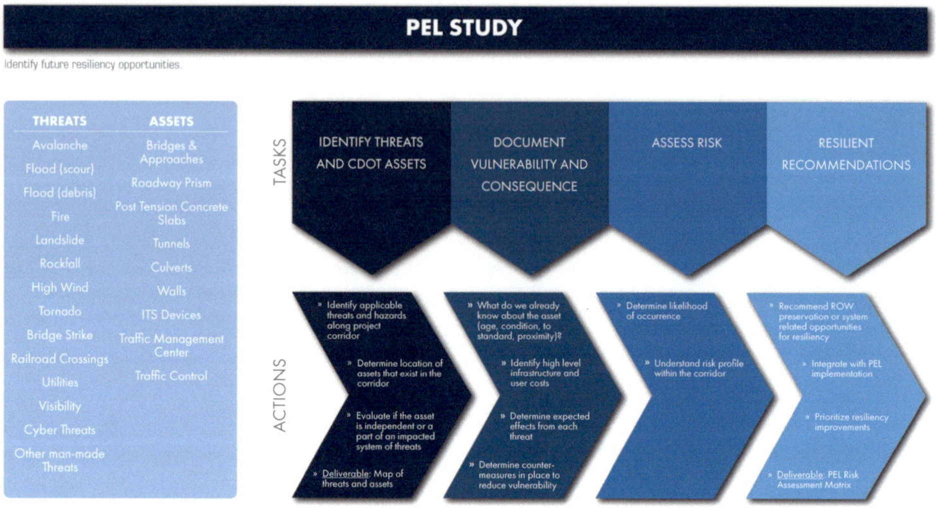

planning perspective of the PEL process (Figure 5). Planning for potential extreme weather or human-caused disasters can help stakeholders identify more resilient infrastructure designs to protect the integrity of the transportation system and maximize their ROI.

The PEL process take a macro look at a project and its broader potential impacts, including the resiliency measures identified during the study. The PEL recommendations use a rough order of magnitude level that will require further analysis to validate the estimated resiliency costs and benefits. The PEL for US 34 is discussed below and further illustrates the process.

US 34 Planning and Environmental Linkages Study

Evaluating Natural Threats

For the 2019 US 34 PEL, flooding and the subsequent consequences to railroads and utility infrastructure were identified as primary threats and impacts for this critical corridor. Floodplain limits and railroad and utility crossings were evaluated in relation to CDOT-owned assets along US 34. CDOT assets include major structures, the roadway, and traffic control items such as signals and intelligent transportation system (ITS) devices. Each area where a potential threat overlapped with an asset was labeled a "risk area." A risk assessment matrix was developed, pairing risk areas with the associated infrastructure and user costs that would be incurred if an event were to occur. Assuming a natural hazard causes the US 34 corridor to shut down, user costs were calculated based on the length of time it takes to travel the detour route using CDOT's Road User Cost Calculation tool. The US 34 PEL study team determined three ranges of user costs: low, moderate, and high.

Infrastructure costs, including the cost to replace a damaged asset, were established using CDOT's Project Cost Planner tool, with standard percentages and en-

Figure 6. **Identified Threats and Resilience Strategies**

Figure 7:

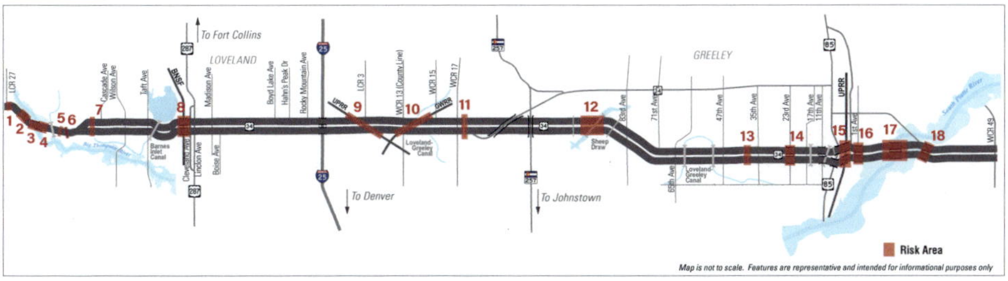

RISK AREA ID	THREAT	PRIORITY	RESILIENCE
1	Flood	High	Ensure structure C-16-DD is built to withstand a 100-year flood event
2	Flood	Moderate	Raise vertical profile of road out of the floodplain or replace with overtopping pavement
3	Flood	Moderate	Ensure structure is built to withstand a 100-year flood event, perform floodplain conveyance improvements to channelize flow and prevent flooding on the north side of the highway (see Risk Area ID #4)
4	Flood	High	Ensure channel and structure C-16-AE is built to convey and withstand a 100-year flood event (see Risk Area ID #3); or raise vertical profile of roadway out of floodplain or reinforce with overtopping pavement
5	Utility Failure	High	Work with utility owner to replace or reinforce existing waterline
6	Utility Failure	High	Work with utility owner to replace or reinforce existing waterline
7	Utility Failure	High	Work with utility owner to replace or reinforce existing waterline
8	Railroad Proximity	Low	N/A
9	GWRR Railroad Proximity	Low	N/A

RISK AREA ID	THREAT	PRIORITY	RESILIENCE
10	GWRR Railroad Proximity	Low	N/A
11	Utility Failure	Low	N/A
12	Flood	Low	N/A
13	Utility Failure	Low	N/A
14	Utility Failure	Low	N/A
15	Railroad Proximity	Moderate	Ensure railroad structure is equipped with acceptable pier protection per AREMA standards
16	Utility Failure	Low	N/A
17	Flood	Low	N/A
18	Flood	Low	N/A

gineering judgment. Three ranges were assigned for the infrastructure costs: low, moderate, and high.

These costs and the asset's vulnerabilities were used to predict consequences that natural or human-caused threats could have on each specific asset (e.g., structure, roadway segment). The team then evaluated the probabilistic risk of occurrence for each natural threat and created a prioritization list of the highest risk areas. The list also weighed the criticality of a roadway segment in relation to a community's social and economic vitality. Recommendations to increase the resiliency of the corridor and reduce risk exposure from each threat were identified for future projects (Figure 6).

Evaluating Threats to Roadway Operations

Because of recent changes largely associated with the state's population growth, CDOT adapted the process of evaluating threats to develop a Land Use Sensitivity Analysis that could predict stresses that would diminish highway operations (functionality) due to traffic increases and land use changes. A flood can physically damage or destroy a section of roadway, cutting off access and disabling the highway's capacity to connect. Unanticipated increased traffic demand can cause the same effect. Even if a highway does not suffer physical damage, congestion from rapid development is a stressor that can overwhelm a corridor and reduce functionality, rendering the corridor almost impassable—and thus unreliable—and can decrease roadway safety.

The purpose of the Land Use Sensitivity Analysis was to enhance the communi-

Figure 7. **Level of Service of Existing Conditions, 2040 No-Build, and 2040 Build**

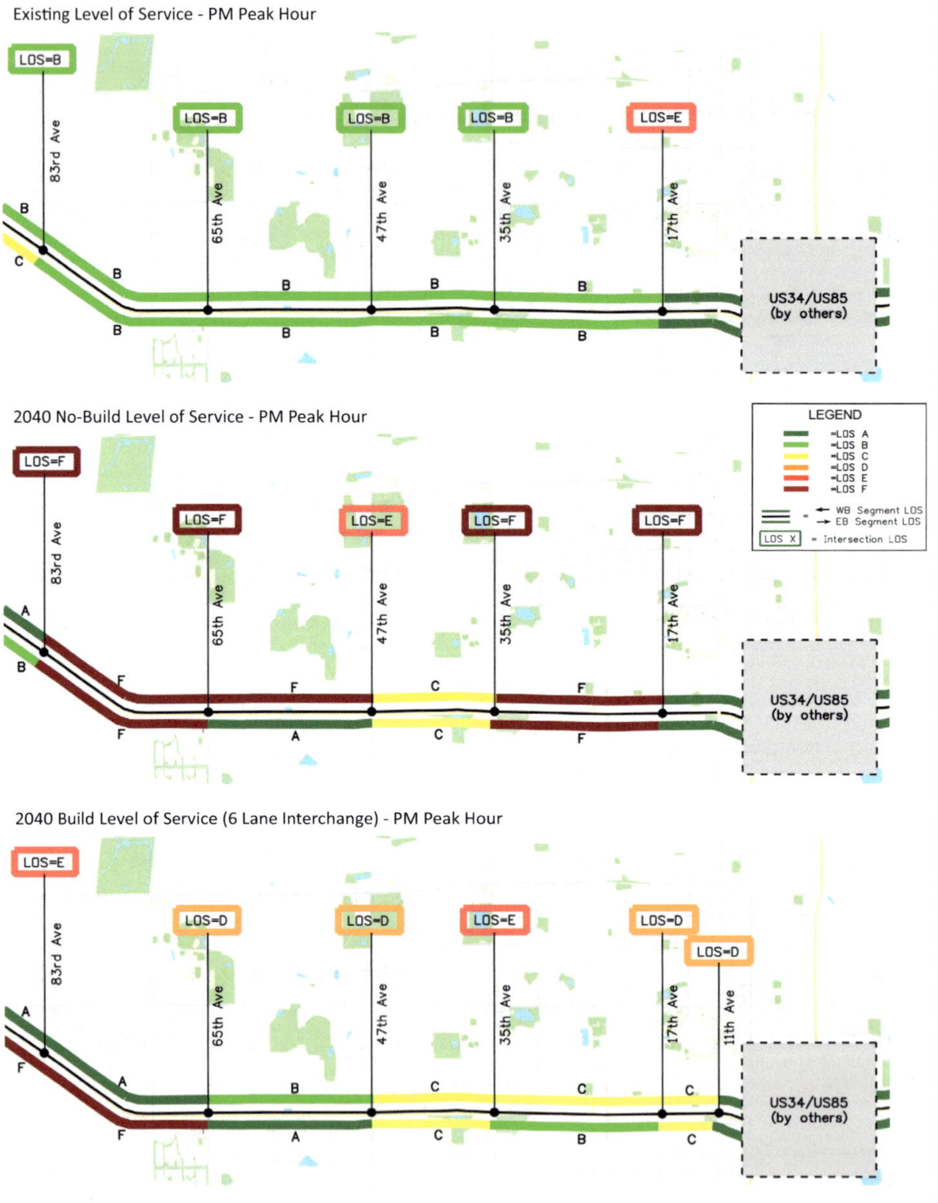

cation between CDOT and local agencies to better understand the operational risks that are present along a highway corridor. If traffic demand at an existing intersection is at or near its available capacity, there is a high risk that the highway will experience functional loss as demand increases in the absence of improvements to address capacity.

By identifying the risks to a highway corridor's functionality, CDOT and local agencies can partner together to address necessary improvements and proactively maintain corridor operations as land use conditions change and traffic demand increases.

Locations along the US 34 corridor that were potentially more sensitive to unanticipated traffic increases were identified by comparing the corridor's levels of service (LOS) under three scenarios using the 2040 Metropolitan Planning Organization (MPO)-modeled conditions (Figure 7):

1. Existing conditions;
2. Conditions without improvements; and
3. 2040-modeled conditions with the PEL-recommended alternatives considered.

The Intersection Capacity Utilization (ICU) rating was also calculated for the existing conditions of an intersection and for the 2040 MPO-modeled conditions without any improvements to identify intersections with the lowest capacity. A list identifying intersections at risk of functionality loss shows CDOT and local agencies where operational issues could arise as demand grows. The initial assessment of the land use sensitivity analysis for US 34 confirmed the PEL recommendations and the initial priority projects.

The intersections were mapped showing the existing and proposed 2040 land use data, including Traffic Analysis Zones (TAZs) and their corresponding population and employment densities, to illustrate the anticipated rate of growth forecast for the corridor at each intersection. Segments of the corridor were characterized as urban or rural, which determined the extents to be included within the map. For example, a development several miles offset from US 34 in a rural condition could have a more significant impact on its operational functionality if there are no alternate routes, whereas route redundancy is typical in an urban environment. The maps will serve as helpful tools as CDOT and local agencies contemplate the potential impacts of a proposed development on the operational functionality of the highway.

As corridor conditions remain dynamic, priorities may be adjusted to meet changing demands. The Land Use Sensitivity Analysis was purposely developed in conjunction with the PEL, but not attached to it, to allow for continued updates to be made as priorities, Metropolitan Planning Organiztion (MPO) growth forecasts, and corridor conditions change. Identifying population growth hubs along US 34 where there are projected, corresponding increases in development helps CDOT and local agencies prioritize roadway improvements to prepare for an increases in traffic volume and integrate multi-modal access. This strategy reduces the future strain on the corridor by increasing the usability of the road for multiple modes and reduces stress during peak traffic hours and safety risks.

Continued Statewide Resiliency Actions

CDOT was first introduced to resiliency as they responded to the challenge to "build back better, stronger, and more resilient" in the wake of the 2013 floods. In the years after, resiliency has expanded into a statewide approach, not only for disaster management, but as a decision-making tool that incorporates and brings resiliency and co-benefits to every aspect of the CDOT system. Over time, resiliency and its many applications will continue to develop and evolve so that Colorado remains vigilant in its efforts to remain adaptable in the face of changing conditions and new challenges, including climate stresses.

The *Colorado Resiliency Playbook* (*Playbook*) is a tool for government agencies, including infrastructure owners, to use to institutionalize resiliency principles, strategies, and practices in their operations. Both the *Framework* and the *Playbook* embody the commitment to resilience prioritized by two gubernatorial administrations since 2013, and build elasticity into the way Colorado does business by making resilience a day-to-day priority. The *Playbook* (Figure 8) is designed to actively integrate resiliency into existing government processes and critical path decision-making in policies, practices, programs, and budgets.

8.3 Quantification, Prioritization, and Decision-Making

The RnR Pilot Program is essential to determine which locations are vulnerable to changing conditions. It is not possible to apply its principles to all locations within the CDOT system. Therefore, CDOT needs a strategy to quantify the cost of incorporating resiliency to determine which areas are in greatest need. Determining which areas have the highest priority helps CDOT decide where funds should be allocated.

Sometimes the most resilient solution is planning for failure during an event or disaster. For example, in Colorado's flood recovery, a resilience solution for numerous roadways involved reinforcing the roadbed with a new riprap base so that, in the event of a similarly severe flood to 2013 that could result in roadway overtopping or washout, only the surface would need to be replaced, as opposed to the entire roadway.

Infrastructure context, including criticality and roadway redundancy, matters. For example, I-25 is a key connector used by many tens of thousands of people every day. Thus, the impacted I-25 corridor segments were built back more resilient after the 2013 floods because the corridor is the backbone for user connectivity and is integral to the state's economic well-being. In contrast, State Highway 7 has an ADT rate of fewer than 1,000 vehicles; moreover, a redundant route is available. As it was a lower criticality corridor with an alternate route available to roadway users, the most effective solution was to make a more limited financial investment to repair 2013 flood damages and plan for corridor failure and rapid repair in a future flood event.

8.3.1 Pursuing Quantifiable Solutions

Evaluating infrastructure within the framework of resiliency is challenging. The reason to implement more resilient systems is to mitigate risk, which is quantifiable. CDOT assesses the annual monetary risk due to the probability of natural threat

Figure 8. **Modules and Description from the Playbook**

impacts and their severity. To do so, CDOT etablishes risk by defining a process that identifies threats to the asset and develops a methodology to consistently characterize each asset and its context, such as the severity of disaster damage as well a criticality and redundancy.

Once those are identified, risk can be quantified as the product of consequences (C) in dollars times the vulnerability (V) of the asset as a probability time the specific threat (T) likelihood as a probability.

$$\text{Risk} = C \times V \times T$$

RnR equations for natural threats can determine the quantifiable and monetary risk of prioritizing or not incorporating resilience improvement into the design approach for a permanent repair project (see Figure 9, next page).

Figure 9. **Resilience from Natural Threats and Risk from Natural Threats**

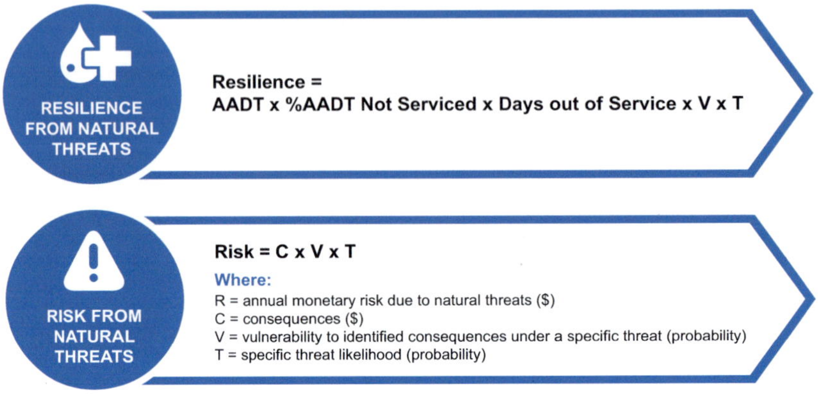

In CDOT's RnR benefit-cost-analysis method (BCA), consequences are defined in terms of as what would happen to the asset if there were a severe disaster event (disaster impact types and severity) combined with the duration the impacts would be experienced. Vulnerability is a measure of how sensitive or responsive the asset is when conditions change. Threat is the probability of the event itself occurring (e.g., a 100-year flood event return rate). Defining risk as a combination of consequence, threat, vulnerability, and threat allows for a qualitative analysis that readily translates into a quantitative BCA, which provides both consistency and rigor as part of the backbone of resilience decision-making.

For example, imagine a house with glass walls abutting a beginner's golf course. The house is worth $1 million. Of every seven golf balls hit by a novice player, the house will be struck by four golf balls. The probability that a golf ball will hit the house with enough force to shatter the glass walls is 1 in 70. Thus, the associated annual monetary risk is:

$$\text{Risk} = \$1{,}000{,}000 \times 4/7 \times 1/70 = \$8{,}163$$

8.3.2 **Prioritization and Decision-Making**

Deciding which locations along a corridor receive resilient infrastructure improvements is calculated based on benefits and costs to avoid and reduce transportation disruptions. Prioritizing decision-making to maximize the returns on investment on the most critical corridors and segments is necessary because it is not feasible to protect everything. As shown in Figure 10, strategic investments are made where they will reduce disruption and yield the greatest benefits to the traveling public. This includes considering resilience investments risk to the physical roadway along with the consequences of transportation disruption to people and communities, the environment, and the regional economy. Quantifying risk using a rigorous method of benefit-cost analysis (such as the equation above) can help state and local agencies

Figure 10. **Natural Threats and Human-caused Threats Along I-70 Corridor**

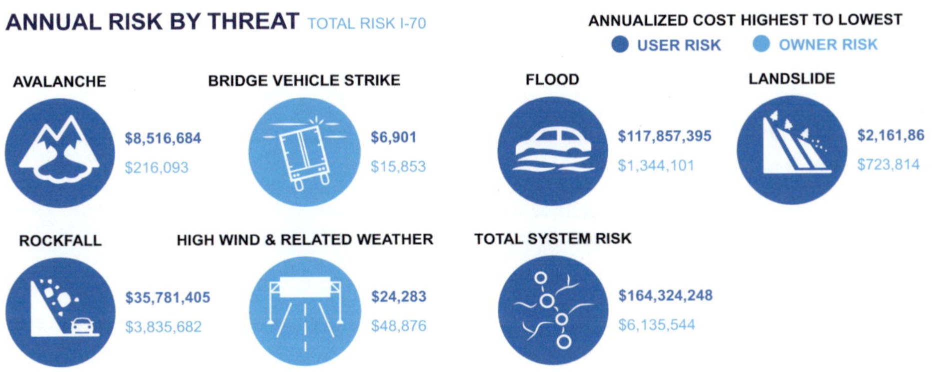

understand its risk exposure and clearly quantify the costs of resilience investments in line with its key priorities. It sheds light on the fact that the least expensive solution in near term is not always the best solution. As importantly, it helps a community put a dollar value on the cost of doing nothing.

The RnR: I-70 Pilot Study used risk quantification to assess potential natural and human-caused threats along a major east-west corridor.

With a monetized understanding of the threats posed to the I-70 corridor, CDOT is positioned to develop cost-effective, resilient solutions on critical corridor segments and structures. Resiliency measures may include new roadway and bridge designs, operational plans, alternate route identification, or improved operations and maintenance.

US 34 A Case Study

This site experienced road damage in the 2013 flood event due to an overflow channel resulting in the complete closure of the road for 17 days. The only alternative route was to be 106 miles in length thus the resilience index determined that this segment of US 34 had high criticality due to the freight value, AADT, and redundancy of the route. Two alternatives, constructing a box culvert or a bridge, were evaluated through the RnR analysis and constructing a bridge that will typically cross dry land provided the highest B/C. The picture of the right shows the bridge under construction from the 2013 event, when a 25-year event in 2015 was experienced. The bridge over a dry channel proved to be the best resilient solution since US 34 maintained usability for corridor users.

8.4 Conclusion

Just like a rubber band, Colorado's infrastructure and communities have been stretched, sometimes to their breaking points. The state's new approach to designing resilient infrastructure gives communities improved capabilities to mitigate risk exposure and bounce forward instead of bouncing back—resilient like a strong and elastic rubber band.

Resilience solutions can attenuate the risk exposure and improve the performance of an individual asset in the face of increasing threat frequency and severity throughout its useful life. Resilience and adaptation to change should influence how new construction is designed and capital expenditures are made. Resilience methodologies can be applied to an owner's entire infrastructure portfolio to improve whole-system performance and even drive down risk exposure within a region. Key factors include an asset's tolerance to risk, resources that are available, and an asset's criticality and degree of utilization. A holistic approach to resilience allows the State not only to reduce risks, but to leverage important co-benefits for environmental sustainability, to support quality of life for people and communities, and to promote economic stability and growth.

While there is no single best approach to resilient design, there is only one way to drive down risks to infrastructure assets, and that is to initiate decisive action and move forward.

- Craft a resilience vision with goals and objectives that are measurable and achievable over time. Be rigorous and consistent when applying selected methodologies.
- Gather key stakeholders, both internal and external.
- Use the time available now, before a shock or climate stressor damages the system's capacity to perform its essential functions, to adopt a definition of resilience and chart your organization's course.
- Be open to adopting new ways of doing business.
- Evaluate diverse approaches to resilient infrastructure design.
- Take stock of how you are doing. Test. Adjust.
- Move forward, again and again.

As you consider ways to apply information about resiliency presented here to your organization's planning, design, construction, operations, and maintenance of its infrastructure assets, the following resources may be useful (see reference section).

- FHWA: *PEL and Corridor Planning: State of the Practice Review of Planning and Environment Linkages Implementation in Corridor Planning*. This is an excellent introduction to PEL.
- Colorado Department of Transportation Action Summary: *A Resource for Pre-*

paring for and Responding to Future Disaster Events, 2013 Flood Event Lessons Learned and Best Practices. This resource walks you through the incident command organization, critical path tasks, and methods CDOT employed during successful 2013 flood response and recovery operations. It provides an operational assessment of what worked and could be implemented by another public infrastructure owner prior to a disaster to shorten downtime and optimize response and recovery. It candidly presents imperfect solutions that CDOT used in subsequent disaster recovery operations.

- National Institute for Standards and Technology (NIST): *Community Resilience Planning Guide for Buildings and Infrastructure Systems*. NIST says the Guide provides "a practical and flexible approach to help all communities improve their resilience by setting priorities and allocating resources to manage risks for their prevailing hazards. Volume I of the Guide describes the six-step planning process (Figure 11) and provides a worked example to illustrate the process. Volume II describes how to characterize the social and economic dimensions of the community, dependencies and cascading consequences, and building and infrastructure performance. Using the Guide can help communities to integrate consistent resilience goals into their comprehensive, economic development, zoning, mitigation, and other local planning activities that impact buildings, public utilities, and other infrastructure systems."

Figure 11. **Six-Step Planning Process for Community Resilience**

Like safety, resilience is never perfect and never complete, but the strength and elasticity designed and built into infrastructure and properly managed will save lives when disaster is at the doorstep.

References

Colorado Department of Transportation: Action Summary: A Resource for Preparing for and Responding to Future Disaster Events, 2013 Flood Event Lessons Learned & Best Practices (July 21, 2015).

Colorado Department of Transportation: Incorporating Resiliency into CDOT.

Colorado Department of Transportation: Policy Directive 1905.0, Building Resilience into Transportation Infrastructure and Operations (November 15, 2018). https://www.codot.gov/programs/planning/documents/plans-projects-reports/projects/resilience_program/policy-directive-pd-1905.0.

Colorado Department of Transportation: Project Cost Planner (Tool) https://drive.google.com/file/d/1wFBgW0shqZWSDZWfszQgLndMzPmUrSBW/view.

Colorado Department of Transportation: Road User Cost Calculation (Tool) https://usercost.codot.gov/.

Colorado Department of Transportation: Statewide Criticality Map.

Colorado Executive Order Declaring Disaster Emergency, 2013. https://www.fmcsa.dot.gov/emergency/colorado-governor-issues-executive-order-declaring-disaster-emergency.

Colorado Resiliency Framework https://sites.google.com/a/state.co.us/coloradounited/resiliency-framework.

Colorado Resiliency Playbook https://drive.google.com/file/d/1GXg16LFW2mxFP4B_xRQ1jxcgSe6xXp4e/view.

Federal Highway Administration (FHWA): Emergency Relief Manual (May 31, 2013) https://www.fhwa.dot.gov/reports/erm/er.pdf.

Federal Emergency Management Agency (FEMA) Presidential Disaster Declaration: Colorado Severe Storms, Flooding, Landslides, and Mudslides (DR-4145)https://www.fema.gov/disaster/4145.

Federal Highway Administration (FHWA): PEL and Corridor Planning State of the Practice: Review of Planning and Environment Linkages Implementation in Corridor Planning https://www.environment.fhwa.dot.gov/env_initiatives/pel/corridor_planning_report_July2014.pdf.

Governor's letter to Colorado Resiliency Framework. May 28, 2015. https://sites.google.com/a/state.co.us/coloradounited/resiliency-framework.

9

The Language of Urgency

Rachel Bannon-Godfrey

9.0 Introduction

The world is waking up to the urgency of addressing climate change. From municipalities to private sector corporations, ambitious goals are being set around carbon emissions from energy sources, supply chains, and operations. However, in many cases, buildings are still the forgotten part of the supply chain. Corporate commitments to large renewable energy procurement deals and R&D into carbon capture technologies are by far a more popular topic. Buildings, providing a basic human need of shelter, will remain the key to survival in the face of increasingly harsh weather conditions that are symptomatic of climate change. The construction of buildings, the materials used in them, and their operational energy are significant contributors to climate change. We must find resilient solutions in which quality construction methods and a careful use of materials allow us to adapt to our worsening climate without further exacerbating the contributing factors. Without mitigation, adaptation will never keep pace.

The evolution of "smart" technologies, from the scale of a phone to an entire city, has the potential to significantly improve everyone's lives. When technologies are applied with this purpose, rather than just adding to the noise of big data, smart buildings offer the promise of a more resilient future. However, energy efficiency and the resulting reduction in greenhouse gas emissions are often seen as side benefits of smart building technologies rather than the drivers. Our phones are still smarter than our buildings.

As a branding exercise the term "green building" implies stability and sensitivity at a time when we need to convey urgency, action, pragmatism. Instead of green and smart, we need to talk about resilient and adaptive. This chapter will focus on the next evolution of "green" buildings, the strategies that are pushing us towards a more resilient future, and the issues we must keep at the front of mind when designing, developing, and operating them.

9.1 The Human Factor

Climate change is a big, scary, complex problem that significantly impacts human life. But it is often hard for people to grasp the impacts of climate change on more than an abstract level. To successfully address this planetary problem, we need to approach it at the human scale with human-centered solutions. Human-centered design approaches have been successful for breaking down otherwise insurmountable problems, such as those faced in the developing world, into relatable components. Breaking significant design challenges down to the scale of their impact on the individual makes problems of great magnitude appear more easily solvable, and the solutions will more likely be grounded in empathy for the user. The building industry as a whole has an opportunity to accelerate the progress of climate change mitigation if we can stay focused on the impact our activities shaping the built environment have at the scale of the individual, and by extension the community.

The United Nations Sustainable Development Goals (UN SDGs)[1] break big problems down to a human scale. They translate the most pressing issues facing the planet into seventeen building blocks, each with quantifiable targets and indicators. By becoming familiar with the metrics and indicators of each of the SDGs, we can translate the global and abstract into the tangible and actionable.

The UN SDGs expand the definition of sustainability beyond energy efficiency and recycled content to include fundamental human rights such as quality education, nutrition, and living wages.

On a smaller scale but along the same lines, the creation of the International WELL Building Institute's WELL Building Standard,[2] a healthy-building rating system focused on the relationship between the built environment and human health and well-being, has produced a similar outcome within the real estate industry. It expanded the definition of a green building to include topics such as the impact of design on mental health, physical agility, food choices, and employment policies such as family leave. Fueled by research into the relationship between indoor environmental conditions and physical and psychological factors ranging from cognitive functioning to workforce engagement, health and well-being by design has become a driving factor in the business case of developers, employers, school districts, and homeowners. The Covid-19 pandemic led to increased interest from building own-

United Nations Sustainable Developmental Goals

ers and tenants in WELL's quantitative and qualitative metrics, adding legitimacy to fundamental strategies that designers have been applying for years to enhance the experience and functionality of spaces.

Up until the pandemic focused the world's attention on the indoor air that we breathe, the surfaces we touch, and the proximity of other people in buildings, many health-focused design strategies were perceived as simply adding cost to a project instead of fundamentally addressing human health needs—mental and physical. Daylight, natural materials and patterns, non-toxic alternatives to common products, a variety of spaces types—some that are bright and welcoming, and others that are calm and private—to suit different psychological needs, opportunities for movement and physical agility during the day—these are examples of how health and well-being are being incorporated into buildings. Designers advocate for these value-added strategies because they know they make people feel better, while learning, working, playing, or just living. A common thread to many of these strategies is the value of comfort—thermal, visual, acoustic, olfactory, ergonomic—and how in business it translates to the return on investment for the building owner, operator, or occupant.

When considering the business case for improving the experience of a building, we must also factor in the increasing impacts of climate change on buildings' ability to maintain safe, let alone comfortable, conditions in the face of increasingly harsh weather. When so many of our data points and financial propositions depend on the benefits of bringing the outdoor environment in, what happens when the outdoor environment is dangerous for our health? Think of the effects of climate change, such as wildfires of increasing frequency and magnitude, creating dangerous air quality conditions and decreased visibility for larger geographic regions, or summer temperatures raising the threshold of the heat index. How can we maintain the benefits to human well-being of natural ventilation, operable windows, daylit spaces, and walking or biking commutes under those conditions? Likewise, as winter storms increase in magnitude and severity, our parks, trails, and outdoor amenity spaces may become unusable for more prolonged periods. More than its effects on amenities and ideal recreation conditions, climate change is creating conditions our building envelopes and mechanical systems may not have been designed for, threatening not just comfort but human survival. We need to consider both the physiological impacts of climate change, such as asthma, heat stress, and food- and water-borne illnesses, and also the mental health impacts of climate disruption, such as depression, anxiety, aggression, and weakened social relationships,[3] and the potential financial toll, whether from medical bills or potential job insecurity, on the individual and their dependents. And this must all be approached through the lens of climate justice, the intersection of the impacts of climate change and human rights, so that no one community is unfairly burdened by the impacts of climate change on health and livelihood. The global Covid-19 pandemic put a spotlight on disparities in public health vulnerabilities and access to resources in many disadvantaged communities. In the case of Covid-19, notably that respiratory vulnerabilities due to poor air quality can be linked to proximity to pollution sources such as power plants, manufacturing plants, and other toxic facilities. In the context of the building industry, mitigation efforts must look at the impacts of design strategies on communities beyond the property boundary of any one site, to also include the impact on people in communities many miles away, such as those living next to the power plant supplying that building with energy or the manufacturing plant supplying the building with materials and furnishings. Consider working with the expertise of community leaders and environmental justice experts, such as the NAACP Environmental and Climate Justice Program, and leveraging industry data such as environmental product declarations and health product declarations from building product manufacturers, as a starting point for mitigating the impacts of climate change on those communities least responsible for the contributing factors.

9.2 From Comfort to Survival

The perspective of architectural and engineering professionals must be included in discussions of public health and emergency preparedness—building construction standards have an impact on human survival in both temperate and harsh climates. In one recent example, a study carried out by Stantec[4] calculated the time it takes for a typical single-family home to cool down to 0°C (32°F) during a power outage, at which point the conditions inside the home would become a health risk—if not before. The freeze time was calculated using computer modeling and simulation of eight homes in Canada's Yukon Territory, built according to four construction standards, and three outdoor conditions. The following parameters were also tracked for each scenario: peak heating demand (kW), outdoor dry-bulb temperature (°C), global solar radiation (W/m2) (sum of direct, diffuse, and reflection radiation), and time of day. Because these homes were not heated at night, data showed the response of the temperature in the homes in free-floating winter conditions. In the simulations, a power failure meant that all heating systems were off and there were no additional internal heat gains beyond those from the human occupants, which represents a worst-case scenario, as all other heat sources had been eliminated. Human presence was factored into the simulations and could, in reality, mitigate heat loss and help maintain higher temperatures. However, these sources cannot, in general, be relied upon. The results were then validated against measured temperature profiles from eight houses in the Yukon.

Each modeling method and set of assumptions had limitations and intrinsic errors, so the results were not expected to match the real-world scenario exactly, but to provide insights on the general applicability and indicators for conditions harmful to human health. The aim of this study was not to make a highly accurate calculation of

Table 1. **Freeze Time (Hrs) per House Construction Type and Outdoor Condition**

	Outdoor Condition[a] During Power Outage		
House Construction	**1. Sustained, extremely low outdoor air temperatures**	**2. Low temperature, rising outdoor air temperature**	**3. Low temperature, dropping outdoor air temperature**
1980s Typical construction for the region	19	29	120
R-2000[b] Constructed to the R-2000 standard	41	120+	120+
SuperGreen Constructed to an EGH[c] rating of 85+	198	198+	198+
Passive House[d] Constructed to Passive House levels	456	456+	456+

a. These scenarios were extrapolated from Canadian Weather for Energy Calculations data in order to use realistic weather patterns. For example, Scenario 1 used weather data from a typical cold spell rather than artificial parameters (i.e., a fixed outdoor temperature). This shows the house's response to realistic weather parameters, which includes outside air temperature and solar radiation.

b. R-2000 is a voluntary standard developed by Natural Resources Canada. Certified R-2000 homes are on average 50% more energy-efficient than typical new homes. (Natural Resources Canada, 2012)

c. EnerGuide for Houses rating, a 0-100 energy performance rating scale for houses developed by the Government of Canada. 0 represents a house that leaks a lot of air, has no insulation, and uses a lot of energy. 100 represents a net-zero energy home, which is airtight, well-insulated, and generates as much energy as it uses through renewable sources (such as solar panels or geothermal).

d. It has not been confirmed in the Passive House Planning Package that the 15 kWh/m^2 limit is really met, or possibly exceeded.

the length of time it takes to reach freezing temperatures, but to determine how rapidly a response to a power failure becomes necessary, from the perspective of both the utility company and medical first responders. An architectural and engineering study focused on health outcomes.

As shown in Table 1, freeze times vary greatly with the construction of the building envelope. In older homes the temperatures can drop to freezing within one day, while in highly insulted, airtight homes such as SuperGreen or Passive House, one can expect the temperature to stay within habitable ranges (although probably uncomfortable) for one or more weeks. The practical consequences of this are significant. They suggest that, particularly for vulnerable populations, a power outage in a standard or sub-standard construction could pose an immediate threat to life and well-being, whereas with Passive House–type shelter, there is time to wait for a winter storm to pass and services to be restored. With the increasing frequency, intensity, and uncertainty of winter storms due to climate change or climate disruption, focusing on the physical and social resiliency of building envelopes is one way in which the building industry can change from bystander to active participant in climate adaptation. When a community's infrastructure is hit with extreme weather conditions, they may not prepared for outcomes that can be catastrophic, as seen in Texas in February 2021 when a polar vortex led to widespread loss of power and running water, resulting in fatalities. Global warming is weakening the polar jet stream that keeps areas of low pressure and cold temperatures around the poles. As a result, cold temperatures are moving south into regions typically not prepared for such conditions, and likewise warmer temperatures are being pushed northwards. These increasingly extreme conditions are putting a strain on infrastructure designed and built for the climate patterns of decades ago, not the increasingly erratic and unstable patterns we are currently experiencing.

The relationship between building construction and human survival is critical in all climate conditions, not just cold climates. As shown by studies of the urban heat island effect, survivability is an urgent issue in high heat conditions, which climate change is exacerbating worldwide. In addition to adding to general physical and mental stress, high temperatures and heat waves are increasingly responsible for premature deaths. According to the Centers for Disease Control and Prevention[5] from 1979 to 2003 extreme heat exposure was responsible for more premature deaths in the United States than all other weather-related hazards combined—hurricanes, lightning, tornadoes, floods, and earthquakes.

A 2018 report by the UK Parliament's Environmental Audit Committee[6] found that climate change may cause the number of heat-related deaths every year in the UK to triple by 2050.

Mitigating heat-related deaths will require a coordinated response at the community level, especially in the face of compounded disasters such as extreme heat during a pandemic, complicating the logistics of providing community cooling centers. Considering the public health impacts of increased temperatures is not just an issue for periodic heat waves. A report from McKinsey[7] estimated that India may be the first place in the world to experience temperature and humidity conditions that exceed the limit of human survivability while resting in the shade. Rising temperatures and humidity levels pose an everyday threat to workers whose jobs require long periods outside, such as construction site workers, and can threaten lives, livelihood,s and entire economies.

Buildings—specifically building materials—play a significant role in the urban heat island effect in which dark, absorbent surfaces such as asphalt and concrete cause air temperatures in urban locations to increase significantly during the day and not cool off as much at night, compared to more rural or vegetated areas. As in the cold climates, the building envelope performance (wall insulation, air-tightness, proportion of glass to solid surface, orientation, shading, and thermally insulating properties of the glass) helps mitigate the impact of climate change on human health. A well-insulated, well-shaded building can act as a buffer from the temperature extremes outside and maintain a level of comfort and survival for occupants for a minimum amount of time until relief or medical assistance can arrive. In addition to architectural details, building commissioning–design stage quality control reviews and preoccupancy testing of building components, systems and controls can have a significant impact on whether a building in operation achieves the long-term performance intended by the design team.

New performance-based standards address the role of the building envelope in achieving overall energy and carbon goals. The Canada Green Building Council's Zero Carbon Building Standard[8] sets targets for the Thermal Energy Demand Intensity (TEDI) of buildings in different climate zones. TEDI is calculated as the annual heat loss from a building's envelope and ventilation, after accounting for all passive heat gains and losses. In addition to emphasizing the thermal autonomy (ability of a space to remain thermally comfortable without mechanical systems) and comfort of the building, the TEDI score indicates where to reduce a building's energy loads before adding renewable energy sources to reach the zero carbon equation. The first building to achieve the CaGBC's Zero Carbon Building design certification was evolv1, a developer-driven speculative office building (see next page). The TEDI score of 24 kWh/m^2 was determined through parametric modeling of thousands of design options. evolv1 includes R-30 walls, an R-40 roof, an R-15 perimeter slab, and a 40% ratio of window area to wall area. The result represented the financial "sweet spot" between investing in the building envelope and its mechanical systems.

The Passive House standard raises the bar for the airtightness of building envelopes, not only for control of heat loss and gain but also for moisture control, another significant health issue in which designers have influence. The University of Northern British Columbia's Wood Innovation Research Laboratory (WIRL) building in Prince George, Canada (see page 233), was not only one of the first Passive House–certified industrial buildings in the world—and in one of the coldest climates—it was also the most airtight Passive House building in North America at time of its completion in July 2018.

The overall energy use of buildings is undeniably a critical factor in their contribution to climate change and adaptive ability. Designers often focus on the role of a building's HVAC (heating, ventilating, and air-conditioning) systems and overall energy use. But most building owners don't want a specific boiler or chiller, they want heat in winter and cooling in summer, for the least cost and zero maintenance headaches. It can be challenging to make the case for a significant investment in more efficient systems. Redirecting our view of buildings—and entire building portfolios—towards a carbon-neutral future (and the importance of details like HVAC

evolv1

Waterloo, Ontario, Canada | 110,000 ft² |
Completed September 2018
Team: Cora Group, Stantec, http://www.coragroup.com/project/evolv1/

evolv1 is a class AAA commercial office building located in the David Johnston Research + Technology Park, within Waterloo's Idea Quarter tech hub. The first building in Canada to be awarded the Canada Green Building Council Zero Carbon Building–Design certification, the building is verified Net Zero Carbon, Net Positive Energy, generating 105% of its annual energy usage and is LEED Platinum certified. From the onset, evolv1 was envisioned as an economically viable prototype for a high-performance office building in a competitive real estate market. Passive and active building systems were iteratively modeled to optimize the value-on-investment, and include a geoexchange system serving a variable refrigerant flow (VRF) HVAC system, triple glazing, transpired solar wall for preheated ventilation, indoor green wall. On-site solar photovoltaics, mounted on the building roof and adjacent solar carport, generate 105% of the annual energy use of the building and contribute to achieving net zero carbon status. In collaboration with the University of Waterloo, the impact of the energy, health, and well-being -focused design strategies on the building users will be studied to inform future iterations of this groundbreaking project.

systems) we must make the connection between the design of a building—materials, systems, orientation, insulation, and shading—and human health.

The UN SDGs revealed the range and magnitude of factors that must be in place to achieve a truly sustainable life. The SDGs translated the ideal of global sustainability into seventeen identifiable goals. In parallel, the health and well-being movement expanded the idea of a "green building" into something more tangible to the person standing in a space. The health and well-being movement is powerful because it feels personal. There is significant attention and a strong economic engine behind the health and well-being movement, and that offers the design industry an opportunity. It is in a crucial position to place design strategies, operational protocols, and organizational policies that address well-being into the context of our changing climate and then to show a path towards, in effect, "futureproofing" them.

Architects, landscape architects, engineers, interior designers—all have an impact on well-being through design and must collectively as an industry have a stronger voice in public health issues, specifically those most impacted by climate change. Like the well-being movement, we need to personalize the impact of climate change on us and in doing so spotlight the critical role our buildings can play in mitigating and adapting to this global crisis.

University of Northern British Columbia Wood Innovation Research Laboratory

Prince George, British Columbia, Canada | 13,000 ft² | Completed July 2018
Team: University of Northern British Columbia, Stantec, Aspect Structural Engineers, IDL

UNBC's Wood Innovation Research Laboratory (WIRL) is the first certified Passive House industrial and higher education building in North America—a particularly impressive feat due to the cold climate in Prince George (average summer temperatures 60°F/16°C, average winter temperatures 21°F/–6°C). Certification was achieved through use of the Passive House Planning Package (PHPP) energy model, blower door testing, and mechanical systems commissioning. The facility features a simple wood structure—conceived as a modest building inspired by passive design principles and wood innovation. The south side of the building is a source of heat, views, and natural light for the office and seminar spaces while external devices shade the facility's windows to reduce unwanted solar gain. The south façade is also where the air intake for the building is located, strategically positioned to draw air from a future adjacent park to provide higher air quality than the road and laneway to the north and east. Because the lab is designed to meet Passive House standards, it uses very little energy for heating while providing healthy amounts of outdoor air for ventilation and is estimated to achieve 90% savings in heating energy costs than an equivalent building constructed per current building code. The Passive House standard sets a maximum heating demand based on the gross floor area of the building, regardless of height. Due to the high-bay lab space, WIRL ratio of interior heated volume to floor area is three times higher than a conventional commercial building. Therefore, the heating demand of WIRL was limited to one-third that of an equivalent commercial building built to the Passive House standard, putting increased pressure on the building envelope to control heat loss.

WIRL achieved an initial blower door test of 0.07 at 50 Pascal pressure, making it one of the most air-tight buildings in North America. Students now have a beautiful, sustainable, and modern facility to conduct research on new uses for wood, creating new jobs and markets in the forestry industry—helping British Columbia secure a sustainable future.

9.3 From Smart to Analytical

Are smart buildings the solution to raising awareness of the impact of climate change on our lives?

While their definition is still nebulous, changing with each new app and sensor that enters the market, two clear trends are emerging. Smart buildings are evolving from an initial focus on energy efficiency to greater automation and personalization of comfort, convenience, and productivity. Building management systems, the software behind the sensors and controls found in buildings, have moved beyond simply presenting data to analyzing the data, identifying trends, and providing recommendations for improvement. Thanks to technologies such as indoor space sensing we know more than ever before how our buildings are being used: which spaces are in demand and which are underutilized, where points of friction arise (e.g., double-booked conference rooms), where lights are on, or where blinds are down. The rise of net-zero leases and sophisticated green leases, which allocate energy use per unit area (e.g., W/sf), can not only tell us energy use by tenant type (e.g., law firms vs. engineering firms), but also give tenants greater clarity for budgeting their energy costs. In addition to shifting building management tasks from schedule-based to condition-based (or data-based), building analytics and forecasting also can open up sources of data to be used to make decisions. How can we bring science back into comfort?

As smart building technology increasingly emphasizes the individual's experience of the building and develops more sophisticated uses of data analytics to determine that experience, is the next logical step for climate change data and forecasting to become an integral part of that experience? Due in large part to the WELL Building Standard's focus on air quality, dashboards and sensors now alert occupants to both indoor and outdoor air quality. How far can the building industry push the level of awareness of the relationship between air quality and health, and in so doing increase understanding of the connection between climate change and health?

We have a long way to go. Many building automation systems don't talk to each other; portfolio managers have to assess data from multiple separate systems and different protocols, depending on when they were installed. Thankfully, the rise of net-zero energy buildings is helping drive awareness of the need for ongoing measurement and verification of building performance.

Though once financially unattainable by most building owners, today net-zero energy (NZE) buildings, districts, and communities are financially realistic for many project types, sizes, and contexts. Increasingly stringent building codes, the decreasing cost of photovoltaics, innovative green financing mechanisms, new frameworks to guide the design process (from benchmarking to completion, such as the Zero Code, and 2030 Palette[9]), and technologies that make NZE buildings smart buildings have all made this possible. In fact, we are now building not just NZE, but net-zero carbon buildings, and the net-zero goal is turning into net-positive. Whatever the details of the mathematical equation (zero or positive, energy or carbon), the most significant lesson learned from Stantec's decade-long portfolio of net-zero and net-positive buildings is that the effort doesn't stop at the ribbon cutting. In fact, the most important phase in a net-zero energy building starts when people move in. And this is where the smart building steps in.

9.4 Post-Occupancy: An Ongoing Relationship

The project team and client relationship doesn't end when the keys are handed over. For NZE buildings the hand-off to the operations team is critical to achieving the required level of energy efficiency to achieve the annual net-zero energy target. With net-zero-ready buildings, this is just as important, because the procurement, sizing, and installation of on-site photovoltaics may still be a while out from initial occupancy, and excess energy use may increase the size (and cost) of the system that is eventually procured.

Rather than a one-time event, consider the project hand-off as an ongoing process. All buildings go through a teething phase in the first year of occupancy, as HVAC systems face each season for the first time and occupants learn how to use the various control switches and appliances. For NZE buildings, this phase is considered "verification" and is critical to determining whether the project will indeed reach net-zero on an annual basis.

The rise of analytical smart building technology means this verification phase can continue beyond the warranty period for the life of the building. In the face of an increasingly erratic and unpredictable climate, the verification phase is critical; temperature conditions, storm frequencies and intensities may already be different than those that occurred during the design phases, and the building systems must be calibrated and tuned properly to achieve their full adaptive/reactive capacity.

9.5 The Language of Commitment

With all this compelling evidence, why are building owners still reluctant to invest in efficient buildings? Two reasons may be a language barrier between industry and clients, and because the science is sometimes obscured by emotion. Clients are often more sensitive to the perceived disruption of the status quo than the very real impacts of climate change that are already being felt today around the world, and will only increase in severity and magnitude the longer we wait to take action. To get over this barrier, it may now be time to reframe the language we use to describe green buildings in general, specifically their energy efficiency. When designers and engineers explain building energy efficiency in the industry jargon of "EUI" (energy use intensity) and "percent better than code," we lose half the audience. If we dive into the nuanced metrics of site energy, source energy, and carbon emissions factors, we are likely to lose a good portion of the other half. Much like the way UN SDGs translate global problems into tangible goals, we need to give building owners and operators tangible, relatable, desirable terminology to get behind. "Efficiency" evokes strategies for cost-cutting, downsizing. Describing energy consumption on a "per-unit-floor-area-per-year" basis sends a message that sustainability is a space-related, fixed-cost issue. It suggests overhead and a revenue burden, instead of an opportunity for revenue growth.

"Energy productivity," on the other hand, evokes opportunities for value creation.

It quantifies a company's economic output for every unit of energy consumed. Instead of conventional metrics of energy consumed per unit area of a maintenance building, data center, or office space, energy productivity is a way to look at, for instance, the number of buses serviced, units of data traffic, or hours billed in a consulting firm per unit of energy consumed. In this approach, productivity, energy performance, and economic performance are captured in one metric.

Energy productivity quantifies the amount of service provided, or useful work produced, per unit of energy. It's an indication of the financial value created by every unit of energy. Instead of the traditional method of selling the benefits of an HVAC system upgrade on the basis of "percent savings in energy consumed per floor area per year," energy productivity can be used to describe the increase in billed consulting hours per unit of energy in an office, higher number of buses cleaned and repaired per unit of energy consumed in a vehicle maintenance facility, or greater number of passengers served per unit of energy consumed in a transit terminal.

This is not to downplay the important role of energy efficiency and building performance in both mitigating and adapting to the climate crisis. To achieve the high-level goals of energy productivity, organizations will likely first need to invest in a full suite of energy-efficiency strategies. Aligning energy use goals with the C-Suite's business metrics—such as product output, worker productivity, or customers served—can give energy efficiency a louder voice and a more defined role in a company's business model. Using the language of a company's specific revenue and services to quantify the benefits of energy efficiency not only makes a more tangible and compelling argument for the investment, but also helps demonstrate that energy efficiency can, and often does, result in a net economic benefit. This is not just an exercise in semantics: reframing the metrics of energy efficiency as the metrics of business investments and revenue growth helps identify opportunities for value creation.

Recent innovations in systems thinking and circular economy approaches have shown that energy efficiency and economic growth are not in opposition but are rather goals that can be achieved simultaneously. Companies can reduce their energy footprint and continue to grow their revenue at the same time. They will do well to move beyond simply looking at energy consumption and utility bills to also consider the impact of typical energy-efficiency strategies on worker health and well-being. A workplace environment that features quality daylight, operable windows, and broader temperature ranges has been shown to boost staff health and well-being, benefiting a company's personnel-related costs. On the logistics side, many strategies aimed at increasing energy efficiency also decrease a company's operating and capital costs, again improving its bottom line. When we factor the holistic—direct and indirect—benefits of efficiency into the equation, energy productivity proves that a company can simultaneously consume less energy and increase revenue. Rather than being mutually exclusive, energy efficiency and company profitability are interdependent.

Energy productivity is not a new concept. The EP100 campaign,[10] a partnership between the °Climate Group and the Alliance to Save Energy, has grown to over 100 member companies across the world, each of which has committed to one or more of three pathways for using energy more productively within defined timeframes:

- Doubling economic output from every unit of energy consumed within 25 years (from a 2005 baseline);
- Cutting energy waste by implementing a smart energy management system within 10 years; and
- Owning, occupying, or developing buildings that operate at net zero carbon by 2030.

According to the Climate Group, if 100 companies doubled their energy productivity by 2030, the potential avoided emissions would be equivalent to taking 37 million cars off the road for a year.

9.6 The 4th Dimension of Carbon—Time

Energy efficiency and a transition to clean energy sources are critical elements in our path towards being carbon neutral, but they are not enough. In addition to the carbon emissions associated with the energy use of our buildings— the operational carbon—we must also look at the emissions associated with the extraction, transportation, and manufacture of the materials used to construct our buildings—the embodied carbon. The total carbon footprint of a building has started to take shape long before opening day, long before flipping the switch on the lights and mechanical system. Per the Global Alliance for Buildings and Construction 2018 report,[11] the three most common building materials—steel, glass, and cement—account for 11% of the total annual global carbon emissions by sector, in addition to the 28% of global emissions from building operations.

Operational carbon is determined by how a building is used; embodied carbon is determined by how the building is built. Designers have a large influence on both, but the spotlight is now turning to our role in embodied carbon. Operational carbon refers to the emissions associated with the energy consumed by a building's operations—heating, cooling, ventilation, lighting, equipment, and appliances. Embodied carbon refers to the emissions associated with the energy consumed in the manufacturing, transport, and installation of the building materials and components. As buildings become more energy-efficient and include on-site renewable energy, decreasing their operational carbon emissions, embodied carbon emissions in buildings become increasingly significant. Embodied carbon is measured by the Global Warming Potential (GWP) of the materials measured in kilograms of CO_2 equivalent (kg CO_2e), calculated from a full cradle-to-grave life-cycle assessment (LCA) of the materials used in the project. LCA software is available to help identify the GWP of different materials and products.

The top three ways to reduce embodied carbon:

1. Reuse existing buildings, structures, and materials before constructing new;
2. Carry out a life-cycle assessment to prioritize which materials and design strategies will make the biggest impact on the overall embodied carbon footprint. A typical ratio of embodied carbon is 50% structure, 30% enclosure, 20% finishes; and
3. Design for structural efficiency to minimize the total volume of materials needed for the structure and substructure.

Operational and embodied carbon are two different measurements from two different time periods in the life of a building, but both must be considered and calculated if we are to meet our climate goals. In the words of Ed Mazria of Architecture 2030,[12] we must recognize the time value of carbon in the building industry.

Consider the timeline of most climate goals—2020, 2030, 2050. Now consider the carbon emissions associated with new construction from this moment forward. As you read this, materials are being extracted, processed, and delivered to building sites, all in preparation for our new building stock that may not be occupied for a few years. Between 2020 and 2030, the operational carbon of new construction accounts for just 28% of the total carbon emissions; embodied carbon the other 72%. Between 2020 and 2050, as the newly constructed buildings are now occupied, lighting, heating and cooling the people within, that ratio will shift to 48% operational carbon, 52% embodied.[13] Only focusing on the carbon footprint of the building operations fails to recognize (and potentially mitigate) half the picture. By adding the fourth dimension of time to the carbon discussion, increasing awareness and accountability of embodied carbon today becomes increasingly clear, and urgent.

9.7 Value Creation

The time is right for green building language that directly addresses the concept of value for building owners, operators, and the public in general. CEOs and CSOs of companies large and small are moving beyond sustainability as a risk-management strategy (*help me protect my assets*) towards setting sustainability goals for their potential value creation (*help me identify opportunities and find value*) and most recently for financial and political market alignment (*help me align with the Paris Agreement, with climate-related financial disclosures*). At the same time, the scope and definition of "value" is changing rapidly across different sectors and geographies. As the building industry continues to raise the bar for excellence in building performance, energy efficiency, and management, there is a growing need to create stronger alliances within the general business community. Alliances are stronger when we all speak the same language, especially one that addresses everyone's metric for success, whether financial or environmental.

In this chapter we have discussed the role of building design in personal and public health, survival even, how buildings fit into the goals for sustainable development at a global scale, the role of smart buildings and net-zero buildings, and the language used to align the goals of the design industry and clients.

Translating the global into the personal and identifying alignment with financial criteria is a pathway towards elevating buildings from green and smart, to resilient and adaptive.

Notes

1 UN Sustainable Development Goals site is available at https://www.un.org/sustainabledevelopment/

2 Available at https://www.wellcertified.com.

3 Clayton, S., C. M. Manning, K. Krygsman, and M. Speiser, (2017). Mental Health and Our Changing Climate: Impacts, Implications, and Guidance. Washington, D.C.: American Psychological Association, and ecoAmerica. Available online at https://www.apa.org/news/press/releases/2017/03/mental-health-climate.pdf.

4 Andrea Frisque, Jeanie Chan, and Innes Hood, "Freezing Time For Yukon Houses," Stantec, 2018.

5 Centers for Disease Control and Prevention, "Climate Change and Extreme Heat: What You Can Do to Prepare," 2016, available online at https://www.cdc.gov/climateandhealth/pubs/extreme-heat-guidebook.pdf.

6 Jocelyn Timperley, "In-depth: How the UK Plans to Adapt to Climate Change," Carbon Brief, July 26, 2018. Available online at https://www.carbonbrief.org/in-depth-how-the-uk-plans-to-adapt-to-climate-change.

7 "Will India get too hot to work?" https://www.mckinsey.com/business-functions/sustainability/our-insights/will-india-get-too-hot-to-work.

8 Canada Green Building Council, available at https://www.cagbc.org/zerocarbon.

9 For more information about Zero Code, see https://zero-code.org/ ; for more information about 2030 Palette, see http://2030palette.org/.

10 Available online at https://www.theclimategroup.org/project/ep100.

11 Available online at https://www.worldgbc.org/sites/default/files/2018%20Global-ABC%20Global%20Status%20Report.pdf.

12 https://architecture2030.org/.

13 "The 2030 Challenge for Embodied Carbon," available online at https://architecture2030.org/2030_challenges/embodied/.

10

Climate Adaptive Stormwater Management: A Living Systems Approach

Amy Seek, April Schneider, and Amanda Ludlow

10.0 Introduction

The most significant design challenge presented by climate change isn't higher seas or stronger waves. It is uncertainty. We face uncertainty about how dramatically climate change will alter weather patterns, how quickly the Antarctic Ice Sheet will melt, how soon the seas will rise, and whether we may ever meaningfully reduce our carbon emissions. Uncertainty causes us to fixate on the damage brought about by the most recent disaster, leaving us vulnerable to risks we can imagine less vividly. Uncertainty pits safety against quality of life, causing us to enlarge pumps, to elevate edges, to build expensive structures that separate us from the water.

Uncertainty can bring about inequitable distribution of protections, and, perhaps most consequential of all, uncertainty can lead to ambivalence and inaction.

Uncertainty is our biggest design challenge, and yet often the systems built to address climate change are fundamentally ill-equipped to deal with uncertainty. Many of these systems remove or alter natural hydrologic or ecologic operations: they get rid of the stormwater, block the storm surge, resist the wave. They mitigate risk for one scenario but magnify it for another. These systems offer no forgiveness beyond a certain contingency; a breach in the system means a community cannot operate normally until a lengthy restoration has been completed.

The shortcomings of such systems have been made evident by New Orleans' efforts to rebuild after Hurricane Katrina; the $14-billion levee repair completed in 2018 is sinking and may not provide protection beyond 2023.[1] An approach that relies solely on resisting natural forces, counterintuitively, *creates* vulnerability; such systems offer a determinate line of defense designed for hazards that cannot be predicted with precision. They offer no opportunity to learn from and adapt through incremental impacts,[2] leaving communities unaware of and unable to organize or innovate to reduce their vulnerability. Further, such systems can generate unforeseen externalities. The armoring of the water's edge in New Orleans and upstream, intended to provide flood control, has contributed to the loss of protective Gulf coastline at a rate of 162 acres per day,[3] leaving the region increasingly vulnerable to storm impacts.

These failures are not due simply to a lack of imagination. Agency standards and permitting processes are designed to approve practices that have been approved before, and new ideas face harsh scrutiny, project delays, and likely denial. New environmental conditions demand the exploring of new responses, and yet, where innovation is possible, the burden of liability rests on the designer—a risk too great for many practitioners. This combination of factors tends to disincentivize innovation and ensure that standard practices continue beyond their obsolescence. And still, it is important that we resist inertia where we can.

Protecting ourselves is necessary, but it alone is not sustainable. It takes a measure of optimism to consider what it means to live *well* amidst an unprecedented degree of uncertainty about an unprecedented level of risk. Can we build coastal safeguards substantial enough to protect us against storm surge without building so high that we make our waterfronts unpleasant places to be? Can we make our neighborhoods more beautiful and habitable as well as resilient? Can we collaborate across site boundaries to act at multiple scales, creating layers of protection and redundancy? Can we work as communities from the ground up to ensure solutions that are effective for all?

These questions, which we have asked ourselves in our practice, have led us to consider the role of "softer" systems as part of a response to new imperatives. While many essays and books have been written now on this subject, the chapter that follows is an appeal specifically to engineers. It details several projects ranging in status from planning to construction to completion that point us to a new, more nuanced approach to climate adaptation, one that features living systems—wetlands, streambanks, coastal marshes—as key components. These systems offer promise as adaptation tools, and they inherently enhance quality of life for our own and for other species. These projects are helping us to develop grounds for better practices in design for climate change, and they suggest an evolved concept of resilience that, in turn, points to new ways of developing our cities to thrive in an uncertain future.

10.1 Living Systems

A project we completed several years ago provides a simple illustration of this shift in thinking. Where the conventional response to persistent flooding of a road would have been to elevate it—a process that would have been repeated as extreme precipitation events increased—a project team that included engineers and ecologists recognized the degraded state of a nearby stream and instead looked to creek restoration as a solution. The restoration created a stable geometry and a flood plain connection compatible with adjacent road and rail infrastructure. The restored system could accommodate higher volume flows without resulting in devastating flooding, erosion of the road, or degradation of the stream itself. Such a project illustrates the value of drawing from the resources of a diverse team. It brings attention to the capacity of natural systems to accept flux. And it speaks to the necessity of flexibility as a feature of adaptation in the face of uncertainty.

Resilience has commonly been defined as an ability to "bounce back" and return to the pre-event state after a disruption, with the measure of a community's resilience being the speed with which it can achieve that return.[4] But as the impacts of climate change become more evident, we are beginning to understand resilience differently. We believe that bouncing back to the status quo is not actually desirable; instead, we should be using disruption as an opportunity to improve and gain feedback on existing systems. This calls for interventions that bend rather than break, allowing for impact without total disruption. It calls for systems that maintain functionality within a range of environmental conditions—in the absence of flood water as well as in their exceedance. Resilience cannot be a static state or an end goal; it must be an ability to evolve and grow in the face of shifting conditions.

Climate adaptations—that is, physical changes implemented to withstand the impacts of climate change—are not necessarily resilient; a wall as an adaptation to rising sea level offers no point of return if breached. What we see in living systems is the potential for a more resilient approach to climate adaptation, because living systems provide functions such as flood storage and wave attenuation while changing over time alongside changing environmental conditions.[5] Living systems such as coastal wetlands have a capacity to self-heal and adapt; they grow in complexity, as when root masses develop and provide stability for the soil, and they evolve with changing conditions: as water levels rise, vegetation seeds itself and moves upland. To optimize the function of such a system, we must provide it with space and autonomy to adapt, and then carefully assess when it requires management.

Flexibility to respond to short-term fluctuations in conditions and adaptability to evolve over time are key aspects of resilient adaptation. Redundancy is also key. A principle that is already common for most engineers, redundancy improves reliability, mitigates risk, and allows for operational continuity. Redundancy can be realized as a set of diverse living systems layered at a range of scales: urban green infrastructure installations, floodplain reconnections, and living shorelines might all contribute to a city's resilience. A layered system would not likely be accomplished in a single contract but would require designers, clients, or agencies to coordinate their work with planned projects with the aim of building such a system.

Living systems solutions are defined not only by the nature of their components but by the scale of analysis necessary to design them. Living systems call for systems thinking—that is, analysis of the interacting systems beyond site boundaries to understand the potential synergistic impacts that an intervention in one system has on the others.[6] Systems thinking means that we consider downstream, watershed-wide, and ecosystem impacts in an effort to provide benefits across multiple systems. Understanding a system's larger context allows us to recognize and capitalize on these relationships.

Living systems are most often accompanied by aesthetic, recreational, or habitat co-benefits. They may provide park space or natural beauty that can be appreciated even during periods when protective functions are not deployed. Although the term co-benefit might make these aspects of a project seem secondary to safety or flood control functions, they contribute to social resilience, to community health, and to quality of life. Design that accommodates a range of uses is necessary in the context of climate uncertainty. It has purpose, function, relevance, and benefit within a wide range of scenarios, ensuring return on investment regardless of future outcomes.

Realizing diverse goals in resiliency projects requires that opportunities be identified early and prioritized by designers, team members, and clients. A living systems design approach calls for multidisciplinary teams, sometimes led by landscape architects and ecologists—professionals trained to think broadly and to design for recreation, habitat creation, and aesthetic value. These professionals can provide a wide scope of analysis, identify multiple design goals, and then assemble a diverse team of experts to achieve the greater project goals. Living systems signify not a softening of approach, but a broadening of perspective.

10.2 **Inland Systems**

10.2.1 **Green Infrastructure**

Increased flooding due to development has led many cities to implement dispersed living systems, in the form of green infrastructure (GI), as a safety buffer for overtaxed grey systems that are cost-prohibitive to upgrade. GI is among the most palatable living systems in cities because these systems suit densely developed urban conditions: they take up little space, are relatively inexpensive to implement piecemeal, and enliven city streets with seasonal interest and shade. The City of

Best Practice Adaptation Approach

- Responsiveness: Allows for incremental changes and learns from them
- Flexibility: Stays functional within a wide range of environmental conditions
- Adaptability: Utilizes natural ecosystem functions optimized for adaptation purposes; values adaptability even in non-living elements
- Redundancy: Creates multiple lines of defense
- Systems analysis: Relies on analysis at multiple scales, including regional scales
- Multiple-benefits: provides cobenefits that give system relevance within range of environmental realities
- Multidisciplinary: requires a range of experts, a breadth of understanding

Philadelphia used this range of benefits to demonstrate the cost-effectiveness of implementing green infrastructure to mitigate its combined sewer overflows (CSOs). The city performed a comprehensive analysis of potential alternatives, including complete sewer separation, large-scale storage, treatment plant expansion, and satellite treatment. The analysis considered affordability, scalability, social benefits, and environmental benefits such as ecosystem restoration and potential energy savings.[7] As a result, in 2011 Philadelphia created the Green City, Clean Waters program, a plan to systematically introduce GI throughout the city.

Defined in practice as a microscale stormwater management tool,[8] GI plays an important role in a living systems framework. It can reduce flooding, improve water quality, protect streams, mitigate CSOs, and recharge groundwater. Philadelphia determined that by capturing the first 1–2 inches of runoff from each precipitation event, green infrastructure significantly reduced outfall volumes from CSOs across the city.[9] By the fall of 2017, the system was diverting more than 3 billion gallons of stormwater from combined sewers each year.

The use of GI for flood mitigation has several advantages for municipalities. The small scale of an individual GI project makes it relatively easy for a city to pilot the latest GI technology while making gradual improvements toward the reduction of CSOs. Because GI does not require intensive master planning, the city can be opportunistic about siting GI projects and implement them in coordination with other street improvements or in locations that are particularly prone to flooding. GI can be more equitable than other solutions, as improvements can be distributed across the city incrementally according to a range of criteria, including social metrics. A large grey infrastructure solution such as sewer separation, by contrast, improves only one drainage shed at a time and requires a greater investment of time and money upfront.

Despite its advantages, the application of GI in many cities has been lackluster. Stormwater management has traditionally been situated within the civil engineer's

Green infrastructure as it's often seen: poorly maintained and absent of original planting

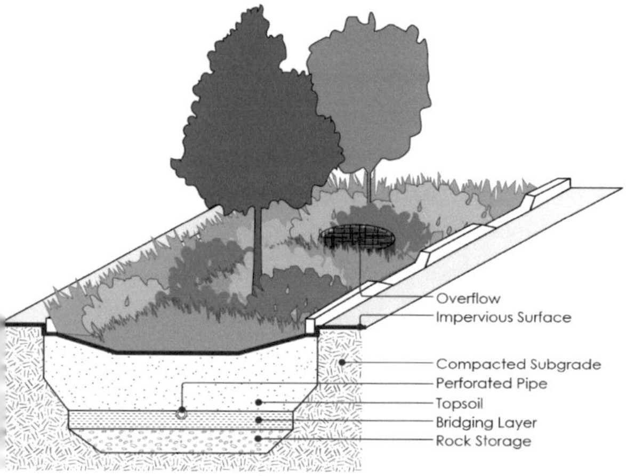

A great urban bioswale often includes a separate filter to absorb pollutants, as street runoff tends to compromise plant health. It incorporates a broad range of species in designed plant communities selected for a site-specific environment and microclimate. Vegetation is densely seeded or planted to minimize weeds and to allow for succession of well-adapted species. Soils above rock storage are graded, providing a bridging layer of medium-size rocks to encourage water to permeate to the storage layer. A well-functioning GI installation uses the absorptive capacity of plants as well as soil to remove water from sewers.

domain, and so engineers are often the primary designers of GI, leading to designs that have failure points outside an engineer's area of expertise. What has been missing from many green infrastructure programs is the recognition of GI as a living system, one in which plants contribute fundamentally to overall system benefits. Also missing from such projects are the professionals who understand the growing conditions necessary to promote plant health and system longevity. Without this knowledge, designers utilize traditional, mulch-based plantings that are low in biodiversity, require significant maintenance, and lack the robustness urban plantings need. They often specify highly permeable soils that are ideal for drainage but provide insufficient time for plants to uptake water. Ideally, GI design teams would include landscape architects, urban planners, ecologists, and horticulturists, resulting in designs that incorporate specific knowledge of appropriate plant species and densities within the harsh conditions of an urban environment.

10.2.2 New York City GI

Beyond stormwater management, green infrastructure can be used to address the increasing danger of extreme heat. Stagnant air, dark pavement, and waste heat from buildings and cars create a microclimate in midtown Manhattan that can be up to 22°F[10] hotter than the surrounding suburbs. Although often overshadowed by flooding, extreme heat is a dangerous weather event that exacerbates other health issues and disproportionately affects a city's most vulnerable residents. Heat is the number one cause of death from extreme weather nationwide,[11] and climate change is intensifying its impact. Record heat was experienced in July 2019 across Europe, where an event that might have occurred once every

1,000 years is expected every 50 to 150 years in the current climate.[12] By 2080, temperatures in New York will sit above 90°F for 24 to 75 days of the year.[13] The intensity of heat waves is also expected to increase, creating a new summertime normal, with mean air temperatures rising to 99°F in New York (from a current baseline of 94.2° F during a heat wave).

A public realm improvement plan for a district in midtown Manhattan is aiming to capitalize on the cooling potential of GI, siting generous GI beds throughout the area. The evapotranspiration process carried out by the vegetation in such beds can reduce ambient temperatures by as much as nine degrees Fahrenheit,[14] making the trees and bioswales deployed specifically for heat mitigation no less critical than the pipes beneath the street. Typically, siting GI in such densely developed spaces is challenging, as a variety of uses compete for limited public right of way both above and below grade. But these GI beds—proposed and funded by a private developer—will utilize site-specific designs that include soil cells, utility realignments, and optimized bed sizes, features that are not accommodated in the City's standard design book. Without such flexibility, among more than 3,000 potential locations for GI projects in other parts of NYC subject to city standards, fewer than 10% have been deemed feasible upon further investigation. Locations may be initially eliminated, based on above-grade conditions—fire hydrants, streetlights, bike racks, public information kiosks, and signage—or community concerns. Many residents do not want swales in front of their homes out of concern that they won't be well-maintained. But most GI sites are deemed unfeasible because of what lies below: bedrock, high groundwater tables, and the utilities that keep our cities functioning day to day. Gas mains, water mains, sanitary sewers, and more all have minimum offsets set by utility companies. Some of these offsets must be maintained, but many are legacy minimums that have become obsolete due to new and emerging technologies. Soil cells, for example, are demonstrably effective at accommodating both utilities and uncompacted soil volumes to support healthy trees. Asking utilities to revisit standards is arduous, but there is desire among many of these agencies to find means to adapt policy to better fit evolving conditions, and as designers we hold influence simply by asking the right questions.

Finding such avenues for siting GI within developed urban contexts is important, as GI provides an effective first layer of flood mitigation, with measurable local benefits that multiply downstream. GI intercepts stormwater, removing it wholly from, or delaying its entry into, municipal systems and streams. It decreases water temperature, velocity, and sediment load where stormwater does enter into streams, reducing erosion that occurs when streams are continuously hit by high volume, high temperature stormwater. Erosion in turn makes streams susceptible to catastrophic flooding; it is often managed by armoring and channelizing streams. Yet such risks can be reduced by doing just the opposite—restoring rivers' natural morphology. In the best-case scenario, GI and stream restoration work together to create a layered living system to reduce flood risk over the long term.

10.2.3 **Stream Management**

In nature, a river's floodplain accommodates the seasonal flux of large volumes of water, but in most rivers throughout the United States this capacity has been compromised. Within urban areas, rivers are confined to structured channels and disconnected from their floodplains. Impermeable surfaces and piped systems convey stormwater to receiving streams at high speeds, scouring channels and destroying vegetation. Even small amounts of impermeable surface within a watershed can have significant impacts on a receiving stream and increase the risk of flooding. With over half the world's population living within 3 km of a river, many communities are made vulnerable by the instability of our rivers in combination with anticipated extreme precipitation.[15]

Restoring a river's natural capacity for flux means, in part, restoring its floodplain. Within urban areas, restoration is often not feasible due to constraints posed by surrounding development, but in these cases, upstream parks and open spaces can become effective relief valves for rivers prone to flooding. We identified such an opportunity while creating a new masterplan for Eisenhower Park, a 330-acre park that sits a few miles upstream of Milford, Connecticut, on the Wepawaug River. The town has a history of severe flooding—floods from a heavy rainstorm in 1982 killed six people and forced 1,300 out of their homes.[16] During that storm, the river overtopped its banks, crashing into Milford's Town Hall and other downtown structures in the process. Flooding has remained an issue in the community as extreme precipitation events have become more common.

Although not a comprehensive flood mitigation solution, creation of a wetland within the park would provide a measure of flood storage to reduce flood impacts downtown. The area designated as wetland was graded and planted, and the manmade dike that had confined the river (a remnant of past gravel mining on site) was removed, allowing water to flow into a 6-acre area and creating a permanent floodplain connection. Regulations are often a hurdle for new approaches, but they can also be leveraged in support of living systems. Here, the local utility funded the project as a wetland mitigation requirement. However, because it was permitted as a mitigation project, it was monitored for five years as part of that requirement but was not assessed for downstream flood reductions. To expand the use of living systems, we must find avenues for monitoring and optimizing these systems for flood mitigation purposes. What is evident from the monitoring that has been conducted is that the wetland is healthy

Eisenhower Park in Milford, CT

and thriving, taking on a periodic flow of water and supporting significant wetland habitat—vital co-benefits to a living system approach.

10.2.4 **Reducing Stream Disruption**

Green infrastructure works alongside stable stream geometry and floodplain connection to reduce flood risk. Stream continuity is another factor in this equation. Dams disrupt ecosystems, interrupt seasonal inundation of floodplains, and cause extreme imbalances in the erosion and deposition processes of a stream, which leads to flood risk. Dams accumulate sediment within their reservoirs, reducing storage capacity and removing sediment from the natural stream deposition processes. They also erode downstream channels by releasing fast-moving water. Whereas it is often assumed that dams provide flood control, only 17% of dams in the United States were built for this purpose, and many dams now pose flood risk due to diminished storage capacity from sedimentation.[17] In many places, more frequent and intense storm events brought about by climate change are introducing higher volumes of water than accounted for in the design of the dams. Flood risk is amplified because significant development has occurred within these floodplains, in part due to the security provided by dams. In most cases, the area that would be inundated in the event of a dam failure is larger than the 100-year storm zone.[18]

The average age of a dam in the United States is 56 years,[19] and, as time passes, uncertainty about potential dam failure increases. With age comes outdated design criteria, material fatigue, and deferred maintenance that compounds with each year that foundations, pumps, valves, and cracks are not repaired. More than 4,000 dams in the United States are rated unsafe because of such deficiencies, but bringing the entire system of 90,000 dams up to current standards would cost approximately $79 billion.[20] Increasing risks and the price tag associated with mitigating dam failure led the U.S. Fish and Wildlife Service to prioritize the removal of five small dams in Connecticut as part of a resilience plan following Hurricane Sandy, utilizing funds from the Disaster Relief Appropriations Act.[21]

One of these, the Pond Lily Dam, was wedged within highly developed retail and residential neighborhoods in New Haven that were subject to frequent flooding. Built in 1794 to power a grist mill, the dam's 6-foot-tall, 200-foot-long structure posed a high likelihood of failure.[22] The Connecticut Fund for the Environment financed its removal in 2015. Investigations into the natural character of the channel prior to the dam's construction suggested that the historic channel would constitute a stable condition for the river, one less prone to flooding, once remov-

Pond Lily Dam Removal in New Haven, CT

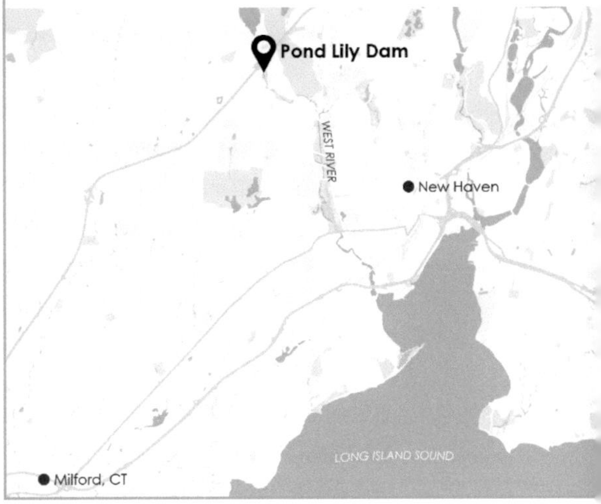

al was complete. The dam's removal created a floodplain that accommodates as much water volume in the event of a flood as the dam reservoir would, without the risk of catastrophic failure. What was once the footprint of the reservoir was transformed into a streamside nature preserve, creating habitat, migratory fish passage, and open space for recreation. The entire restoration process was accompanied by events and activities, such as educational tours and volunteer plantings, for a very engaged community, an example of the social resilience potential of living systems projects.

10.2.5 **Water Quality Impacts**

Often it is the complexity of quantifying the impacts of a living system—inherently less predictable and measurable than a conventional system—that makes it difficult for a municipality to invest funds for a living systems approach. In the Chesapeake Bay watershed, it took the demands of federal regulation for protection of the bay to overcome this complexity, which ultimately led to the implementation of stream restoration projects throughout the watershed. The Chesapeake Bay is a federally designated "impaired" ecosystem. High levels of nitrogen, phosphorous, and sediment entering the bay through runoff have resulted in algae blooms and destroyed habitat essential to the economy and culture of the region.[23] In 2010, having failed to meet federal goals for pollution reduction, the watershed was placed under strict limits to be met by 2028. Storm outfalls flowing into the bay are subject to regulated maximums (total maximum daily loads, or TMDLs) with stringent waste load allocations (WLAs) for phosphorous, nitrogen, and sediment. Localities planned projects to reduce nutrient loads, but they struggled to find enough opportunities for projects to achieve the required reductions. Many projects also proved financially unfeasible, especially considering the maintenance burden they would entail.

An expert panel convened by the Chesapeake Bay Program led a search for innovative solutions and found that stabilization of stream banks could directly prevent sediment and associated nutrients from entering downstream waterbodies, resulting in highly efficient reductions in pollution.[24] Per pound of nutrient reduction, stream restorations were shown to be the most cost-effective nutrient reduction method.[25] While conventional best management practices such as detention ponds or hydrodynamic structures cost upwards of $20,000–$30,000 per pound of phosphorus reduced, stream restorations cost less than $800 per pound.[26, 27]

With this knowledge, we took a unique approach to a flood mitigation project for our client, the Virginia Department of Transportation (VDOT), one of the many stakeholders assigned pollution reduction targets through the Chesapeake Bay TMDL, to address both localized inland flooding and earn valuable and much-needed credits for downstream pollutant reductions. At Lithia Road, in a rural area within the Federal Emergency Management Agency (FEMA) 100-year floodplain near the town of Buchanan, Back Creek was flooding as many as 30 times per year. Heavy rains would cause the creek to overtop its banks and spread across the adjacent road, causing a maintenance burden for the DOT and threatening the nearby rail corridor. Alternatives such as elevating the road out of the flood plain or armoring the stream channel were considered by VDOT, but they proved cost prohibitive. A more promising solution was the restoration of the stream itself.

In its condition at the time of the project, the stream flowed into the area of concern at a fast speed and spread into a braided channel that migrated around the flood plain, causing excessive erosion and pushing additional sediment downstream at the point where it reconnected with a more concentrated channel. The stream's condition was highly mobile, and if nothing was done it would continue to migrate and threaten the road and rail corridor.

As VDOT was among the "dischargers" being monitored by the EPA, a living systems approach would not only stabilize the stream and resolve the flooding issue but also provide credits towards its TMDL requirements. The quantity of sediment and nutrients credited to a project for purposes of the Chesapeake Bay TMDL is determined using the Chesapeake Bay Program's Expert Panel guidance document, which describes the practice, recommends removal rates, and outlines qualifying criteria.[28] Using these criteria, we estimated a 94-pound annual reduction in phosphorous, a quantity that is significant given the very low concentrations of phosphorous in native soil.[29] In fact, the restoration produced twice the annual reductions anticipated, significantly reduced VDOT maintenance, and produced a 95% reduction in annual flooding events. While monitoring of living systems is often not funded, we believe that implementation of living systems will increase as their benefits are understood and quantified. Here, the effect of such quantification is demonstrated—it has led to the implementation of hundreds of stream restorations throughout Virginia.

A Stream Restoration in Buchanan, VA

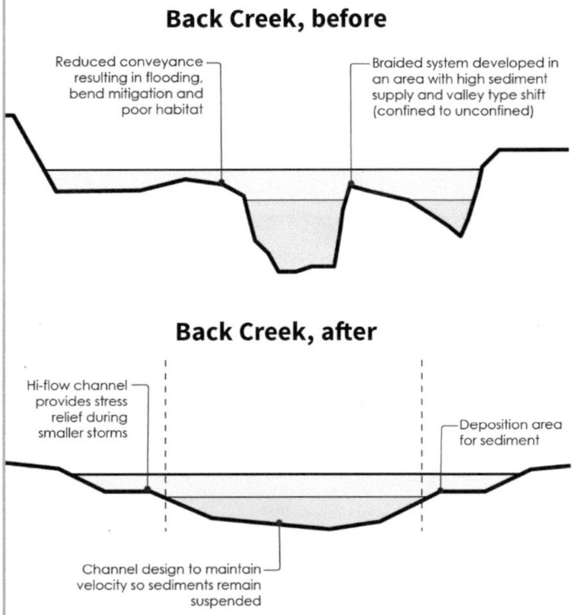

A braided stream's restoration to a stable channel with floodplain connection.

10.3 **Coastal Systems**

10.3.1 **Northeast Coastal Systems**

Chesapeake Bay's TMDL program illustrates the direct relationship between inland stormwater management and the health of coastal habitat. Dunes and coastal wetlands, the living systems that naturally occupy the coast, depend on replenishment of healthy sediment from inland waters. They provide a vital first line of defense against storm surge and sea level rise for coastal communities, where 39% of the United States' population resides. A team of researchers from UC Santa Cruz, Risk Management Solutions, and the Nature Conservancy estimated that coastal wetlands prevented $625 million in property damages during Hurricane Sandy—despite the degraded condition and reduced size of many wetlands along the coast.[30] It found that on average, such systems could reduce property damages by 20%.

10.3.2 **Restoring a Critical Line of Defense**

The $38-million restoration of Prime Hook National Wildlife Refuge, a 10,000-acre sanctuary along the Delaware Bay, speaks to the value of healthy coastal habitat. The Refuge had fallen into a degraded condition after a period in the 1980s when tidal marsh was converted into freshwater marsh to attract certain birds and wildlife, and the altered hydrology proved difficult to maintain. The dune system that kept inland waters fresh was continually breached by storms, and saltwater intrusion compromised inland freshwater habitat, killing off vegetation and leaving the landscape vulnerable to impacts. When Hurricane Sandy hit in 2012, it put much of the refuge under water, resulting in the loss of 4,000 acres of wetland, eroded beaches, and ecologically "dead" zones. Restoration of the Refuge meant restoring the natural hydrology to create a system better equipped to deal with sea level rise and future impacts. The United States Fish and Wildlife Service, with funding from the Hurricane Sandy Disaster Relief Act, set out to do just that.

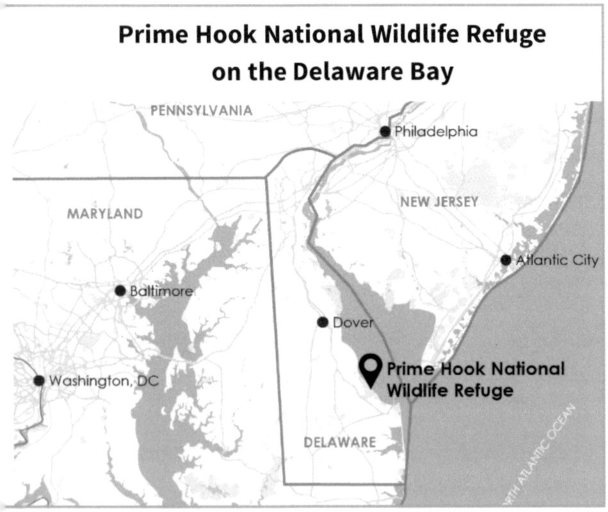

As part of the restoration design team, we evaluated the physical impact of Hurricane Sandy and estimated the sand volume needed to build a protective beach barrier system. We conducted topographic and hydrographic surveys and developed hydrodynamic and salinity models of the Refuge and adjacent bay region, including morphology, salinity, and constituent transport within the hydrodynamic modeling framework. Understanding the natural hydrology of the Refuge made it possible to determine the ideal marsh configuration for restoration efforts and, in conjunction with the coastal engineering analysis, to recommend an approach.

The dredging of over 100,000 linear feet of channels helped to recreate historic circulation and ensure healthy salinity levels and water movement through the marshland. Dredged material deposited along the sides of these channels supports the development of mud flats. The 2-mile barrier dune protects interior marshland and controls inflows of saltwater, which enter the site primarily through smaller channels north and south of the Refuge.

The restoration was not intended to provide protection against the next Sandy but to create a stable system within a dynamic context—one that would strengthen and evolve over time. The system's resilience comes entirely from healthy plant material and natural hydrology. Here, neither resilience nor ecological habitat can be regarded merely as co-benefits of one another. Research increasingly shows that a healthy coastal habitat *creates* resilience, and such a project is an inextricable combination of the two.

Restoration of Prime Hook National Wildlife Refuge

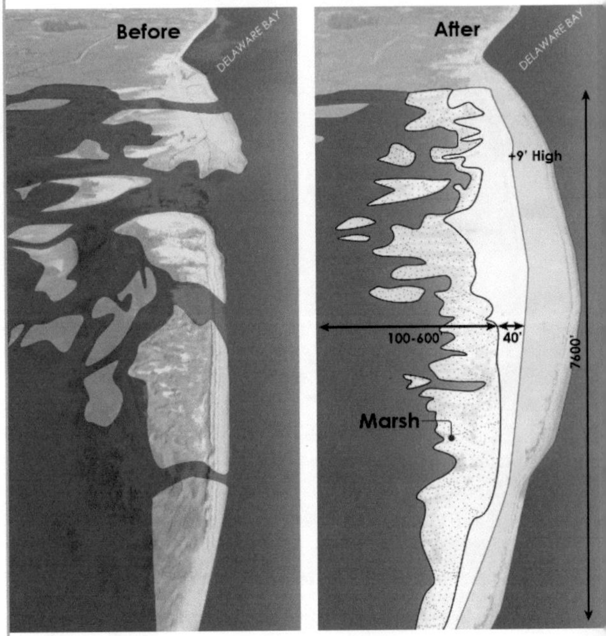

Restoration of a barrier dune helps to control influx of ocean water, protecting marsh ecosystem inland. Channels were dredged to recreate healthy circulation of the marsh system.

10.3.3 **Implementing a Hybrid Approach**

On Gerritsen Creek in New York City, the need for a no-maintenance landscape in a location highly vulnerable to wave impacts led to the development of a hybrid living system to stabilize a shoreline. In the early 1900s, Charles Downing Lay designed the entire area around Gerritsen Creek as an intensely programmed park. Known as Marine Park, the project would have obliterated the Gerritsen Tidal Marsh to create an 1800-acre play space, bigger than Central and Prospect Parks combined.[31] In a concept plan that illustrates the limited perceived value of salt marsh at the time, the creek would be hardened into a two-mile-long water feature to host yachts, model boats, racing sculls, and canoes. The filling of the creek for the construction of the park began in 1933. That same year, a change in government and the appointment of Robert Moses as Parks Commissioner put an end to these plans. While Moses intended to build a park of some kind, it was never realized, and the marsh was left largely untouched. Over time, the marsh became a nonhazardous municipal waste dump, and White Island, which sits in the middle of what would have been the main

White Island Shoreline Restoration in Brooklyn, New York City

canal of Lay's park scheme, became a landfill. Over time, erosion from wave action had begun to expose the underlying waste.

Construction of a nearby 227-acre affordable housing development, Gateway Estates, returned attention to White Island in 2006. As a mitigation requirement for the 56 acres of maritime grassland habitat that the development would destroy, White Island would be restored as a habitat for the Henslow's Sparrow, a once-common grassland species that has fallen into decline due to coastal development.

Restoring its habitat on the island required stabilizing the shoreline to prevent waste from entering the surrounding waterways. Several factors determined the selection of stabilization options, including slope, the unsuitability of cutting into existing edges due to the presence of waste or of building out into the water, and the erosive forces of the relentless wave action. But a key criterion for the selection of shoreline stabilization approaches was longevity. The island would not provide public access and would not be maintained; therefore, any chosen shoreline treatment would need to be self-sustaining.

The design team recognized that vegetation would provide an effective and sustainable form of slope protection that would also improve drainage by removing moisture from the soil, thereby reducing erosive impacts on the slopes. Over time the plantings would create a living mat that would further bind the shoreline together, but the effectiveness of vegetation alone would be limited when exposed to persistent or concentrated wave action. The establishment period for vegetation can leave new installations vulnerable to storms and foraging wildlife. To maximize both habitat and long-term stability of the shoreline, a range of hybrid techniques incorporating hardened structures and bioengineering solutions were applied according to the specific conditions along each segment of shoreline. Where possible, in flat areas with dense, mature, and native vegetation, the shoreline was preserved to save cost and to capitalize on the resistance to erosion provided by established plants. Where possible, salt

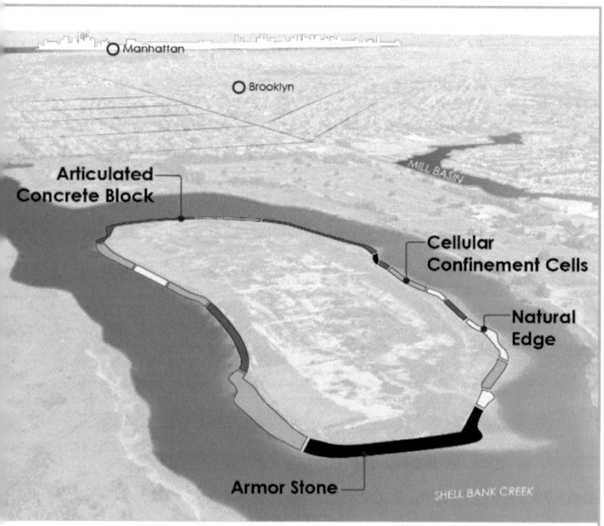

Varying shoreline conditions call for different restoration approaches.

marsh was introduced around the island to provide additional protection and habitat.

Construction of the island's shoreline was completed in August 2012. On October 29, 2012, Hurricane Sandy hit the New York City area, producing a 15-foot storm surge on the island. Even immediately after the storm, there was no evidence of erosion or vegetation loss, and the island is thriving as bird habitat. While a naturalized shoreline may have over time been able to withstand a powerful storm, the use of a redundant system here prevented a potential loss of investment.

Hurricane Sandy was more destructive, however, on the shores of Staten Island. Tottenville sits on the south shore of Staten Island facing Raritan Bay. The shape of the harbor intensifies waves as they move toward the low-lying community. Over time, waves have eroded Tottenville's beaches and coastal bluffs, threatening public property, infrastructure, natural resources, and homes. Peak storm tides during Hurricane Sandy reached 16 feet on Staten Island, and waves up to 6 feet scoured the shoreline.

The unique geography of the area has made it the subject of study as well as the recipient of funding for climate adaptation projects since Hurricane Sandy. Among these is the Tottenville Shoreline Protection Project. Originally conceived as part of the New York Rising Community Reconstruction Program, the initial concept was a series of stone-core, sand-capped dunes that would attenuate wave energy and prevent the intrusion of sea level rise (up to 30 inches) into interior neighborhoods. However, modeling showed that the natural sediment supply in this area would not be sufficient to replenish dune sand over the long term, and many areas slated for permanent dunes were not wide enough for an installation with the desired level of protection. Given the inland constraint of an established residential community, constructing a larger dune meant locating it further out to-

Hybrid Approaches to Shoreline Restoration

Armor stone (D50=36in) was installed to protect the steepest and most severely eroded shoreline where consistent wave action required hard solutions.

Articulating Concrete Block (ACB) systems were installed along slopes >2:1. ACB consists of a matrix of individual concrete blocks placed together to form an erosion-resistant blanket. The term "articulating" refers to the ability of the matrix to conform to minor changes in the subgrade, while remaining interconnected with cables.

Cellular Confinement Systems (CCS) were installed along slopes >3:1. CCS, commonly referred to as "geocells", consist of high-density polyethylene (HDPE) strips bonded together to form a honeycomb matrix that can be backfilled with soil and vegetated.

Tottenville Shoreline on Staten Island, New York City

Tottenville Shoreline Protections

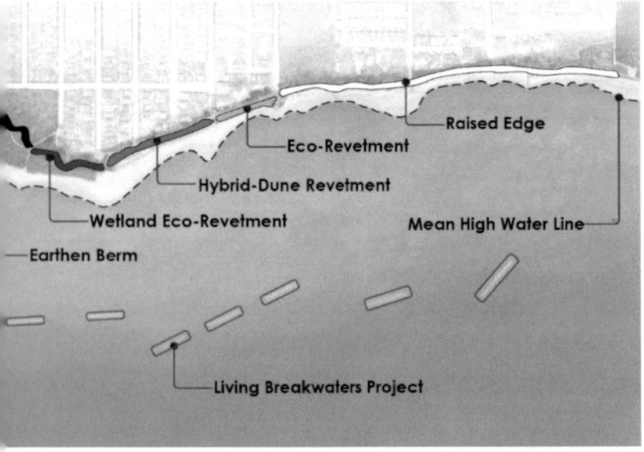

Coordinated coastal protections are responsive to changing conditions along the shoreline.

wards the water and placing earthen fill below the mean high-water line (MHW), a measure that would require extensive permitting and had the potential to significantly delay the project.

As a result, the concept was revised to ensure that the continuous raised structure was located entirely above the MHW line. This was a major driver for the design and one with which many coastal projects must contend. It presents a choice between designing a more future-facing system that may take many months to permit, if it is permitted at all, or designing something that can be built sooner. With this regulatory uncertainty and sense of urgency to prepare for the next storm, the design of the mile-long structure evolved to include different treatments that respond to site-specific ecologies and exposure to wave action. Along with coastal protection, the design provides recreational trails and an accessible pathway along the full length of the beach, even at high tide, when the shore is currently impassable.

Many of the regulations meant to protect the environment can present obstacles for new and innovative approaches to resilience. While regulatory agencies are well equipped to evaluate engineered systems, the protective capacity of living infrastructure—such as dunes and ecological systems below the MHW line—has not been as comprehensively quantified or documented. Evolving these processes will become more important for designers, researchers, and regulators as the use of living systems has the promise to increase. Legislation introduced in Congress in 2019 acknowledged the effectiveness of living shorelines for increasing coastal resilience. The Living Shorelines Act proposes new federal funding for shorelines that incorporate natural infrastructure.[32] It recognizes the protections natural systems can provide, as well as their long-term co-benefits. The Act, if passed, will authorize $50 million per year for construction and monitoring of living shorelines. While this figure pales in comparison to the level of funding needed to create such infrastructure, it could form the basis for a national program through which the value of such systems can be more precisely quantified and in turn validate further use of them.

10.3.4 **Louisiana Coastal Systems**

The coastline in Louisiana, like all wetland coastlines, erodes naturally as tides

Shoreline Protection Techniques

Earthen Berm

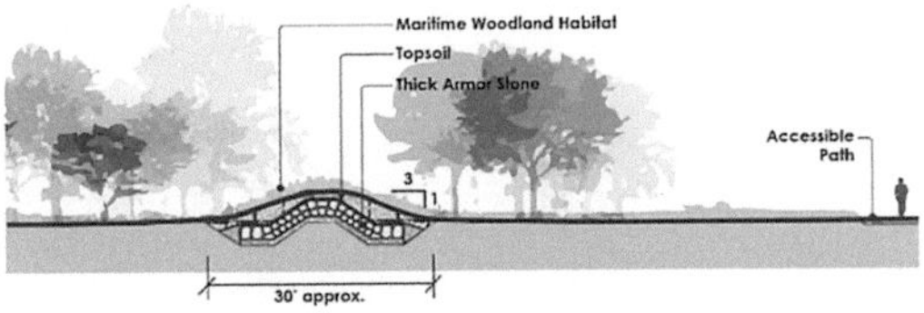

1 A wide berm with shallow slopes and a minimum of 18″ of topsoil to facilitate root growth of woody plants. Runs through an existing woodland.

Wetland Eco-Revetment

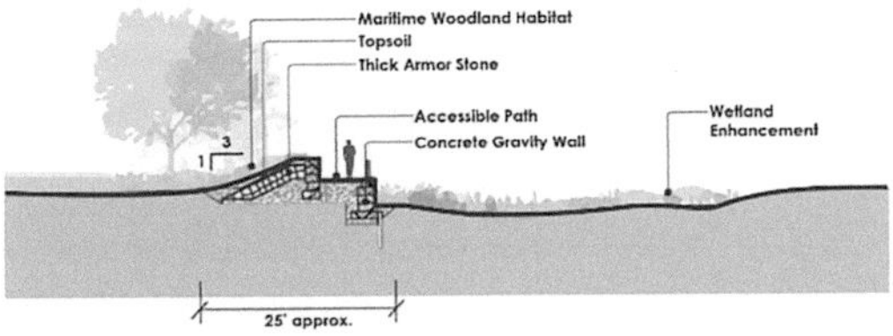

2 A hybrid structure that is part gravity wall and part vegetated revetment. Maintains a minimal footprint to make space for an existing wetland.

Hybrid Dune Revetment

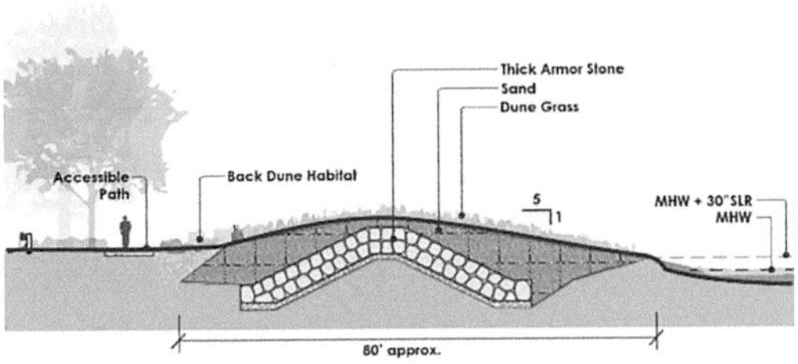

3 A stone core, sand-capped dune planted with native beach grasses. Located where sufficient beach space exists. Sits inland to minimize risk of erosion.

Surf Ave Eco-Revetment

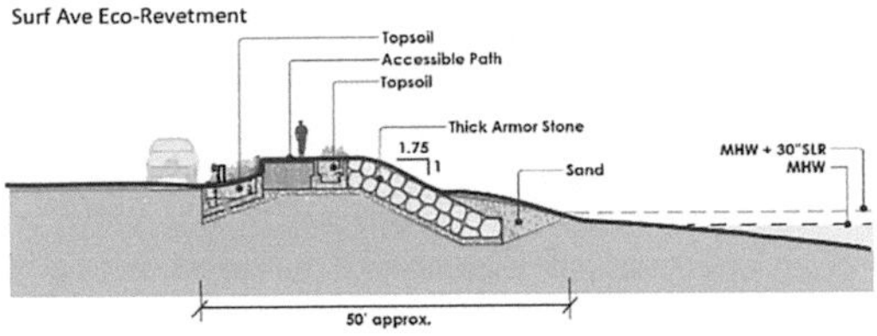

4 The most robust shoreline structure, including a thick layer of topsoil for root mass growth and a larger stone core. Located in areas subject to the largest hydrodynamic loading and the least beach width to accommodate an intervention.

Raised Edge

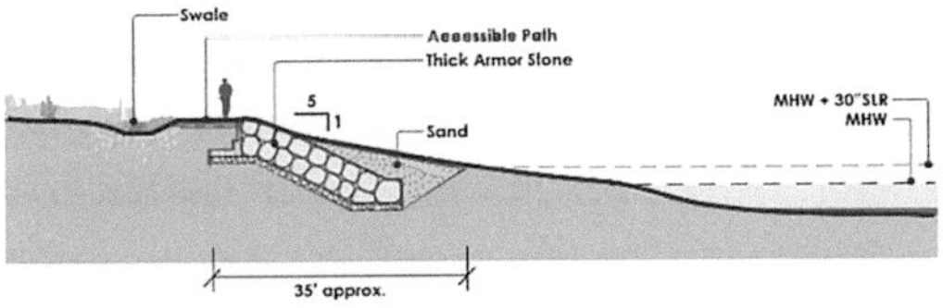

5 The section of the shore that is not subject to moderate wave action. Inland community is already at a higher elevation. Design minimizes regrading, continues the path alignment along the shore, and improves water quality by introducing a drainage swale.

come and go and waves lap the shore. This process is accompanied by the flow of sediments through inland streams and rivers, which are deposited in wetlands as the water flows out to the sea. This cycle—sediment out, sediment in, also known as the fluvial process—perpetually builds and unbuilds the shoreline, and it is how the low-lying areas of the Mississippi delta were formed over 7,000 years.[33] Sand, silt, and clay particles traveled hundreds of miles across the Midwest—from Wisconsin

Mid-Breton, Near the Louisiana Coast on the Gulf of Mexico

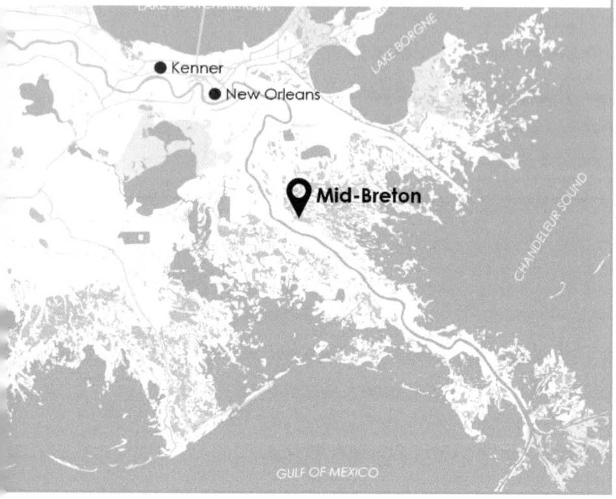

and Iowa to Arkansas and Mississippi—until the riverbed widened and the flow slowed enough to drop them out of suspension. Once settled, another process took hold. Marsh plants and diverse animal species took up residence, creating an ecosystem in this brackish terrain that bridges land and sea.

In a healthy ecosystem, the fluvial process is balanced, with roughly equal amounts of accretion and erosion. When one or more of these processes is interrupted, the imbalance causes ripple effects that may not be immediately evident. Over the last 200 years, more than 3,500 miles of levees have been built along the Mississippi River system,[34] constraining the Mississippi and its tributaries and limiting release of water into the Gulf of Mexico through a select number of locations while causing severe sediment deprivation in others. Dredging and canal building by the oil and gas industries channelize freshwater flows and induce saltwater intrusion, which kills freshwater marshes and contributes to land loss.[35] Without plant roots to hold soil in place, the land erodes at an even faster rate, resulting in loss of land and habitat.

Impacts from climate change—namely, sea level rise and increasing storm frequency—threaten to hasten the loss. The loss of land and habitat is not just disconcerting: wetlands and salt marshes have important functions for mitigating the more intense storms that will result from climate change. Coastal wetlands attenuate wave energy and have been shown to reduce wave height by up to 18%.[36] Where wetlands and marshes still exist, they are often severely fragmented, which reduces their ability to perform functions that are critical defenses for flood mitigation and storm surge.[37] The irony is that conventional approaches to flooding from storm surge—levees built along the river—are the very infrastructure that have weakened natural defenses. By eliminating the low-lying, fan-like geometry of floodplains that causes water to slow, sediments remain suspended in water until they are far into the Gulf instead of dropping in the Delta and counterbalancing erosion. As a result of this and other factors such as upstream dams, invasive species, subsidence, and sea level rise, Louisiana has lost more than 2,000 square miles of coastal wetlands since 1930.[38] The Breton Basin alone has lost more than 700 square miles[39] of land since the 1930s, representing one of the highest land loss rates in the world.

10.3.5 **Rebalancing Natural Processes**

When faced with the issue of land loss, a more conventional approach might have been to dredge a borrow site and apply that fill in the delta, a short-term solution that would eventually require further dredging and filling. A relatively new approach to righting the unintended consequences of river levees, however, is sediment diversion, wherein a structure allows a portion of the river to flow to where it might have gone without levees. This process replaces eroded coasts with sediment deposits, thereby allowing a dynamic living system to establish.

One of the first projects to be designed for this specific purpose is the Mid-Breton Sediment Diversion. The Mid-Breton diversion structure is being designed 69 miles from the Louisiana coast to divert 35,000 cubic feet of water per second, a fraction of the overall Mississippi River flow, to the Breton Delta. Our design team used historic records and hydrologic models to optimize the location of the diversion and to anticipate the volume of sediment that could be transferred into the delta. Modeling also helped determine where sediment will be deposited, allowing designers and planners to visualize the growing coastline in 50 years and beyond. This modeling is crucial for understanding potential impacts and risk reduction.

The approach allows a system to build itself. This is radically different from building hardened systems that aim for complete control. In this case, a sediment transport system formerly disrupted by levees will begin to passively build salt marshes and wetlands. The innovation is in the design intent. By simply allowing a process to unfold and not designing for the specific end product, we are providing opportunity for many potential futures. The exact plant and animal species that will eventually make up the marshes at the bottom of the delta are not specified—those will evolve as the land rebuilds.

Salt marshes are especially applicable to climate adaptation, not only because they absorb wave energy, but also because they can move and adapt to sea level rise. If given the space, a salt marsh will migrate inland as seas rise. They also continue to accumulate sediment even as water rises, replenishing the coastline.[40] This is why land use policy is a critical component of any climate adaptation plan. Policies that allow for a future with reconnected wetlands, space for salt marsh migration, and undeveloped flood plains are necessary. It is in this evolution that living systems have the most potential to address the key concern that plagues all climate adaptation projects: uncertainty.

10.3.6 **A Community-Focused Design Approach**

Further inland, the city of New Orleans is also taking steps to reduce risk and uncertainty. For over a hundred years, New Orleans has been pumping water out of the city at rates that increased as technology improved. While this created the dry land that allowed the city to expand, it also caused much of the city to sink below sea level, making it dependent for its survival on its system of walls and levees. Today it is home to some of the largest and most sophisticated pump systems in the world, and yet, facing over 60 inches of annual rainfall and growing risks associated with climate change, sea level rise, and subsurface soil subsidence, New Orleans is now investing in adaptive, layered living systems that account for both existing and future demands.

New Orleans is a city marked by social and economic inequity, where minority and lower income residents tend to live in neighborhoods more vulnerable to flooding, subsidence, and infrastructure failure. Before Hurricane Katrina, the city did not have a strong tradition of urban planning and citizen engagement, and water management adhered to the tenets of "convey and dispose." In the ten years after Hurricane Katrina, a series of planning initiatives laid the foundation for using a living systems approach to building resilience in New Orleans in response to precisely those social justice imperatives.

Initial recovery planning introduced the idea of using green open space to manage flooding, but the "green dot plan" was controversial because construction moratoriums, relocations, and storm basins published on a planning diagram were located in predominantly African American and low-income neighborhoods, where communities were in large part committed to rebuilding. As a result, subsequent recovery planning, culminating in the 2006 Unified New Orleans Plan, gave priority to robust community engagement in the planning process. A series of efforts continued to build community support and create holistic and inclusive concepts of resilience, including commitments to community quality of life and to living systems flood management practices.

Specifically, New Orleans adopted an approach to water management called "Living with Water," which is characterized by stormwater infiltration-based practices and strategic, multi-benefit water storage systems to reduce the City's reliance on pumps and levees while providing a variety of social, economic, and environmental benefits—the "Triple Bottom Line." In 2015, the city participated in the US Department of Housing and Urban Development's (HUD) National Disaster Resilience Competition (NDRC) with a proposal to create the city's first Resilience District. It was awarded over $141 million through the NDRC, which will supplement previously allocated investments from FEMA's Hazard Mitigation Grant Program.

The Gentilly Resilience District in New Orleans, Louisiana

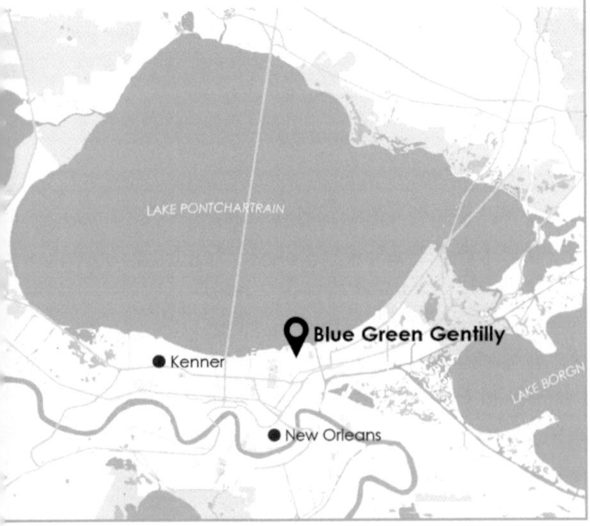

The Blue & Green Corridors project is just one of the projects to receive this funding and to begin to realize the Living with Water approach—reversing the impacts of pumping water out of the city in favor of natural hydrologic processes. The project will create a network of canals, recreational parks, and community spaces along eight linear miles of public right-of-way within the Gentilly Resilience District. Along the streets slated as "blue corridors," the wide neutral grounds between vehicle travel lanes will be transformed into open channel canals to receive and manage runoff and relieve stress on the pumping system. Along the streets slated as "green corridors," GI practices such as bioswales and permeable pavement will help to recharge groundwater and reduce subsidence. Several lots left vacant by Katrina and subsequently purchased by the city will be converted to floodable parks, providing recreational space as well as additional flood storage. Early in the design process, our team engaged with the community to determine what kinds of amenities and programming it might desire along the corridors—from seating and lighting to playgrounds, safe bike lanes, and outdoor fitness equipment. Desired amenities will be located throughout the corridors, alongside canals and plantings, with canopy trees to beautify the neighborhood, keep canal water cool, and reduce the heat island effect. Plans to reduce traffic lanes and make streets narrower were introduced to decrease impervious cover, calm traffic, and allow safe cycling and walking.

The city has other open canals, most of which are engineered structures with steep, lawn-covered side slopes, visible outfalls, and no water quality protocols. The desire for blue corridors to provide a visual amenity—something people would choose to walk alongside—and serve a potential recreational function means that water quality, smell, insect-borne health risk and nuisance, and appearance must be considered. The most efficient and low-maintenance way to accomplish these aesthetic and water-quality goals while providing significant flood storage is through the creation of a living water body, in this case, a linear wetland that can remove contaminants through natural processes such as sedimentation, filtration, sorption, plant uptake, and microbial breakdown. The transformation from a bioretention canal to a wetland means stabilizing the side slopes with vegetation, providing a varied terrain to support high and low marsh habitat, and creating backwater pretreatment areas where stormwater initially enters the system. This design approach will minimize maintenance, one of the community's

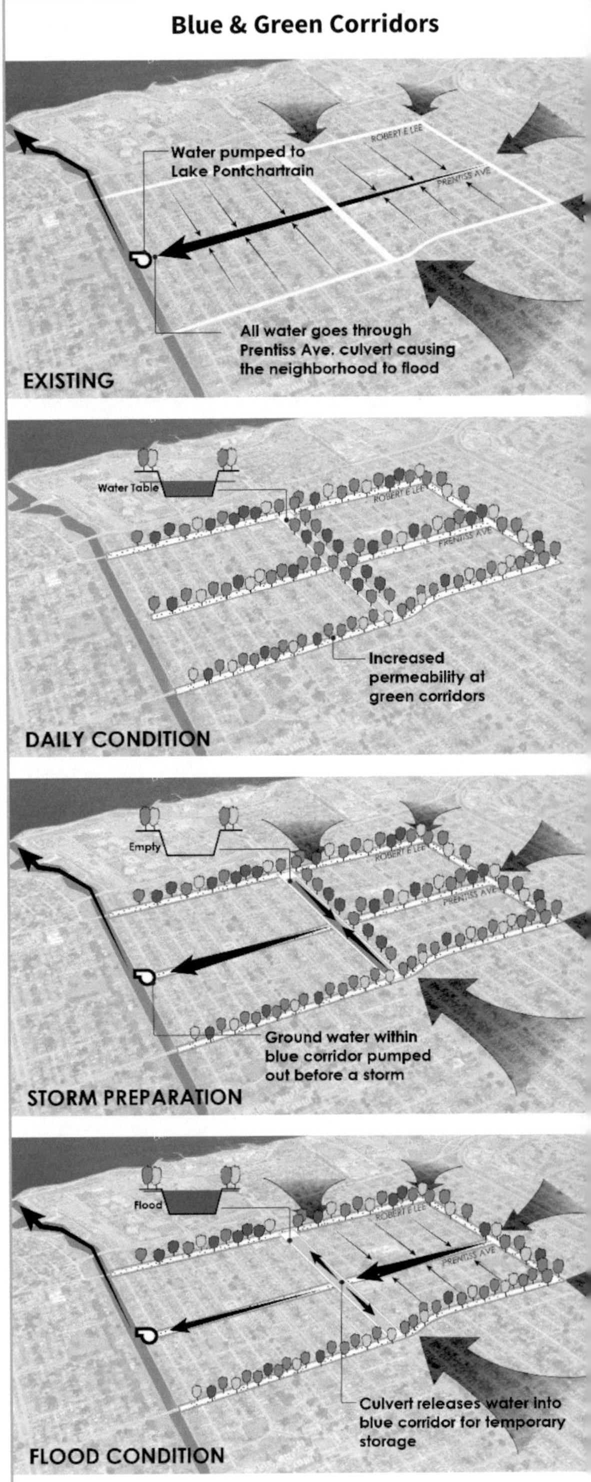

Multi-benefit infrastructure performs as flood mitigation during a storm; a neighborhood park the rest of the year.

primary concerns, which will involve invasive species control, edge mowing, and growth management. The end result will be a beautiful living water body that will attract wildlife and draw people to sit, play, exercise, gather, and walk.

The programming and flood storage goals of the project competed for square footage throughout the design process, but the demand for a balance between them underscores the evolution of thinking about resilience that has occurred in New Orleans. Resilience does not originate in flood mitigation alone but is also nurtured through spaces for communities to come together and thrive.

Hydraulic models show that the blue and green corridors will mitigate flooding from average events, not storm surge or severe events. But these areas constitute an important part of a larger resilience system that serves as an example of a new form of multi-scale, layered resilience planning. These communities will experience future storms, but the purpose of resilience is to reduce the impact of disruption and commerce and to ensure that recovery will be swift following storm events. These communities will be better prepared for the next storm.

10.4 **Conclusion**

As practitioners and consultants, our work is often reactionary, with project approaches largely defined by clients and funding. Increasingly, clients and politicians are helping to define more innovative and proactive projects, projects that afford us the opportunity to capitalize on and further understand the potential of a living systems approach. But despite growing client ambition and an increasingly supportive political climate, there are challenges ahead for implementing living systems. Understanding the new parameters created by climate change is one of them. Throughout this chapter, we have referred to optimized locations for infrastructure and ideal conditions; in many cases, we refer to the "natural" condition as the target. Similarly, low-impact development standards refer to the pre-development condition as the appropriate design criteria and benchmark. But the unpredictable nature of climate change calls for a new understanding of what the "right" design criteria should be. Given that we face new conditions altogether, resilient design cannot be a matter of merely returning to a previous condition; instead, it should rely on evolving science and leverage project research to gain a better understanding of the forces at work to allow for the emergence of new ecologies and new standards. We must shift design criteria from historical benchmarks to more future-facing factors.

10.4.1 **Empowering Clients with Better Planning Tools**

One way to develop better design criteria is to develop more nuanced models for analyzing risks at the planning stage. In the context of flooding, many of our conventional tools are inadequate. FEMA maps do not provide the resolution necessary to understand the real vulnerability of problem areas, and our models have long been based on a simple calculation of impervious and pervious surfaces within a catchment area and a given storm size. The new models we are developing, however, are more context-specific and detailed. They represent spatial variability in flow and timing. They consider different development and rainfall scenarios, different long-term plans, and future uncertainty. They look at back-to-back storm events—key because of the dynamic nature and timing of storms—the spatial distribution of rainfall, and the effects saturated soils and storm systems have on flood severity.

The Automated Mitigation Project Identifier (AMPI) is a risk-assessment model that correlates flood data with building stock data, as well as proposed project data, to assist municipalities in prioritizing projects to minimize damage in a range of scenarios. It also tests implications of potential solutions, such as raising or removing structures, and produces cost/benefit analyses that incorporate social, financial, and environmental impacts. Where clients need more than refined data to generate project support and funding, there are modeling tools that turn to the power of images to spur action. Such models utilize video game visualization technology alongside FEMA Risk MAP data to allow a client or community group to virtually walk through a neighborhood to see the anticipated impacts of a storm event. As communities negotiate the balance between priorities of protection with quality of life, more sensitive models can offer more accurate risk profiles on which to base decisions. More precision allows us to generate more nuanced solutions—layered and decentralized systems that are right-sized for the risks. They help us to identify opportunities for incorporating living systems into comprehensive resilience plans.

New tools must be complemented by ongoing scientific research, keeping pace with the changing conditions brought about by climate change and the corresponding changes in our understanding of resilience. But even with the most advanced models and visualizations, implementing living systems remains challenging. The spatial requirements for living systems are often at direct odds with dense and valuable developments that are concentrated along shorelines. Retreat from the coast is one strategy for creating space for living systems and removing human settlement from at-risk locations. Retreat is a slow and piecemeal process, with the real estate market in many places making buyouts—the primary tool for achieving it—impractical.[41] Discussions about retreat must be transparent and grounded in equity and community-based planning. Changing course in a developed world is complicated, even given accurate data and political consensus, but implementation of living systems in the near term points to retreat in the long term. These efforts should be coordinated.

Finally, while we argue that a living systems approach is ideal for confronting uncertainty, resistance to the use of living systems is most often rooted in uncertainty, namely, uncertainty about these systems themselves. Regulations are written to support measurable and proven technologies, and while we have seen willingness in regulatory agencies to consider new approaches, the litigiousness of the design profession puts the burden on designers to carry the risks associated with them. The potential for error is magnified when combined with the uncertainty of climate change. We engineers are trained to build a factor of safety into our designs; the responsibility for public safety we bear calls for conservative, measurable, tried-and-true solutions. The advancement of a living systems approach requires increased monitoring of the living systems that have been built and the sharing of that data among design professionals so that clients can understand the full monetary and risk-reduction values of such systems.

What we describe above is both an approach—one based on systems analysis and community engagement—and a physical intervention—one that nurtures natural processes, accommodates adaptation, and prioritizes aesthetic and community benefits along with protection. The projects we have used to illustrate this approach do not do so perfectly, but they offer techniques and tools that help us formulate a new direction. Living systems are promising as adaptation tools in part because they adapt to changing conditions—sometimes by changing themselves. We believe that as designers we have a responsibility to do the same. The urgency of the climate crisis requires designers to be proactive in developing project approaches and advocating innovative solutions. We must monitor completed projects and share knowledge generously. In the face of uncertainty, the power of designers to envision a future, to expand the imagination of our clients and stakeholders, and to generate a constituency in support of action is critically important. We must not undervalue our role nor underestimate our responsibility in realizing a livable future.

Notes

* The authors would like to acknowledge Stantec colleagues who provided insights into their project work: Bernadette Callahan, Phillip Champagne, Michael Chelminski, Dan Grandal, Ashley Hall, Josh Running, Gary Sorge, and Jeff Tabar.

1 Frank, T., "After a $14-Billion Upgrade, New Orleans Levees Are Sinking," Policy & Ethics, E&E News, last modified April 11, 2019, https://www.scientificamerican.com/article/after-a-14-billion-upgrade-new-orleans-levees-are-sinking/.

2 Wenger, C. 2017, "The Oak or the Reed: How Resilience Theories Are Translated into Disaster Management Policies," *Ecology and Society 22*(3):18. https://doi.org/10.5751/ES-09491-220318.

3 Craig, N. J., et al., "Land Loss in Coastal Louisiana (U.S.A.)," Center for Wetland Resources, Louisiana State University, Baton Rouge, Louisiana, USA. https://doi.org/10.1007/BF01867025, https://link.springer.com/article/10.1007/BF01867025.

4 Bogardi, J. and A. Fekete, 2018, "Disaster-Related Resilience as Ability and Process: A Concept Guiding the Analysis of Response Behavior Before, During and After Extreme Events," *American Journal of Climate Change*, 7, 54-78. doi: 10.4236/ajcc.2018.71006.

5 Narayan, S., et al., "The Value of Coastal Wetlands for Flood Damage Reduction in the Northeastern USA," *Nature, Scientific Reports*, August 31, 2017, online at https://www.nature.com/articles/s41598-017-09269-z.

Scientific evidence shows that many coastal habitats, including mangroves, coral reefs, and salt marshes, help protect coastlines by absorbing wave energy and reducing flooding and erosion. Under favorable conditions, these nature-based defenses can keep pace with changing environmental conditions and rising sea levels. In contrast, gray infrastructure, such as sea walls and levees, can be costly, rarely keeps up with environmental change, and often damages nearby shorelines and habitats.

6 Elmqvist, T., J. Siri, E. Andersson, et al., "Urban Tinkering," *Sustainability Science.* 2018; 13:1549–1564. Published online 2018 Aug 6. https://doi.org/10.1007/s11625-018-0611-0.

7 "A Triple Bottom Line Assessment of Traditional and Green Infrastructure Options for Controlling CSO Events in Philadelphia's Watersheds," Stratus Consulting, accessed August 21, 2019, https://www.epa.gov/sites/production/files/2015-10/documents/gi_philadelphia_bottomline.pdf.

8 "What Is Green Infrastructure?" Green Infrastructure, United States Environmental Protection Agency, accessed August 21, 2019, https://www.epa.gov/green-infrastructure/what-green-infrastructure.

9 To measure success, Philadelphia uses a "greened acre" metric. With units of an acre-inch, one greened acre represents the management of one inch of stormwater runoff from one acre of impervious land. In other words, one greened acre is equivalent to the management of 27,158 gallons of stormwater runoff during a storm event. In the fall of 2017, the City surpassed 1,000 greened acres, which was the equivalent of more than 3 billion gallons of stormwater diverted from combined sewers annually.

10 "Heat Island Effect," Environmental Topics, United States Environmental Protection Agency, accessed August 21, 2019, https://www.epa.gov/heat-islands

11 "Weather Fatalities 2018," Weather Related Fatality and Injury Statistics, National Weather Service, accessed August 21, 2019, https://www.weather.gov/hazstat/.

12 Robert Vautard, Olivier Boucher, et al., "Human contribution to the record-breaking July 2019 heat wave in Western Europe." *Environ. Res. Lett.* 15 (2020) 094077.

13 Gonzalez et al., "New York City Panel on Climate Change 2019 Report Chapter 2: New Methods for Assessing Extreme Temperatures, Heavy Downpours, and Drought," Advancing Tools and Methods for Flexible Adaptation Pathways and Science Policy Integration, *Annals of The New York Academy of Sciences*, (2019): pp 37, 2.3.2.2, DOI: https://doi.org/10.1111/nyas.14007

14 "Heat Island Effect," Environmental Topics, United States Environmental Protection Agency, accessed August 21, 2019, https://www.epa.gov/heat-islands/using-trees-and-vegetation-reduce-heat-islands#2

15 Matti Kummu, Hans de Moel, Philip J. Ward, Olli Varis, "How Close Do We Live to Water? A Global Analysis of Population Distance to Freshwater Bodies," PLoS One. 2011; 6(6): e20578. Published online 2011 Jun 8. DOI: 10.1371/journal.pone.0020578. https://www.ncbi.nlm.nih.gov/pmc/articles/PMC3110782/.

16 Huard, J., "A Devastating Rainstorm Sunday Dumped Nearly 10 Inches of…," UPI Archives, United Press International, last modified June 6, 1982, https://www.upi.com/Archives/1982/06/06/A-devastating-rainstorm-Sunday-dumped-nearly-10-inches-of/7300014849288/.

17 Annandale, G.W., "88 Reservoir Sedimentation," John Wiley & Sons, Ltd., accessed August 21, 2019, https://doi.org/10.1002/0470848944.hsa086, https://onlinelibrary.wiley.com/doi/abs/10.1002/0470848944.hsa086

18 Federal Emergency Management Agency, "Living With Dams: Know Your Risks," FEMA P-956, p 12, (2013), https://www.fema.gov/media-library-data/20130726-1845-25045-7939/fema_p_956_living_with_dams.pdf.

19 "Summary," *2017 Infrastructure Report Card Dams*, American Society of Civil Engineers, p 1, 2019, https://www.infrastructurereportcard.org/wp-content/uploads/2017/01/Dams-Final.pdf.

20 Task Committee of the Association of State Dam Safety Officials, "The Cost of Rehabilitating Our Nation's Dams," Association of State Dam Safety Officials, (2019 Update). https://damsafety-prod.s3.amazonaws.com/s3fs-public/files/Cost%20of%20Rehab%20Report-2019%20Update.pdf.

21 The project was mostly financed by the U.S. Fish and Wildlife Service, which provided more than $600,000 through the Disaster Relief Appropriations Act. That law took effect after Hurricane Sandy swept through the Northeast in 2012.

22 "Environmental Assessment: Pond Lily Dam Removal Project," U.S. Fish and Wildlife Service, Southern New England New York Bight Coastal Program, p 3, (2015), https://www.fws.gov/r5snep/pdf/Pond_Lily_EA_05282015%20Final.pdf.

23 "The Economic Importance of the Bay," What We Have to Lose, Issues, Chesapeake Bay Foundation, accessed August 21, 2019, https://www.cbf.org/issues/what-we-have-to-lose/economic-importance-of-the-bay/index.html.

24 The Water Quality Goal Implementation Team, "Protocol for the Development, Review, and Approval of Loading and Effectiveness Estimates for Nutrient and Sediment Controls in the Chesapeake Bay Watershed Model," Chesapeake Bay Program, (2015). https://www.chesapeakebay.net/channel_files/22798/cbp_bmp_expert_panel_protocol_wqgit_approved_7.13.15.pdf.

25 Robert Wieland, Doug Parker, Will Gans, Andrew Martin, "Costs and Cost Efficiencies for Some Nutrient Reduction Practices in Maryland," NOAA Chesapeake Bay Office and Maryland Department of Natural Resources, (2009). http://www.envtn.org/water-quality-trading/state-programs/Final_BMP_Cost.pdf.

26 "Cost-Effectiveness Study of Urban Stormwater BMPs in the James River Basin," The Center for Watershed Protection, accessed August 21, 2019, https://thejamesriver.org/wp-content/uploads/2016/05/JRA-Cost-effective-Full-Report-June-update.pdf.

27 "Investigating the Cost-Effectiveness of Nutrient Credit Use as an Option for VDOT Stormwater Permitting Requirements," Virginia Center for Transportation, Innovation & Research, accessed August 21, 2019, http://www.virginiadot.org/vtrc/main/online_reports/pdf/15-r9.pdf.

28 Projects are credited using four primary protocols. Most of a stream restoration project's credit comes from the prevented mass of sediment and nutrients that is annually lost from eroding streambanks. This figure is estimated using a predictive model that considers susceptibility to erosion and dimensions of the stream, and it may be verified with bank pin measurements – a method of installing pins within a streambank to measure subsequent changes in bank form. Other protocols account for reductions achieved through nutrient processing by denitrification and nutrient absorption through floodplain reconnection. Regenerative practices used on intermittent and perennial streams to slow flows and increase floodplain interaction are also credited as

a stream restoration practice. http://chesapeakestormwater.net/wp-content/uploads/dlm_uploads/2013/10/stream-restoration-short-version.pdf.

29 94 pounds of phosphorous, 104 pounds of nitrogen, and 61,812 pounds of sediment.

30 Narayan, S., M.W. Beck, P. Wilson, C. Thomas, A. Guerrero, C. Shepard, B.G. Reguero, G. Franco, C.J. Ingram, and D. Trespalacios, "Coastal Wetlands and Flood Damage Reduction: Using Risk Industry-based Models to Assess Natural Defenses in the Northeastern USA," Lloyd's Tercentenary Research Foundation, London, (2016). https://www.nceas.ucsb.edu/files/Coastal%20Wetlands%20and%20Flood%20Damage%20Reduction%20Report.pdf

31 Thomas J. Campanella, "Playground of the Century: A political and Design History of New York City's Greatest Unbuilt Park," *Journal of the Society of Architectural Historians*, Vol. 72 No. 2, June 2013; (pp. 189-204) DOI: 10.1525/jsah.2013.72.2.189. https://jsah.ucpress.edu/content/72/2/189.full.

32 Federal Government of the United States of America "Living Shorelines Act," The National Oceanic and Atmospheric Administration, (2018). https://www.harris.senate.gov/imo/media/doc/ROS18816.pdf.

33 Day, D.W., et al. "Pattern and process of land loss in the Mississippi Delta: A Spatial and temporal analysis of wetland habitat change," Estuaries, accessed August 21, 2019, https://doi.org/10.2307/1353136, https://link.springer.com/article/10.2307/1353136.

34 Alexander, J.S., et al., 2012, "A Brief History and Summary of the Effects of River Engineering and Dams on the Mississippi River System and Delta," U.S. Geological Survey Circular 1375, available online at https://pubs.usgs.gov/circ/1375/C1375.pdf.

35 "Louisiana Wetlands Tattered by Industrial Canals, Not Just River Levees," Sustainability, *Scientific American*, last modified December 1, 2013, https://www.scientificamerican.com/article/carey-louisiana-wetlands-tattered-by-industrial-canals/.

36 Iris Moller, Matthias Kudella, Franziska Rupprecht, Tom Spencer, et al., "Wave attenuation over coastal salt marshes under storm surge conditions," *Nature Geoscience 7*, 727–731 (2014). https://www.nature.com/articles/ngeo2251.

37 Hinkel, J. et al., "Chapter 4: Sea Level Rise and Implications for Low Lying Islands, Coasts and Communities," IPCC SR Ocean Cryosphere, accessed August 21, 2019, https://www.ipcc.ch/srocc/download-report/.

38 "Land Loss," Our Coastal Crisis, Restore the Mississippi River Delta, accessed August 21, 2019, http://mississippiriverdelta.org/our-coastal-crisis/land-loss/.

39 "Mid-Barataria and Mid-Breton Sediment Diversions," Coastal Protection and Restoration Authority, http://coastal.la.gov/wp-content/uploads/2017/09/FAQs_MidBarataria-and-MidBreton-Sediment-Diversions-FINAL-web.pdf.

40 Christine C. Shepard, Caitlin M. Crain, Michael W. Beck, "The Protective Role of Coastal Marshes: A Systematic Review and Meta-analysis," Plos One, (2011). https://journals.plos.org/plosone/article?id=10.1371/journal.pone.0027374.

41 Brogan, J.K., F.Z. McGee, D.H. Morris, and D. Tobias, (2019, June 20). *Post-Disaster Buyout Programs: Insights for Adaptation Planning in the New York Region* [Conference presentation]. Managed Retreat Conference, New York, NY, United States. https://adaptation.ei.columbia.edu/content/thursday-june-20.

11

Looking Ahead: Designing a Resilient Future

Luce Bassetti and Elizabeth J. Bradford

11.0 Beyond Climate Change Resilience

As used elsewhere in this book, the term "resilience" is frequently applied in relation to climate change impacts. From engineering firms to designers, project owners, stakeholders, and the public, people often consider a resilient project to be one that is resilient only to the impact of a limited set of climate-related threats and hazards. These threats most often include sea-level rise, global warming, increasing CO_2 levels, changes in precipitation, and extreme weather events exacerbated by climate change

Figure 1. **Climate Change-Related Threats and Hazards to the Built Environment**

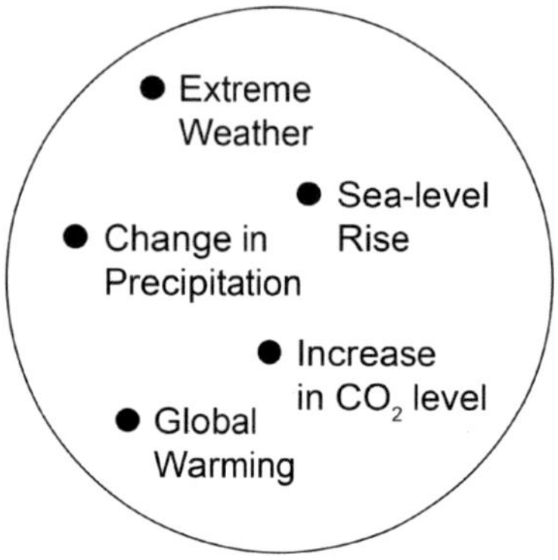

Typical threats and hazards considered in a climate vulnerability assessment of the built environment.

(Figure 1). Similarly, risk assessments are often limited to evaluating climate change-related vulnerabilities to the built environment.

However, as the science of resilience continues to progress, three key themes have emerged. First, the industry is challenged to design systems that are resilient to a broader range of threats and hazards, such as cyber threats, terrorism, infrastructure failure or collapse, and population shifts.[1] Second, there is an increasing understanding of the interconnectedness of existing and future systems. This interconnectedness highlights the relationship between climate change impacts and additional hazards and threats,[2] and shows that the resulting vulnerabilities and risks do not necessarily remain constant over time and can increase exponentially when a tipping point is reached.[3] Third, sustainability and resilience are intrinsically linked around addressing impacts to systems across their full lifecycle. The industry is therefore challenging practitioners to better leverage resilience and sustainability best practices to define and manage a wider array of impacts and develop more holistic and cost-effective solutions that address multiple challenges for a broader range of stakeholders.

11.0.1 **The Expanding Definition of Resilience**

It is increasingly clear that our planet and its systems are vulnerable to a growing range of naturally occurring hazards and human-induced threats. In this chapter, we define resilience in the broadest terms—the capacity of a system to return to an original or desired state after experiencing stress or change,[4] regardless of what generates the stress or change. A risk evaluation should therefore study the full spectrum of threats and hazards (Figure 2) that would disrupt or damage the system under consideration (the vulnerability). It should also assess whether there is the probability (the risk) of a threat or hazard exploiting a vulnerability and what the associated impact and consequences would be.[5] Critical considerations in defining vulnerabilities are whether the threat or hazard is an acute shock (short-term duration and lower predictability) or chronic stressor (longer-term period with higher predictability),[6] and which existing and future systems would be affected (Figure 2) (that is, Infrastructure and Buildings, Environmental and Ecological, Economic and Financial, and Social and Political).

For example, consider that a coastal city's ports, airports, and wastewater treatment plants are often located at or near sea level. Traditionally, a vulnerability assessment may only evaluate climate-related threats and hazards such as sea-level rise

Figure 2. **Threats and Hazards to Natural and Built Systems Across Life-cycle Phases**

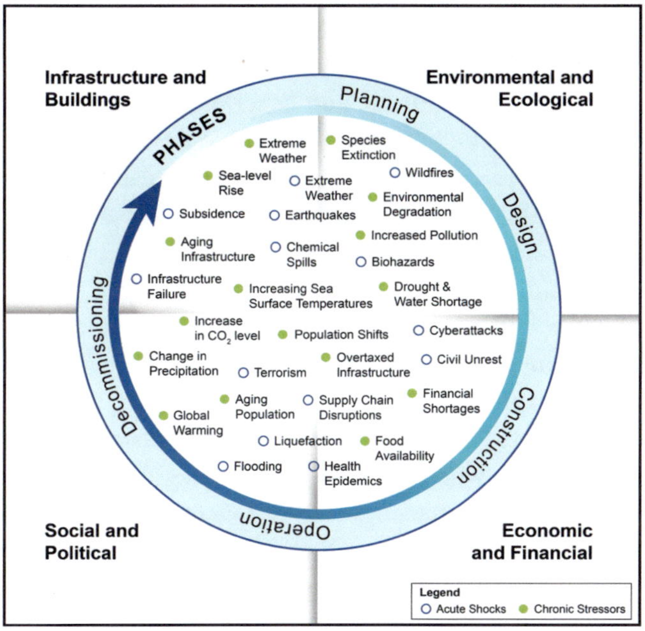

A comprehensive resilience and sustainability assessment evaluates (1) potential impacts of a new or existing system on other systems that might generate threats or hazards, (2) impacts of existing threats and hazards on existing and future systems, (3) impacts across various life-cycle phases, (4) impacts on various natural and built systems, and (5) the relationship of threats and hazards to each other, including possible tipping points.

and extreme weather events (Figure 1). However, Infrastructure and Building systems may also be vulnerable to terrorism, cyberattacks, population shifts, infrastructure failure, aging infrastructure, and financial shortages (Figure 2). Additionally, the chronic stressors (i.e., sea-level rise, population shifts, financial shortages, and aging infrastructure) create different vulnerabilities and risks to understand and mitigate than the acute shocks (i.e., terrorism, extreme weather, and infrastructure failure).

11.0.2 **Interconnectedness and Interdependence of Threats and Hazards**

In addition to, or because of, the growing awareness of a broader range of hazards and threats, there is an increasing understanding of the interconnectedness and interdependence of existing and future systems with near- and long-term threats and hazards.[7] For example, awareness of the relationship between climate change impacts and other types of hazards and threats, such as health epidemics, supply chain disruptions, and food availability, is growing.[8] It is estimated that climate change could drive more than 140 million people to migrate within sub-Saharan Africa, Latin America, and South Asia by mid-century. Moreover, by 2045, desertification will displace an additional 135 million people globally.[9] Climate change-related migration will set off a wide range of impacts, including rapid urbanization, depopulation, water shortages, civil unrest, and shifting demands on the natural and built environment that will likely result in infrastructure stress and environmental degradation.

Work by the United Nations highlights the social dimension of climate change, increasingly linking the emotive and political dimensions to resilience and sustainability. It is estimated that the world's 52 island nations emit less than one percent of global greenhouse gases, yet they will suffer disproportionately from climate change as the magnitude and frequency of climate-related hazards increase, adversely affecting multiple sectors, from tourism, agriculture, and fisheries to energy, freshwater, health, and infrastructure.[10] The United Nations Sustainable Development Goals framework has begun to illuminate the connection between sustainability and resilience, and between climate change, the environment, and social justice issues.[11] This has, in turn, resulted in a better understanding of the concept of climate justice—"that no group of people should disproportionately bear the burden of climate impacts or the costs of mitigation and adaptation"[12]—and the inclusion of climate justice in sustainability and resilience planning, linking human rights and development to achieve a human-centered approach.

The most concerning example of the interconnectedness and interdependence of threats and hazards is tipping points, "when small shifts in human pressures or environmental conditions bring about large, sometimes abrupt changes in a system."[13] Tipping points exacerbate threats and hazards, creating new or higher vulnerabilities and risks. For example, consider the multiple-stressor impacts of sewage to coral reefs (increased pollution), which are well underway. Increases in non-sewage stressors such as temperature (global warming), overfishing (food availability), coastal development (population growth), ocean acidification (environmental degradation), agricultural runoff (increased pollution), unsustainable tourism (overtaxed infrastructure), shipping (environmental degradation), and disease (health epidemics)[14] will increase the synergistic impacts, potentially accelerating environmental degradation and species extinction to a point where recovery is not possible.[15] The loss of a small island nation's coral reef would be profound and would affect existing and future environmental and ecological, economic and financial, and social and political systems (see Figure 2).[16]

11.0.3 **The Intersections of Resilience and Sustainability**

Sustainability is the ability of a system to endure through time and to meet "the needs of the present without compromising the ability of future generations to meet their own needs."[17] If one studies Figure 2 through the lens of one's organization and its technical experts, it quickly becomes apparent that sustainability and resilience are intrinsically linked. However, even seasoned practitioners struggle to clearly state where these two disciplines intersect and how they are distinct from each other. Organizations and infrastructure owners are equally challenged to define, identify, and align these two themes.

Simply put, both resilience and sustainability can be understood by how they impact existing and future systems and the synergies they generate across all life-cycle phases of the systems (Figure 2). Whereas a resilience assessment would require evaluating the impact of a threat and hazard on existing and future systems, sustainability practitioners evaluate the impact of a new or existing system on other systems that would generate a threat or hazard. As demonstrated in Figure 3, both approaches aim to (1) define the impacts quantitatively or qualitatively, as appropriate; (2) identify minimization and mitigation strategies for those impacts; and (3) offset remaining impacts where possible.

11.0.4 Leveraging Intersections to Manage Impact

Industry leaders are beginning to recognize that resilience and sustainability are complementary practices that function best when collaborating in a holistic design that can help alleviate a threat or repair the effects of one. As such, the industry is moving away from focusing on how best to parse which part of a project belongs to sustainability practitioners and which belongs to resilience practitioners. Thus, rather than a resilience team developing risk evaluations and resilience plans independent of the sustainability team's sustainability assessments and plans, organizations are moving toward forming departments of resilience *and* sustainability, conducting resilience *and* sustainability assessments, and developing sustainability *and* resilience plans that evaluate (Figure 3) the following:

Figure 3. **Impact Management Approach**

Sustainability and resilience disciplines both seek to (1) define impacts quantitatively or qualitatively, (2) identify minimization and mitigation strategies for impacts, and (3) offset remaining impacts.

- A broader range of impacts, including (1) potential impacts of a new or existing system on other systems that might generate threats or hazards, and (2) impacts of existing threats and hazards on existing and future systems;
- Impacts across various life-cycle phases, including planning, design, construction, operation, and decommissioning;
- Impacts on various natural and built systems, including Infrastructure and Buildings, Environmental and Ecological, Social and Political, and Economic and Financial; and
- The relationship of threats and hazards to each other, including possible tipping points.

As a result, rather than designing a resilient project to address a single problem and highlighting the sustainability benefits, the industry is asking practitioners to better define and manage a wider range of impacts while offering a more holistic solution that can demonstrate triple bottom-line cost savings.

11.1 Integrating Independently Evolving Sustainability and Resilience Journeys

Confronting climate change calls for a holistic and collaborative approach that supports innovative, cost-effective solutions and addresses multiple impacts while offering co-benefits to a variety of stakeholders. This can be done by assigning resilience and sustainability managers to establish connections across disciplines and functions and use well-established principles and tools to define the vision, set targets, and achieve their climate change goals.

11.1.1 Trends in Sustainability and Resilience Across Architecture and Engineering Disciplines

Sustainability and resilience have evolved as discrete practices that are often seen as management or strategic consulting services, or as solutions-focused disciplines such as natural infrastructure or biosolids. Meanwhile, public demand for resilient and sustainable infrastructure has increased in recent decades,[18] leading state, local, and national governments to respond with mandates and guidance for implementation.[19] As a result, many separate architecture, engineering, and design disciplines have begun to develop more sustainable and resilient solutions to address various impacts (Figure 4).

This evolution has resulted in a variety of subdisciplines that are driving innovative, sustainable, and resilient solutions. For example, within the category of en-

Figure 4. **Trends in Sustainability and Resilience Across Architecture and Engineering Disciplines**

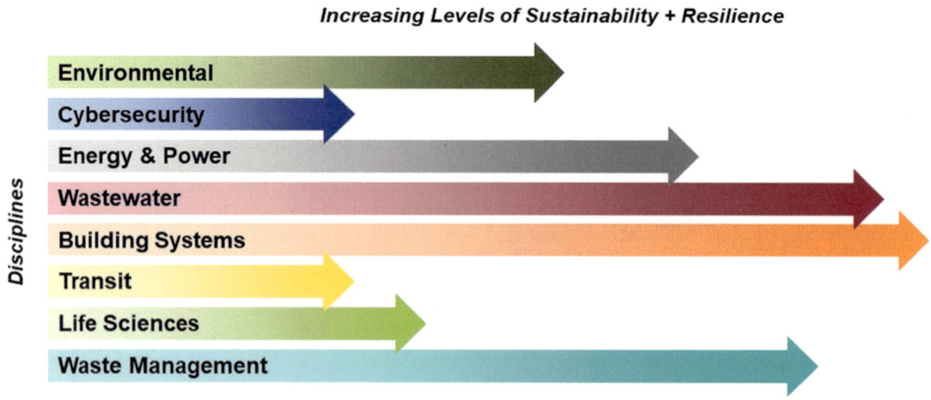

Technical experts across a range of disciplines are each on their own sustainability and resilience journey. Note: Disciplines and progress bars are examples only. As such disciplines are not meant to be comprehensive, and progress bars have been randomly drawn and do not imply true trends for each discipline.

ergy are subdisciplines such as conservation, optimization, renewables, and storage, while transit includes congestion reduction, system reliability, and electric vehicles. These seemingly siloed innovations are leading us toward a convergence of disciplines. By engaging a broader range of disciplines in assessing impacts and developing the mitigation, minimization, and offset strategies, the enabling conditions are set for innovative and multifaceted solutions.

11.1.2 Connecting Sustainability and Resilience to Business and Assurance Functions

While business and assurance functions (for example, asset management, capital investment, communications, compliance, human resources, information management, marketing, procurement, quality management, and risk management) may be audited in a traditional sustainability or resilience assessment, team members from these groups are seldom actively engaged in solutions. However, these functions are also on a sustainability and resilience journey—adopting best practices and developing software and tools to support operational, environmental, social, and economic priorities. By actively incorporating an organization's business and assurance functions, sustainability or resilience managers can use the well-established principles and tools of these disciplines to set priorities. Thus, they can provide stakeholders and investors with data-verified proof to start removing existing financial barriers that delay the implementation of resilient and sustainable design.[20] For example:

- **Asset Management.** Asset management is the coordinated activity of an organization to realize value from assets by balancing asset performance, maintenance, capital investment cost, operating expenditures, and short-term benefits.[21] At its core, asset management focuses sustainability and resilience assessments on a broad range of threats and hazards. Connecting them to asset prioritization systems enhances investment decisions and establishes the economic case for more sustainable and resilient design.[22]
- **Risk Management.** Historically, risk management has addressed both organizational risks and project or program risks (that is, funding shortfalls, natural hazards, and client and stakeholder interactions). And while these risks overlap with some of the threats and hazards evaluated in a sustainability and resilience assessment (Figure 2), they often omit many others that sustainability and resilience practitioners routinely consider. By working together, risk managers and sustainability and resilience assessors can create a more comprehensive risk profile that can address a broader spectrum of threats and hazards.
- **Information Management.** While information management systems are currently being used to track sustainability and resilience metrics, there is room to automate this process better and to link goals set during design to document performance across the full lifecycle.[23] For example, supervisory control and data acquisition (SCADA) and building information modeling (BIM) are used to

track operational performance against goals and include key sustainability and resilience metrics. Integrating these performance-monitoring tools with program management information systems (PMIS) would enable automated performance tracking of sustainability and resilience goals across the full lifecycle.

11.1.3 Leveraging a Sustainability and Resilience Leadership Structure to Deliver Holistic Outcomes

Identifying connection points between threats and hazards, impacts, systems, and phases, and facilitating intersections between disciplines and functions does not just happen.[24] Organizations, programs, and projects need a champion who pushes the sustainability and resilience priorities, identifies and leads the team, and fosters collaboration. A sustainability and resilience manager (S&R manager), sitting at the right hand of organization, program, or project leadership, does that and more.

The S&R manager facilitates, champions, coaches, and connects the team to translate and implement the resilience and sustainability vision throughout organizational departments, program functions, or project tasks.[25] Additionally, the S&R manager provides inspirational leadership, prompts and challenges the team, builds the business case for resilient and sustainable outcomes, and reports achievements and benefits.

An effective S&R manager leads as follows:[26]

- Fosters collaboration to help the team work across lifecycle phases and systems;
- Facilitates intersections among disciplines and functions (Figure 5) to define the vision, set targets, and drive forward innovation to achieve goals;
- Champions and coaches the whole team to harness existing momentum and set enabling conditions for innovative solutions;

Figure 5. **Facilitating Intersections Between Disciplines and Functions**

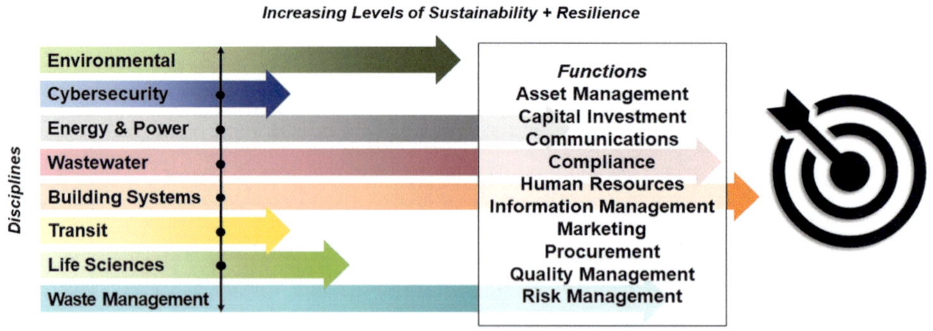

S&R managers champion and coach the whole team, facilitate connections across disciplines and functions to define the vision, set targets, and drive forward innovation to achieve goals. Note: Disciplines and progress bars are examples only. As such, the lists are not meant to be comprehensive, and progress bars have been randomly drawn and do not imply true trends for each discipline.

- Supports and works with the traditional roles of project manager, design manager, procurement manager, and others;
- Is a resource to deliver on emerging sustainability and resilience imperatives;
- Is embedded in project delivery rather than merely being an extra layer of workshops, checks, or reports; and
- Inspires and leads the team to respond to the challenge of "What do we need to do differently?"

11.2 Encouraging Multifaceted Solutions

A whole-system approach that challenges practitioners to move beyond discipline-centered silos will foster broader solutions with co-benefits (ancillary or secondary benefits received beyond the project) that remove the existing financial barriers inhibiting expansion of resilient design. A robust performance management framework allows for collaboration and sets the conditions for multifaceted, multi-benefit solutions.

11.2.1 Implementing a Program Management Approach to Sustainability and Resilience

For an organization, program, or project to integrate resilience and sustainability throughout, a robust performance management framework should be developed. The Observe-Strategize-Plan-Do-Check-Adjust method is an iterative and scalable six-step approach, facilitated by the S&R manager, for creating a framework that supports continuous improvement and evolution of the sustainability and resilience strategy of any organization, program, or project (Figure 6). The six steps are:

1. **Observe: Evaluate the Impacts.** The goal of this step is to establish a holistic view of the current and future impacts on the organization, program, or project. This phase includes a discipline- and function-mapping exercise that identifies key partners who can fully account for, quantify and qualify, and address impacts across lifecycle phases of natural and built systems (Figure 7, next page).
2. **Strategize: Define the Strategy.** During this step, the S&R manager collaborates with the team to set the sustainable and resilience vision, applying structured,

Figure 6 . **Observe-Strategize-Plan-Do-Check-Adjust Method**

An iterative and scalable six-step management approach, facilitated by the S&R manager, for creating a framework that supports continuous improvement and evolution of the sustainability and resilience strategy of any organization, program, or project.

Figure 7. **Sustainability and Resilience Planning Model**

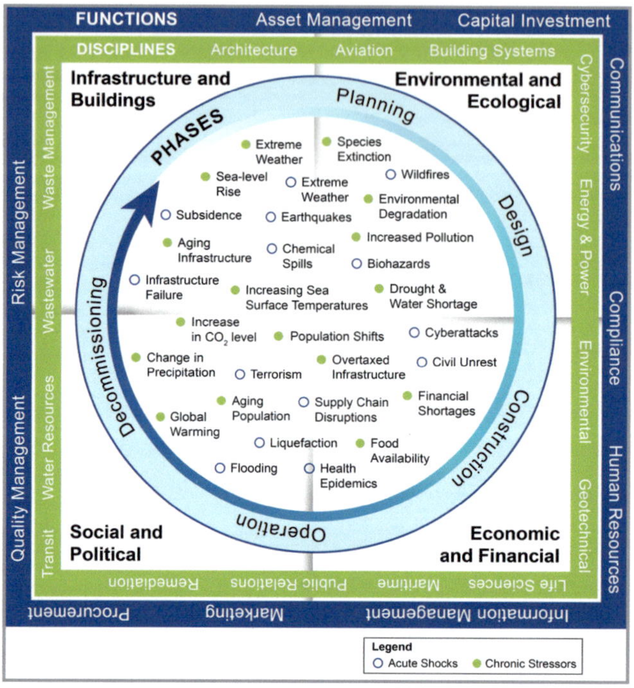

This model facilitates a discipline- and function-mapping exercise that illuminates key partners in completing a comprehensive resilience and sustainability assessment that (1) fully accounts for threats and hazards; and (2) quantifies, qualifies, and addresses impacts across life-cycle phases and natural and built systems. This mapping exercise will highlight points of intersection and illuminate multi-functional, multi-benefit solutions and solution sets.

strategic thinking that builds an integrated, holistic view of the objective, and takes the team beyond business as usual to set the challenge for the organization's, program's, or project's positive legacy.

3. **Plan: Develop an Action Plan.** Meaningful and measurable goals are critical to achieving the vision. Therefore, during this step, the team selects focus areas (for example, energy and greenhouse gas emissions; climate change adaptation; materials, waste management, and recycling; and construction impacts). For each focus area, goals, objectives, and key performance indicators should be established and action plans and initiatives identified and prioritized using a business-case approach. Prioritization models used by asset management practitioners may be helpful here.

4. **Do: Implement the Plan.** During this step, the S&R manager works with the team to develop policies and procedures for implementing the action plans and initiatives, the timeline, and the budget. This includes using existing tools (such as tools developed for information management and asset management), creating new tools as needed, developing a sustainability and resilience management plan, and promoting knowledge transfer across the organization, program, or project.

5. **Check: Monitor the Plan.** During this step, the team should develop evalua-

tion plans to monitor performance against targets. This can be done with tools such as SCADA and BIM and by integrating performance-monitoring requirements with the existing PMIS.

6. **Adjust: Update the Plan.** During this step, the team should evaluate performance against goals, redefine the program baseline, and identify areas for improvement based on past performance while taking into account multiple changing variables. The updated plan should focus on the present but also factor in future uncertainties.

The results of this collaborative effort should be described in a sustainability and resilience management plan or similar document. The action plan and implementation timeline contained therein would outline discrete deliverables such as climate vulnerability assessments, design guidelines, and waste management plans.

The key to success is choosing processes (procedures, tools, and training materials) that are fully integrated in the business management, program management, or project management framework, as appropriate. This includes working with the PMIS team to automate data collection and dashboard development that supports sustainability and resilience goals. It also includes coordinating with risk management, quality management, and health and safety assurance leaders, and integrating best sustainability and resilience practices in the policies, plans, and processes of those functions as appropriate.

11.2.2 Leveraging Interdisciplinary Intersections to Address Economic Roadblocks

A multifaceted risk profile—where risks may be related to, or independent of, climate change—requires a holistic approach from the outset in a project, including innovative, cost-effective solutions that address multiple vulnerabilities while offering co-benefits to a broad range of stakeholders. Unfortunately, as the full range of vulnerabilities and risks are quantified, the cost of the risks increases rapidly, reducing the probability of a sustainable and resilient design becoming a reality.[27] However, by actively engaging experts in an organization, program, or project, teams can apply well-established principles and tools from other disciplines to support priority-setting and give stakeholders and investors data-verified performance proofs to start breaking down financial barriers that impede the expansion of resilient design.

11.3 Summary and a Look Ahead

This chapter focused on how a changing world has modified our perception of what resilience and sustainability are and how to strengthen both concepts by better understanding their interactions, intersections, and overlaps. The S&R manager's role is to connect disciplines and functions to develop a multifaceted impact profile that is quantified and qualified, then work with the team to minimize, mitage, and offset those impacts, by leveraging well-established principles and tools where possible. A robust performance management framework will facilitate a whole-system approach that challenges practitioners to move beyond discipline-centered silos and will foster broader-reaching solutions. This will set the enabling conditions for multifaceted, multi-benefit solutions that increase the probability of holistically resilient and sustainable designs becoming reality.

Of course, innovative solutions and thought leadership will continue to be developed in silos. But it is when we connect thought leaders and innovators that we magnify progress across disciplines. And it is at this point of collaboration that emerging impacts of concern are brought to light. It is said that innovation happens at the intersection of two disciplines, but world-class innovation happens at multi-discipline intersections, and that is what future conditions demand of us.

References

Bradford, Elizabeth. 2021. Breaking Down Silos: Trends in Resilient & Sustainable Infrastructure Planning. Energy Exchange: Connecting the Future. Virtual Conference. August 2, 2021.

Bradford, Elizabeth J., Christopher Corbin, Aaron Hutchins, Mary Jo Kealy, Elizabeth C. Smith, Jonathon Weier. "Making the Case for Increasing Public and Private Investments in Wastewater Treatment: The Importance of Integrating Effects on Natural Capital." 24th Annual Caribbean Water and Wastewater Association Conference and Exhibition. Miami, Florida. October, 2015.

Bradford, Elizabeth J., Michael J. Sarullo, Catherine Zammit. 2017. "Integrating Envision Principles in Program Management for Decision Support." American Society of Civil Engineers International Conference on Sustainable Infrastructure. Brooklyn, New York. October 2017.

Burke, L., K. Reytar, M. Spalding, and A. Perry. 2011. *Reefs at Risk Revisited.* World Resources Institute, Washington, DC. http://www.wri.org/sites/default/files/pdf/reefs_at_risk_revisited.pdf.

Cooley, H., E. Moore, M. Heberger, and L. Allen. 2012. *Social Vulnerability to Climate Change in California*. Pacific Institute. Accessed October 2019. http://www.energy.ca.gov/2012publications/CEC-500-2012-013/CEC-500-2012-013.pdf.

Fritz, Alison and Elizabeth Bradford. *Breaking Down Silos: Trends in Resilient & Sustainable Infrastructure Planning.* American Society of Civil Engineers International Conference for Sustainable Infrastructure. Los Angeles, California. November 2019.

Institute for Sustainable Infrastructure and Zofnass Program for Sustainable Infrastructure. 2018. *Envision Sustainable Infrastructure Framework Version 3*. Washington, D.C.

Institute of Asset Management. *Asset Management – an Anatomy*. Vol. 3, 2015.

Jacobs Engineering Group, Inc. 2016. *Delivering Better Outcomes (Innovative, Sustainable, and Smart)*.

Ocean Tipping Points. 2015. Project Overview. http://oceantippingpoints.org/project-overview.

Rigaud, Kanta K., Alex de Sherbinin, Bryan R. Jones, Jonas Sebastian Bergmann, Viviane Wei Chen Clement, Kayly Jennifer Ober, Jacob Schewe, Susana Adamo, Brent McCusker, Silke Heuser, and Amelia Midgley. 2018. Vol. 2. http://documents.worldbank.org/curated/en/846391522306665751/Main-report.

Solomon, J.D., and Daniel Vallero. 2016. "Communicating Risk and Resiliency: Special Consideration for Rare Events." *The CIP Report*. Center for Infrastructure Protection and Homeland Security, George Mason University.

United Nations Environment Programme. 2014. *Emerging Issues for Small Island Developing States*. Accessed October 2019. http://www.indiaenvironmentportal.org.in/content/393686/emerging-issues-for-small-island-developing-states/.

United Nations. 2019. *Sustainable Development Goals*. Accessed October 2019. https://www.un.org/sustainabledevelopment/sustainable-development-goals/.

World Business Council for Sustainable Development. 2015. *The Business Case for Natural Infrastructure*. https://www.naturalinfrastructureforbusiness.org/wp-content/uploads/2016/02/WBCSD_BusinessCase_jan2016.pdf.

World Commission on Environment and Development. 1987. *Our Common Future*. Oxford: Oxford University Press.

World Economic Forum. 2019. *The Global Risks Report 2019 14th Edition*. Geneva: World Economic Forum. http://www3.weforum.org/docs/WEF_Global_Risks_Report_2019.pdf.

Notes

1 Institute for Sustainable Infrastructure and Zofnass Program for Sustainable Infrastructure [ISI and Zofnass]. 2018. *Envision Sustainable Infrastructure Framework Version 3*. Washington, DC.); World Economic Forum. 2019. *The Global Risks Report 2019, 14th Edition*. Geneva: World Economic Forum. http://www3.weforum.org/docs/WEF_Global_Risks_Report_2019.pdf.

2 World Economic Forum. 2019. *The Global Risks Report 2019*.

3 Burke, L., K. Reytar, M. Spalding, and A. Perry. 2011. *Reefs at Risk Revisited*. World Resources Institute, Washington, DC. http://www.wri.org/sites/default/files/pdf/reefs_at_risk_revisited.pdf; Bradford, Elizabeth J., Christopher Corbin, Aaron Hutchins, Mary Jo Kealy, Elizabeth C. Smith, Jonathon Weier. 2015. "Making the Case for Increasing Public and Private Investments in Wastewater Treatment: The Importance of Integrating Effects on Natural Capital." 24th Annual Caribbean Water and Wastewater Association Conference and Exhibition. Miami, Florida. October, 2015.

4 Solomon, J.D., and Daniel Vallero. 2016. "Communicating Risk and Resiliency: Special Consideration for Rare Events." *The CIP Report*. Center for Infrastructure Protection and Homeland Security, George Mason University.

5 ISI and Zofnass. 2018.

6 *Ibid*.

7 United Nations Environment Programme (UNEP). 2014. *Emerging Issues for Small Island Developing States.* Accessed October 2019. http://www.indiaenvironmentportal.org.in/content/393686/emerging-issues-for-small-island-developing-states/.

8 World Economic Forum. 2019. *The Global Risks Report 2019.*

9 Rigaud, Kanta K., Alex de Sherbinin, Bryan R. Jones, Jonas Sebastian Bergmann, Viviane Wei Chen Clement, Kayly Jennifer Ober, Jacob Schewe, Susana Adamo, Brent McCusker, Silke Heuser, and Amelia Midgley. 2018. Vol. 2. http://documents.worldbank.org/curated/en/846391522306665751/Main-report.

10 Ocean Tipping Points. 2015. Project Overview. http://oceantippingpoints.org/project-overview.

11 Burke et al., 2011, *Reefs at Risk Revisited*; Bradford et al. 2015, "Making the Case for Increasing Public and Private Investments in Wastewater Treatment"; *italics* represent threats and hazards outlined on Figure 2.

12 Ocean Tipping Points. 2015.

13 UNEP, 2014, *Emerging Issues for Small Island Developing States*; Bradford et al. 2015.

14 United Nations Environment Programme. 2014. *Emerging issues for Small Island Developing States.* Accessed October 2019. http://www.indiaenvironmentportal.org.in/content/393686/emerging-issues-for-small-island-developing-states/.

15 United Nations. 2019. *Sustainable Development Goals.* Accessed October 2019. https://www.un.org/sustainabledevelopment/sustainable-development-goals/.

16 Cooley, H., E. Moore, M. Heberger, and L. Allen. 2012. *Social Vulnerability to Climate Change in California.* Pacific Institute. Accessed October 2019. http://www.energy.ca.gov/2012publications/CEC-500-2012-013/CEC-500-2012-013.pdf.

17 World Commission on Environment and Development. 1987. *Our Common Future.* Oxford: Oxford University Press, p. 27.

18 Bradford, Elizabeth J., Michael J. Sarullo, Catherine Zammit. 2017. "Integrating Envision Principles in Program Management for Decision Support." American Society of Civil Engineers International Conference on Sustainable Infrastructure. Brooklyn, New York. October 2017.

19 World Business Council for Sustainable Development. 2015. *The Business Case for Natural Infrastructure.* https://www.naturalinfrastructureforbusiness.org/wp-content/uploads/2016/02/WBCSD_BusinessCase_jan2016.pdf.

20 Fritz, Alison, and Elizabeth Bradford. 2019. *Breaking Down Silos: Trends in Resilient & Sustainable Infrastructure Planning.* American Society of Civil Engineers International Conference for Sustainable Infrastructure. Los Angeles, California. November.

21 Institute of Asset Management. *Asset Management—An Anatomy*, vol. 3, 2015.

22 Fritz and Bradford. 2019, *Breaking Down Silos.*

23 *Ibid.*

24 Jacobs Engineering Group, Inc. 2016. *Delivering Better Outcomes (Innovative, Sustainable, and Smart).*

25 Jacobs Engineering Group, Inc. 2016. *Delivering Better Outcomes.*

26 *Ibid.*

27 Fritz and Bradford. 2019, *Breaking Down Silos.*

PART IV

Case Studies

CASE STUDY 1

Whitney Museum Case Study

Amy Hwang

Introduction

On October 29, 2012, Superstorm Sandy hit the New York metropolitan area with a record-breaking storm surge at high tide, bringing with it 12 inches of precipitation. It resulted in a 6-day power outage in lower Manhattan, widespread coastal flooding, damage to hundreds of buildings, and dozens of deaths despite mandatory evacuation of those areas. In the intervening years since the storm, New York City created the Mayor's Office of Resiliency, embarked on resiliency studies for various neighborhoods, and continues to study the effect of climate change on the city via the New York City Panel on Climate Change.[1]

Figure 1. View of building from west with mobile flood barrier trial deployment

The impact of Sandy on the new Whitney Museum building, which was under construction at the time of the storm, is relevant in the resiliency discussion as a unique case study on the implications to the structural design when minimum code requirements based on historical data are used without considering the future escalation of extreme events.

Key Takeaways

- In accordance with the 2008 NYC building code at the time of design, the new Whitney building, adjacent to the Hudson River, had been designed to the 100-year flood level based on 2007 FIRM data. In 2012, Superstorm Sandy exceeded the design flood level by 4 feet while the building was under construction.
- The Whitney engaged flood mitigation consultants and a research lab from Germany to determine a new design water level based on historic flood data and statistical analysis. Together with the design team, they settled on a new design flood level 9 feet higher than the original requirement and which accounts for the possibility of even higher levels.
- The new flood protection system consisted of flood gates, flood doors, and mobile flood barriers in order to accommodate the structure already in place, as well as preserve Renzo Piano's original architectural design.
- The flood mitigation strategy accommodates future rising flood levels and serves as an example of designing for future risk rather than present-day code minimums.

Original Building Design

The new Whitney Museum building is located on the western edge of Manhattan separated from the Hudson River by only the West Side Highway. New buildings in high-risk flood zones are typically designed using the FEMA Flood Insurance Rate Map (FIRM) flood level with a 1% chance of yearly exceedance—in other words, a flooding event that has an average recurrence interval of 100 years. The Whitney site was located partially in a Flood Hazard A Zone but outside the area of Moderate Wave Action. Therefore, it was designed using the 100-year flood level based on 2007 FIRM data of 8.9 ft. NAVD88 (referring to the North American Vertical Datum of 1988, issued by the National Geodetic Survey in 1991 and used as a standard reference for vertical measurements), with no additional high-velocity wave or impact considerations, per the 2008 NYC Building Code.[2]

The first floor of the museum was set above the design flood level at 11.65 ft. NAVD88 to permit the architect, Renzo Piano, to use pre-tensioned cable glass curtain walls and hang precast concrete façade panels at the lobby—elements that would be damaged if floating debris or waves were to impact them. The 30-ft. deep cellar, housing the museum's main mechanical equipment, was designed to be dry-floodproofed.

At the recommendation of the geotechnical engineers, URS, the 100-year flood level was used in place of the design groundwater level. Per the building code, the load factor applied to groundwater pressure (H) for strength design purposes is 1.6 while the load factor on flood loads (Fa) in non-coastal A Zones is 1.0. Therefore, the foundation walls, concrete mat slab, and caisson foundations were all conservatively designed to withstand the hydrostatic water pressure from a 100-year flood level with 1.6 load factor, propitiously providing reserve capacity. It is worth noting that had the building been in a coastal A Zone (within the area of Moderate Wave Action), the code-mandated design load factor would have been 2.0.

Flood Mitigation

When Superstorm Sandy hit in 2012, the Whitney was midway through construction, with the steel superstructure topped out to the roof and the concrete slabs following close behind. The storm surge inundated the entire site, submerging the entire cellar and destroying all the large mechanical equipment that had been installed early. The storm surge measured at the site was approximately 12.95 ft NAVD88, exceeding the FEMA FIRM 100-year flood level at the time by 4 feet and the FEMA FIRM 500-year flood level by 3 feet—higher than any United States Geological Survey (USGS)-reported surge level to date in the area.[3]

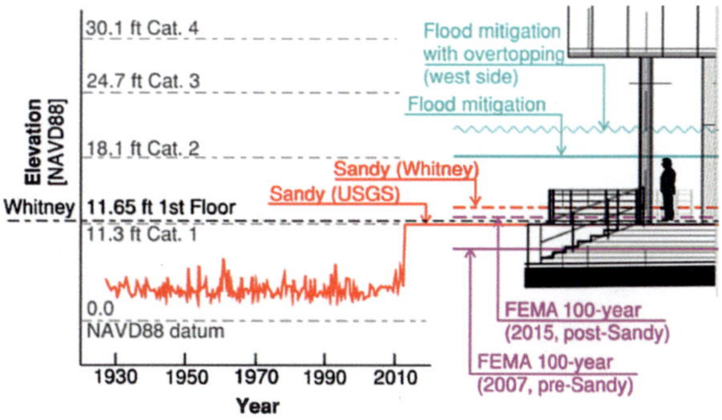

Figure 2. Composite of historical data per Franzius Institut report and Whitney lobby section, NAVD88 Datum.

Flood Mitigation Design Phase

Unwilling to risk damage to their art collections and operations once occupied, and unable to wait for updated FEMA data, the Whitney took the proactive step of engaging WTM Engineers, structural engineers with expertise in flood mitigation based in Hamburg, Germany, a city that has adapted to accommodate frequent flooding. Teamed with the research arm at the Ludwig-Franzius Institute for Hydraulic, Estuarine and Coastal Engineering at the Leibniz University in Hannover in Germany, the consultants were tasked with proposing flood mitigation solutions appropriate for the site and building.

The Ludwig-Franzius Institute carried out extensive scientific studies based on available design methods developed in Germany and historic data from measurements taken at the Battery (about 3 miles downstream from the site) to determine future design water levels. Due to time demands of the stalled construction site awaiting resolution of the flood study, a location-specific hydro-numerical model, which would have given a more accurate understanding of the expected levels at the Whitney location itself, was not performed. The results of the studies were reviewed with the design team and a design water level (DWL) of 18.15 ft NAVD88 was settled upon. This elevation, resembling the predicted level for a Category 2 hurricane, was primarily constrained by the capacity limit of the in-place structure. Silman had determined the reserve capacity of each affected structural and foundation element—including the caissons and tie-downs—without load factors and identified that the foundation walls and first floor framing were the limiting elements. This constraining elevation was judged to be reasonable considering the widespread damage to the entire neighborhood with a flood level of this magnitude. Using this DWL, WTM Engineers then developed the flood protection strategies for implementation by the design team, while Silman determined the impact on and necessary strengthening for the building structure.

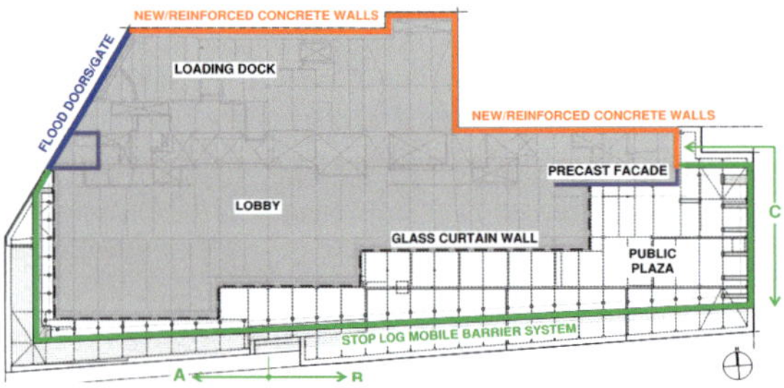

Figure 3. Flood mitigation scheme diagram

Figure 4. 27 ft-long Loading Dock flood door

Floodproofing Measures

Generally, floodproofing measures can be either permanent or temporary. As the impact on the architectural design had to be limited, temporary or hidden systems were chosen. Three different methods were used around the site to work around the physical constraints and different detailing conditions (see Figure 3).

Due to the lower elevation of the entrances on the west side of the building, which would result in more frequent flooding of these areas, flood doors that could be operated quickly and easily were the preferred option for loading dock and staff entrances (Figure 4). The existing steel columns behind the precast concrete panels were encased in concrete piers, and the steel columns and overhead beams were reinforced to support the significant load from the flood doors (Figure 5). Kryton KIM hydrophilic integral waterproofing concrete admixture was used due to limited access for traditional waterproofing.

Along the northern perimeter, the building abuts the neighboring buildings. The concrete masonry unit exterior infill walls were already in place, and space was constrained since the adjacent loading dock and egress stair widths could not be reduced. The solution was to add 6-inch concrete walls between the steel columns up to the second-floor steel using flowable concrete poured in lifts (Figure 6). The steel was reinforced for the eccentric loading. Where the wall is blocked by an adjacent building, no debris impact load was considered.

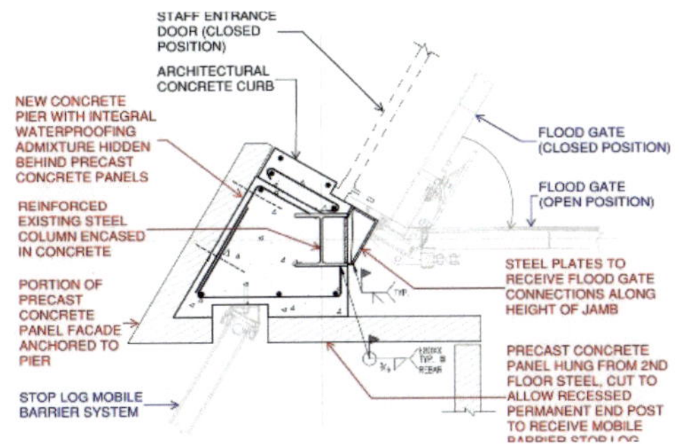

Figure 5. Plan detail at flood door attachment to structure

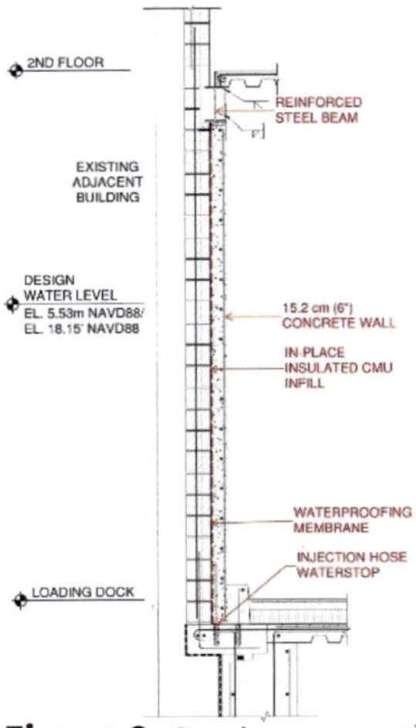

Figure 6. Section at north wall

For the higher and more visible public plaza area, various temporary options were evaluated. Based on its location in the dense Meatpacking District and adjacent to the West Side Highway, the system was expected to withstand impact by floating debris (cars, etc.) despite not being required by the code for non-coastal A Flood Zones. Also, a watertight, robust connection to the overall building waterproofing system was critical. Therefore, systems like Aquafence or Quick Dam were not deemed appropriate for this application. Instead, a custom aluminum stop log system consisting of vertical posts and horizontal infill logs was selected. These would only be deployed when early storm warnings deemed it necessary. Standard products could not be used since the structural work was mostly completed, requiring the posts to be spaced further apart than in a typical application in order to transfer overturning loads to the in-place first floor steel beams. A custom solution also allowed custom details that would fit in with the architecture.

Along the perimeter of the public plaza, the stop log system (SLS) was split into three sections: A (west and southwest), B (southeast) and C (east)—see Figure 3. The maximum height of these sections was based on two considerations. On one hand, the building had to be protected for the established DWL plus wave action, which is mostly expected on the west side of the building located closest to the Hudson River. On the other hand, this DWL is close to the maximum water level the as-built structure could withstand—meaning that if higher levels should occur, a quick flooding of the building is required to ensure the overall stability of the building. In this case, the SLS would function as a weir and the structure would be able to withstand a higher future

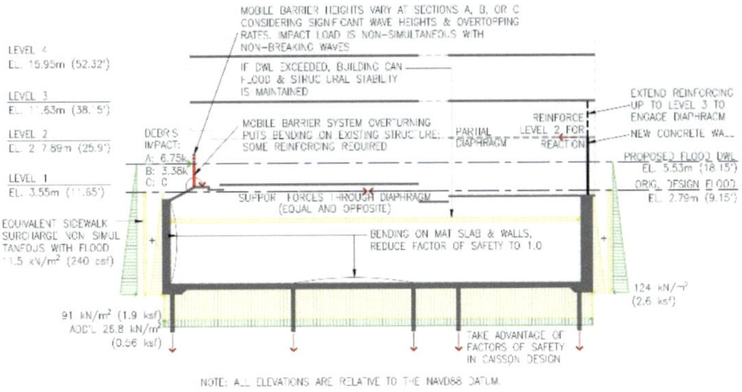

Figure 7. Flood load diagrams showing load assumptions

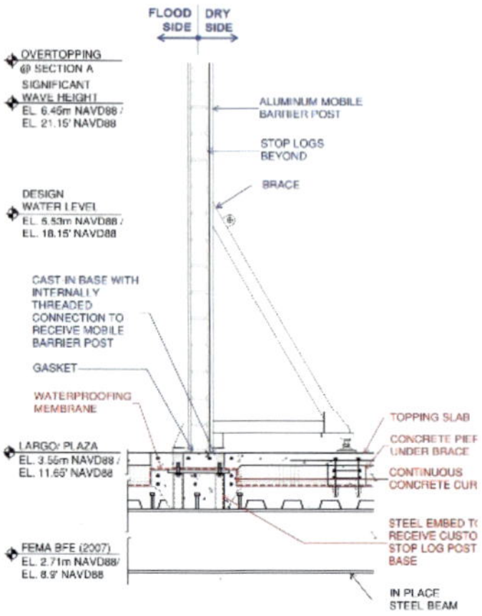

Figure 8. Structural support for stop log system at section A

Figure 9. Trial deployment of mobile barrier posts without logs

Figure 10. Preliminary FIRM map (Dec 2013) of area, NAVD88 datum.

DWL should it be needed (Figure 7). Studies for different elevations of sections A, B, and C relating maximum overtopping rates to the outer water level at complete flooding of the cellar were performed and the ideal section heights were determined iteratively. Section A—the closest to the river—would extend up to the DWL plus the significant (non-breaking) wave height and the additional overtopping height to control the water entry rate, and would be designed for the impact of a large debris object (Figure 8). Section B—facing the narrow Gansevoort Street—would extend to the DWL plus the significant wave height, and would be designed for the impact of a smaller object since the approach speed of a piece of debris is limited by the street width. Section C—the farthest from the river—would extend to only the DWL with no debris impact potential.

To provide a continuous watertight tub and to transfer the flood loads to the structural steel framing, a series of hidden concrete beams, piers, and custom embedded steel inserts were added underneath the public plaza topping slab along the entire 460-ft. mobile barrier system perimeter (Figures 8 and 9). The sloped topping slab and structural slab required extensive reworking and coordination to accommodate the change in drainage created by the addition of the concrete barrier within this interstitial space, and in order to stay within allowances for the topping weight.

Additionally, all points of water entry were reviewed and strengthened.

Current Code and Future Implications

In 2015, NYC formally adopted the Preliminary FIRM maps dated December 2013 into the 2014 NYC Building Code. The updated version increased the 100-year base flood level for the western portion of the Whitney site to 11 ft NAVD88, still landward of the Limit of Moderate Wave Action line (Figure 10). The new base flood elevation (BFE) is 2.1 ft above the previous level, but the addition of 1 foot of required freeboard height per the clarified code requirements would bring the design flood elevation (DFE) to 12 ft NAVD88.[4] This is comparable to the site-reported Sandy surge level. However, by placing it outside the Area of Moderate Wave Action, were it to be designed now, the building would still be categorized in non-coastal Flood Zone A, where high-velocity waves and impact from floating debris are not required to be considered. For a further point of comparison, if the current non-mandatory NYC Climate Resiliency Guidelines issued by the Mayor's Office of Resiliency were also applied to the museum, as a noncritical facility with an expected 100-year life span, another 3 feet of height would be recommended for future sea level rise,[5] yielding a DFE of 15 ft NAVD88.

New Preliminary FIRM maps were originally expected in 2021 and were anticipated to include the effects of the "growing risk of climate change and sea level rise."[6] They were expected to reflect the work presented in the 2015 NPCC Report, which projected the effect of climate change on the NYC area through 2100, reporting increasing annual precipitation, accelerating sea level rise in the years to come, and thus increased coastal flooding as a result.[7] No new timeline has been suggested in the subsequent 2019 NPCC Report.[8] Given that the area of inundation is expected to increase for the 100-year floodplain in coming years, the Limit of Moderate Wave Action will likely shift inland in the future. If it does, the requirements for increased wave effects and debris impact that was accounted for in the flood mitigation retrofit would then become code-mandated for the area, and the load factor on the flood loads would double. Buildings that were designed satisfying present-day building codes would find their flood protection schemes outdated.

Conclusions

By taking advantage of reserve capacity, the Whitney Museum structure was able to accommodate a flood retrofit that far exceeds current minimum code requirements. Moreover, by electing to retrofit to a standard that considers future increased risk of flooding with a flood mitigation strategy deployed continuously around the perimeter, the Whitney Museum is well prepared to meet the expected increase in coastal resiliency demand irrespective of the future direction of the FIRM maps – an option that building owners, developers, managers, and designers of buildings situated on the border of high-risk flood zones should perhaps also consider in order to stay resilient in the years to come.

Notes

1 City of New York Mayor's Office of Recovery & Resiliency, "OneNYC 2018 Progress Report," City of New York, New York, 2018, pp. 82–89.

2 NYC Department of Buildings, 2008 New York City Building Code, New York: City of New York, 2008, Appendix G.

3 Franzius-Institute, "Scientific assessment of the flood risk profile and future design storm surge levels of Gansevoort project of Whitney Museum of American Art, NYC, USA Phase 1: Empirical FAAM design analysis," Franzius-Institute, Hannover, Germany, 2013.

4 NYC Department of Buildings, 2014 New York City Building Code, New York: City of New York, 2014, Appendix G.

5 City of New York Mayor's Office of Resiliency, "Climate Resiliency Design Guidelines," Sept. 2020.

6 City of New York Mayor's Office of Recovery & Resiliency, "OneNYC 2018 Progress Report."

7 New York City Panel on Climate Change (NPCC), "New York City Panel on Climate Change 2015 Report Conclusions and Recommendations," *Annals of the New York Academy of Sciences*, New York, 2015, pp. 107–108.

8 New York City Panel on Climate Change (NPCC), "New York City Panel on Climate Change 2019 Report Conclusions and Recommendations," *Annals of the New York Academy of Sciences*, New York, 2019

CASE STUDY 2

US 34–Big Thompson Canyon Flood Recovery Project

Kelly C. Maiorana

Case Study

The lifecycle of the Big Thompson Canyon Flood Recovery Project began in 2013 when floodwaters rapidly rose along the North Front Range and flows within the Big Thompson River became dangerously high. With little warning, water careened down the steep walls of the US 34 canyon, causing the river to overtake its banks and claim roads and bridges through a ferocious show of nature's force. Thousands of residents were stranded without adequate supplies of food, water, and fuel, and one of the largest civilian airlifts in U.S. history was conducted, second only to Hurricane Katrina (Figure 1, next page).

Governor John Hickenlooper charged the Colorado Department of Transportation (CDOT) with two critical challenges. First, to stabilize the canyon for emergency use and to restore essential traffic by Thanksgiving of 2013, and second, to "build back stronger." The roads were opened to essential traffic on November 26, less than 3 months after the shock of the flood. Answering the second challenge introduced CDOT and its partners to the concept of resiliency and started the State of Colorado down a path of incorporating resiliency into everyday operations.

Figure 1. The 2013 flood damage took out entire portions of the US 34 roadway, leaving residents stranded without access to clean water or food.

Resiliency occurs in a cycle of four stages:

1. Readiness or preparedness for the event;
2. Incident management, which includes the event itself and life-saving activities, such as the airlift evacuation that followed the 2013 floods;
3. Exigent emergency repairs to stabilize conditions and restore emergency access and essential transportation; and
4. Planning and design that involves evaluating risk and resilience (RnR), considers alternative levels of protection to meet critical needs, and is feasible, affordable, and practical to implement.

Completing the fourth stage sets the community's readiness for the next event, thus starting the cycle over again. Implementing resiliency across a region, a project portfolio, or a single project reduces vulnerability to risks and in so doing reduces disaster impacts and increases the return on resilience investments. In addition, resilient infrastructure enables communities to rebound faster and adjust to the new post-event conditions more quickly when a disaster happens.

There is perhaps no better example of the resilience lifecycle than the US 34–Big Thompson Canyon in Loveland, Colorado, because in our lifetimes we have seen two revolutions of the cycle. The following case study on the resilient reconstruction of the US 34 includes background information on the canyon, its flood history, and CDOT's stakeholder process. The goal of the stakeholder process is to produce durable benefits for people and communities, to enhance environmental sustainability, and to restart the economy of Colorado's North Front Range.

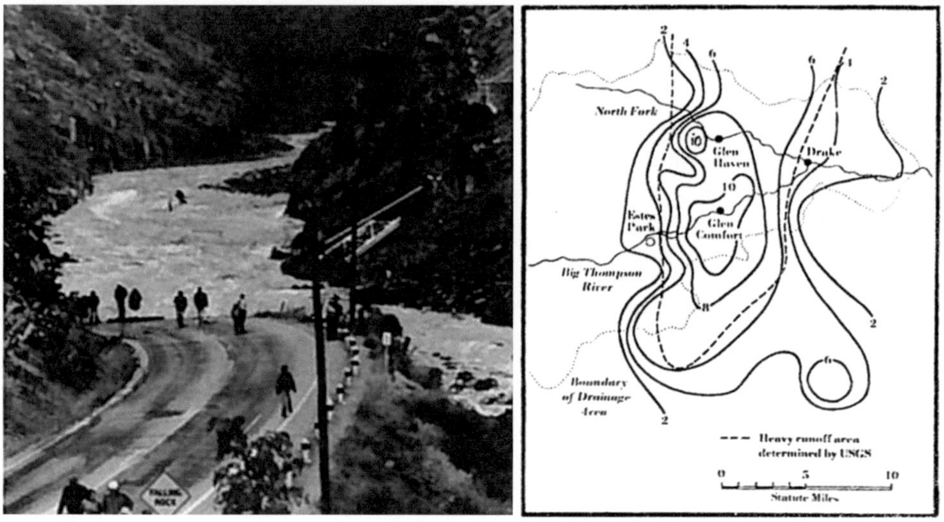

Figure 2. A photo showing a washed-out section of US 34 resulting from the 1976 flood and a map showing the rainfall and heavy runoff area. In some areas it rained more than 10 inches in the span of 4 hours.

The 1976 Flood

The 2013 flood was not the first major flood event the canyon has seen. On the eve of Colorado's Centennial, July 31, 1976, the Big Thompson Flood became the deadliest flash flood in Colorado's recorded history. Cool air coming from Canada converged with moist air from the southwest, creating a thunderstorm directly over Big Thompson Canyon, stalled by the high altitude of the terrain. Beginning at 6:30 p.m., 2 to 4 inches of rain fell per hour. In some areas (see Figure 2) over 10 inches of rain fell during a four-hour period, which is over two-thirds of Colorado's annual precipitation. By 9:00 p.m., the river continued to swell, and US 34 through the canyon began to wash out. Cars and trailers were swept into the river, piling up behind bridges and eventually causing them to fail. The resulting surge of water swept houses and motels into the river.

Due to the canyon's direct access to Estes Park, one of two gateways to Rocky Mountain National Park, the summer season attracted tourists from across the country and the world. Approximately 2,000 tourists, residents, and campers were in the canyon on the evening of the flood in 1976. A total of 144 people perished in the flood, and 150 more were injured.

Those who survived the flood shared their accounts of the night with Dr. Eve Gruntfest, then a graduate student at the University of Colorado at Boulder. One family was lucky enough to escape the flood waters by climbing the mountainside in 1976, and they remained connected to the canyon through the 2013 flood. Lynda Wright narrated an account of her family's escape from the 1976 flood in the spring following the event. She and her late husband, their two daughters (then 9 and 5 years old), and her husband's parents all managed to avoid the torrents of floodwater by climbing the hillside behind their cabin named "Linger Longer." The cabin was built in 1912 and continues to serve the family as a summer retreat, miraculously sustaining only reparable damages during both the 1976 and 2013 floods. In the 1976 flood, over 300 homes and 50 businesses were completely destroyed.

Figure 3. Climb to Safety signs were implemented in canyons throughout Colorado as a result of Dr. Gruntfest's research.

The 1976 flood caused about $35 million in direct damage ($145 million in 2013 dollars). The recovery funds totaled $56 million ($230 million in 2013 dollars). The highway was rebuilt in the third and fourth stages of the resilience life-cycle. When rebuilding US 34, the major priorities during the 1976 recovery were floodplain management and managing the human impacts of the flood, such as providing mental health services, improving emergency notification systems, using family grant programs and Small Business Administration (SBA) loans, and more. During the rebuilding of homes, techniques of floodplain management were used to reduce the "future hazard potential." Additionally, US 34 disaster recovery objectives included highway safety, constructing a road with the ability to withstand a "100-year" (rate of return) flood event, maintaining the scenic drive, and designing canyon bridges to avoid clogs and damage by water-borne debris.

During construction, water from the river was diverted to Pinewood Lake, and it was not until June 1977 that water was released back into the Big Thompson River. Twenty-three new bridges were constructed to serve residential communities. Following the recommended recovery objectives, the bridges were designed to be pushed off

Figure 4. A portion of US 34 that was destroyed by flood waters.

their footings and break up or sink in the event of another flood. The section of the canyon known as "The Narrows" was elevated using concrete slabs and steel beams to try and prevent the undercutting that eroded the roadbed during prior floods.

After the 1976 repairs, US 34 remained close to its original alignment, but the roadway was designed to "prevent inundation of the highway during the 100-year flood event." In addition to these engineering design decisions, Larimer County adopted floodplain regulations that restricted construction in the floodway and required structures in the fringe to be raised above a 100-year flood elevation. The roadway infrastructure was built to be capable of handling the stress of an event that at the time was considered a 100-year flood. Resiliency may not have been the concept at the forefront of transportation planning during this time, and yet a large effort was made to protect the roadway and community to reduce losses in the event of future floods

2013 Flood

With a more prepared roadway after the 1976 US 34 rebuilding, the community was tested again in September 2013 as extreme flood waters washed away homes, cars, and sections of US 34 through the Big Thompson Canyon between the city of Loveland and the town of Estes Park. Total event rainfall reached up to 20 inches in some areas. Eight lives were lost in the 2013 flood. Aerial and ground evacuations were necessary to safely remove residents in the canyon who were effectively trapped—stranded after the flooding destroyed access roads, preventing them from leaving and emergency services from entering.

Statewide, the seven-day rain event resulted in approximately $750 million in damages to federally supported "on-system" roadway facilities, caused 486 miles of state highways to close, damaged over 200 bridges and culverts, required the removal of 120,000 cubic yards of debris, and caused hundreds of millions of dollars in damage to local roadways maintained by cities and towns. In total, US 34 through Big Thompson Canyon suffered 23 miles of severe roadway damages, including 8.5 miles of full, impassable roadway loss, and 2,963 post-flood evacuations, 2,256 of which were aerial evacuations. The flooding left US 34 road users through the canyon with limited or no access to the canyon for 2.5 months, despite Herculean efforts working night and day on emergency repairs.

The Colorado Department of Transportation (CDOT) initiated a program for emergency response (ER) to reopen the highway to traffic by the end of 2013 to fulfill a promise made to the citizens of Colorado by then-Governor John Hickenlooper. The ER effort in the canyon cost $38,728,001 and included removing and replacing overtopped asphalt with temporary paving surface, replacing undermined shoulders, removing debris, rebuilding embankments and embankment slopes, placing temporary riprap, and replacing guardrail to restore access for residents and businesses through the canyon. All roads were open on Thursday, November 26, 2013, a testament to the quick response of CDOT and its team of emergency repair workers.

2013 Permanent Repairs

The US 34 Big Thompson Canyon Flood Recovery Permanent Repairs Project followed the ER response with the goal of making long-term improvements to US 34 by permanently repairing the flood damage and providing a safe and resilient highway for the future. After seeing many of the same sections of roadway from the 1976 flood get washed away by the floodwaters in 2013, CDOT decided to take a new approach to permanent repairs, based on improvements in data, accurate modeling software, and engineering science. A first-of-its-kind hydrology study spanning flood-affected basins of the South Platte River was conducted to provide updated flow rates and flood frequencies to inform the state's flood recovery design efforts. For Big Thompson Canyon, the 2013 flood event was characterized as a "100 year" event with "500 year" event surges for some segments of US 34. This information, combined with using complex hydraulic modeling software not available during the 1976 flood recovery, was a priority to ensure the design team was working with the best available information.

Figure 5. Residents stranded in the canyon without access had to be airlifted out post-flood.

In order to achieve the goal for a resilient corridor, CDOT partnered with contractor Kiewit and prime design consultant Jacobs Engineering through the alternative delivery method of Construction Manager/General Contractor (CM/GC). The team engaged in innovative design engineering, contracting, and construction. As well, there was extensive stakeholder coordination and partnership to help define and build CDOT's vision of resiliency for the US 34 Big Thompson Canyon.

The first goal of the 2013 permanent repairs was to fulfill Governor Hickenlooper's assurance to Coloradans that all highways damaged by the flood would be "built back better than they were before."

Defining the Resilient Vision

It became apparent to the project team that flood-proofing the highway was not realistic. With the highway sharing the narrow canyon with the Big Thompson River, using more stringent design standards was not feasible, and even the estimated cost to provide full, 100-year event protection for the corridor far surpassed the available program budget. CDOT and the project team began to consider where the need for infrastructure was most critical in order to maximize the benefits while minimizing the monetary costs. They began to consider how they could make the US 34 corridor truly resilient.

The answer to this question revealed itself as the team looked back at the 2013 ER phase: almost 3,000 evacuations were conducted, the majority of them airlift, and for 2.5 months residents were without access in or out of the canyon despite round-the-clock efforts. Rebuilding a more resilient US 34 highway would mean connecting as many people as possible to an available emergency access route out of the canyon after a disaster. It would mean reducing emergency response time to hours or days, not months. And specifically, it would mean finding ways to provide—for as many residents as possible—a drivable surface at least 15 feet wide that would remain intact after an event similar to the 2013 flood.

Resiliency Innovations

In order to find ways to maximize a resilient infrastructure within the project budget, the project management created a "challenge team." This team consisted of construction experts from CDOT, Kiewit, and Jacobs, as well as other public- and private-sector partners not involved with the project on a day-to-day basis. They offered objective ideas on how the scope of work could be completed and meet the resiliency goals, given the budget and other constraints.

The challenge team focused on realistic resiliency measures that could ensure that a 15-foot-wide section of drivable roadway surface remained through most of the canyon for emergency access after a similar future flood event, thus making the roadway more prepared for future floods. The challenge team developed several innovative solutions for damage-prone areas that would have otherwise been too costly or schedule-prohibitive to implement.

With limited available right-of-way and access, environmental concerns, economic impact considerations, numerous stakeholders, and the use of federal dollars, the project team was constantly challenged to do things in new and better ways to minimize impacts and maximize results.

Hydrology data from the two floods was reviewed to identify damaged road overlaps from the 1976 and 2013 events. Resiliency measures were implemented in those overlapping damage areas to improve long-term public safety.

They included:

- Using a technique called soil-cement mixing, which involved excavating deep beneath the surface to retrieve natural bedrock material and mixing it with concrete. Crews used this mixture to create a 15-foot wide section of man-made bedrock beneath the roadway, which would ensure that the river could not erode the highway from below. Using native materials allowed crews to mix it in place, reducing hauling costs while providing a stronger, more resilient roadway base. Soil-cement mixing also eliminated the need to construct major roadway embankment protections, reducing construction impact to the river.
- In order to move the roadway alignment farther from the river and place a minimum 15-foot-wide section onto solid bedrock, rock faces were cut into, where feasible. More than 75% of the excavated material removed during drill

[Left] Figure 6. Trucks pouring concrete into an excavated section to be mixed with natural material and adhere to the canyon bedrock, forming a 15-foot drivable surface. **[Right] Figure 7.** Equipment working to realign the highway onto bedrock through a severely damaged section of highway. By shifting the road onto bedrock, the road is protected from future floods and the floodplain is expanded.

and shoot operations was processed and recycled into four different sizes of riprap and crushed road base material.

- The project excavated a five-mile stretch of US 34 and installed layers of riprap from 10 feet below the riverbed to the roadway surface, creating a more solid roadway foundation. In areas of extreme scour and potential damage, partially grouted riprap, or matrix riprap, was used. Matrix riprap partially grouts smaller-diameter riprap sections, forming larger masses that behave like 6- to 8-foot boulders that would be costly to manipulate and could not be created onsite. These large riprap sections were placed under the roadway prism to reinforce it and to secure a 15-foot corridor for post-flood access, minimizing the construction impact to the river, and allowing for increased floodplain conveyance in the event of another historic flood.
- Two new bridges built in the canyon's horseshoe curve—the most severely damaged section of roadway in both the 1976 and 2013 floods—shifted the roadway to the inside of the bend, combining the roadway and river as a system and allowing room for the river to flow along a more natural floodplain through the canyon.

Figures 8 and 9. Instead of constructing expensive protections for the highway in its existing alignment on the outside bend, the project team removed the highway from the threat of the river and gave the floodplain the room it needs.

Navigating Funding Eligibility

The permanent repair effort on US 34 through Big Thompson Canyon had a $280 million program budget in Federal Highways Administrations Emergency Relief (FHWA ER) funds specially appropriated by Congress. The funding came with tight eligibility constraints that limited the use of funds to the reconstruction and protection of the federal asset, the US 34 roadway. Eligibility requirements are necessary to ensure funds are appropriately used while allowing for occasional innovative scope elements.

Flood recovery funding through the FHWA ER program can be legally used in three ways: First, a replace-in-kind option that puts back what was historically there; second, an option that replaces infrastructure and brings outdated elements to a current accepted standard; and third, as a betterment, one that uses a Risk and Resiliency (RnR) benefit-cost analysis to compare the initial cost of an improvement against the cost in damages anticipated to the US 34 roadway if that improvement were not constructed.

Building the highway segments with resiliency required the consideration of the river, residents, and the canyon landscape in addition to the highway. This holistic approach sometimes required use of funds for scope not normally considered within the purview of the federal funds. The CDOT team worked with its FHWA partners to use the RnR analysis to prove that the resiliency scope elements were eligible under the federal program.

For example, RnR analyses were used to prove the eligibility of upsizing five local access structures crossing the Big Thompson River. While the structures were not federal assets, upsizing them to 100-year capacity proved to be less expensive than the cost in damages that the highway would potentially incur if the structure were built with its pre-flood conveyance.

Figures 10 and 11. Local access structures like this one were reconstructed with increased flow conveyance, allowing for better passage of debris during floods and reducing the risk of damage to all assets in the canyon.

[Left] Figure 12. Riparian vegetation making a comeback as the project's rehabilitation work gives the Big Thompson River a chance to heal from the flood. **[Right] Figure 13.** A project goal was to maintain the aesthetic condition of the canyon; many of the resiliencies and protections are buried beneath the roadway.

River and Stakeholder Partners

In conjunction with the restoration of US 34, the team also reshaped the river channel and developed long-lasting resiliency measures within the river so US 34 and the Big Thompson River can function as a system, rather than clash with each other, in the event of a future flood.

A key aspect to the success of this project was stakeholder coordination, as there were more than a dozen federal, state, and local agencies with a vested interest in the project. Monthly stakeholder meetings allowed CDOT, Kiewit, and Jacobs Engineering to meet with representatives from the U.S. Forest Service, Bureau of Reclamation, Army Corps of Engineers, U.S. Fish and Wildlife Service, Colorado Parks and Wildlife, Larimer County, the Town of Estes Park, and the City of Loveland.

The project team also worked closely with representatives from the Big Thompson Watershed Coalition, a collaborative, nonprofit organization formed in the wake of the 2013 flood, to restore the Big Thompson River corridor. This group used local resources and grant programs to make additional stream restorations beyond the scope of the emergency relief funds. The coalition brought in crews under a separate contract to help reshape and restore the river channel in various locations.

The efficient coordination between the numerous agencies and organizations in the project helped canyon residents who endured stress and disruption in September 2013 return to their normal routines.

Permitted Closure

Not unlike the recovery effort after the 1976 flood, the project team was sensitive to the area's tourism industry, including the 3 million annual visitors to Rocky Mountain National Park, many of whom use US 34 and Big Thompson Canyon. The team developed a construction approach to complete most of the work during two off-season closures, limiting traffic to only canyon residents from October 2016 to

May 2017 and from October 2017 to May 2018. Crews worked 7 days a week, day and night, to reopen the canyon before Memorial Day in 2017 and 2018, for summertime visitors.

To manage traffic volumes during these wintertime closures, a one-of-a-kind vehicle permitting process allowed canyon residents to travel through the canyon at specific times each day while detouring through traffic around the canyon. To distinguish canyon residents from those using US 34 for through travel, rear-view mirror placards were distributed to canyon residents, business owners, and emergency service providers to help identify vehicles allowed to travel through the canyon at specified times.

The off-season closures allowed a high-quality final product, but the downside to restricting access to the canyon was the fact that more than 1,000 year-round canyon residents rely on US 34 to get to their homes, and another 6,000 commuters use the highway daily. Working with these stakeholders required an active outreach program and a traffic management approach never previously attempted by CDOT. The project team supplemented the permitted travel times with a Resident Scheduling Service (RSS) that enabled residents to schedule escorted, safe passage through major work zones outside the permitted times.

As work activities would allow, Kiewit opened access for residents beyond the permitted travel times, often on weekends and during extended holiday periods like Thanksgiving weekend and the week between Christmas and New Year's Day.

In total, more than 11,000 calls were fielded through a project hotline, more than 4,000 emails responded to, and 3,600 vehicle permits accommodated 5,095 RSS requests, and nearly 500 parcel packages were delivered to residents.

Figure 14. The permanent repairs to US 34 brought a return to normalcy for residents and commuters alike.

The US 34 Big Thompson Canyon Flood Recovery Project was completed on time and within the $280 million in federal emergency relief funds. The project installed 100-year protection for a 15-ft minimum drivable surface for almost all the 23-mile section of US 34. When the next flood strikes, there will be at least a 15-foot wide section of road remaining throughout almost all the canyon—allowing for emergency services to get into the canyon and for residents to get out.

The concept of rebuilding a roadway with more post-disaster resiliencies that was developed through the 2013 flood recovery effort is something that can be translated to other emergency repair projects—available funds can be managed and applied effectively to achieve the goal. For US 34, the project team's mindset shifted from one of flood-proofing the road, as was done after the 1976 flood, to one focused on maximizing emergency access in and out of the canyon.

Following the success of this project, the flood recovery program, and the Colorado Resiliency Framework, CDOT, and the State of Colorado are incorporating this focus on resiliency into all phases of a project's lifecycle, from planning and design to construction. By identifying risks and building resiliency into standard processes, the state can make transportation improvements, minimize risk, and create durable and adaptable systems for users. In the face of future natural disasters, the goal is no longer to repair transportation infrastructure by attempting to make it flood-proof, in this case, but rather to incorporate resilient strategies that have the adaptive capacity to best withstand the shock of natural disasters.

Sources

"Big Thompson Flood of 1976." *The Archive, The Denver Post*, July 31, 2012, http://blogs.denverpost.com/library/2012/07/31/big-thompson-flood-disaster-colorado-1976/2795/.

"Colorado Floods: Death Toll Climbs as Hundreds Still Stuck." *The Denver Post*, Sept. 15, 2013, available online at https://www.denverpost.com/2013/09/15/colorado-floods-death-toll-climbs-as-hundreds-still-stuck/.

Duggan, Kevin. "One Year Later: Recovering after Rivers Rage." *Coloradoan*, 8 Sept. 2014, available online at https://www.coloradoan.com/story/news/local/2014/09/05/september-flood-anniversary-colorado/15151647/.

Gruntfest, Eve C. "What People Did During the Big Thompson Flood," (Issue 32 of Natural Hazard Research Working Paper), Boulder, Colorado: Institute of Behavioral Science, University of Colorado Boulder, 1977.

Jarrett, Robert D., and John E. Costa. "Big Thompson Flood, Colorado—Thirty Years Later." U.S. Geological Survey, July 2006, available online at https://pubs.usgs.gov/fs/2006/3095/pdf/FS06-3095_508.pdf.

Muller, Larry, and Patrick F. Mulhern. "1976 Big Thompson Flood and Recovery Planning," American Society of Civil Engineers, October, 1977.

Pohl, Jason. "40 Years Later: Scores Killed in 1976 Big Thompson Flood." *Coloradoan*, 31 July 2016, https://www.coloradoan.com/story/news/2016/07/29/big-thompson-flood-killed-scores/87524858/.

CASE STUDY 3

From Molasses to Sea Level Rise: Boston's North End

Julie Eaton Ernst

Introduction

On the morning of March 2, 2018, the tide overtopped the existing seawalls along Boston Harbor and flooded areas of Downtown Boston, including Langone Park and Puopolo Playground in Boston's historic North End. The Harborwalk along the water's edge was submerged, and the tide lapped up the granite steps, flowed into the park, and extended into the outfield of the Little League baseball field. That tide was the third highest in Boston's history (El. 15.63 ft. BCB, 9.17 ft. NAVD88), and it was just shy of Boston's pinnacle tide crest of El. 16.12 ft. BCB (9.66 ft. NAVD88), which resulted in similar flooding only two months earlier on January 4, 2018. While the recent flooding from the back-to-back 2018 winter nor'easters brought public atten-

Figure 1. Flooding on March 2, 2018. Courtesy Weston & Sampson.

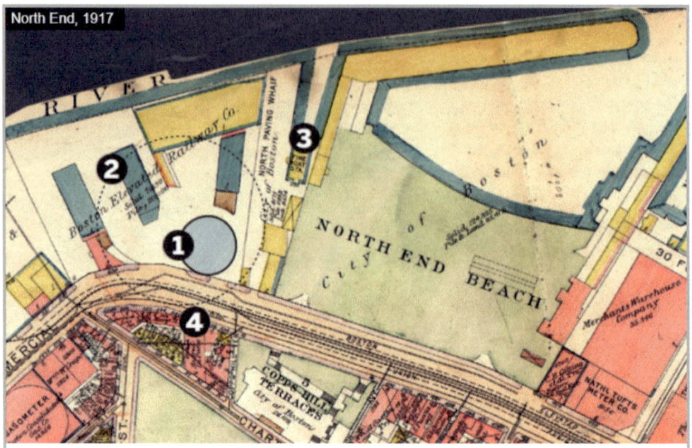

Figure 2. Historic map showing location of the molasses tank. Courtesy Boston Public Library.

Figure 3. Great Molasses Flood damage. Courtesy *Boston Globe* archives.

tion to the threat of climate change and flooding in Downtown Boston, the small 4.5-acre waterfront park in Boston's North End is no stranger to famous floods.

Almost a century before the ocean crested the seawalls, a darker tide propelled Boston into the spotlight. On January 15, 1919, a distillery tank containing 2.3 million gallons of molasses exploded in the North End and caused a wave of molasses roughly 15 feet high and 160 feet wide to surge through the neighborhood into the Harbor—through today's Langone Park and Puopolo Playground. Everything within a two-block area was crushed, flattened, or swept away by the nearly 26 million pounds of molasses. The flood took the lives of 21 people, injured more than 150 others, and resulted in property damage totaling nearly $100 million in today's dollars. Following the accident there was outcry for answers and an investigation, which ultimately found that a structural failure of the tank caused the explosion and the owner of the tank, U.S. Industrial Alcohol (USIA), at fault. The investigation exposed that USIA had not consulted an engineer or architect to design the tank, perform inspections, or provide testing, and the resulting molasses distillery tank was constructed poorly.

In the aftermath of the investigation, the City of Boston and Commonwealth of Massachusetts passed laws to establish accountability in construction, including requirements that engineers and/or architects sign and prepare relevant calculations on building plans. This practice soon became a requirement nationwide, and eventually led to the professional certification of engineers and architects. Engineering and construction standards have continued to grow over the past 100 years to increase accountability and reduce risk to public health and safety.

Engineering best practices must continue to evolve to mitigate the threats of climate change. In 2016, the Boston Research Advisory Group (BRAG) developed climate projections for sea level rise and storm surge, increased extreme precipitation, and extreme temperatures. Since the Great Molasses Flood, sea levels have risen approximately nine inches relative to land. Based on varying carbon emission scenarios, that rise is projected to increase between roughly 1.5 to 4 feet in the next 50 years, with greater uncertainty at the end of the century. As sea levels continue to rise, the likelihood of flooding, like that seen during the 2018 nor'easters, will increase from the current 1% annual chance (or 100-year flood event) to a more regular occurrence.

In response to the risk posed by climate change, the City of Boston instituted "Climate Ready Boston" and tasked the initiative with updating climate projections, preparing vulnerability assessments, and creating neighborhood-scale resilience plans to protect and prepare communities for climate change impacts. The City of Boston identified a planning horizon of 50 years (2070) with an estimated 40 inches of sea level rise as a target for flood protection. Through the Climate Ready Boston initiative, conceptual flood barriers and alignments were prepared for neighborhoods, including East Boston, Charlestown, South Boston, Downtown Boston, and Dorchester. With the growing number of conceptual solutions and urgency for action following the 2018 nor'easters, the City of Boston proactively identified the need to establish a framework for climate-resilient design standards and guidelines and incorporate resilience into current and future projects.

In 2018, the Boston Public Works Department published the *Climate Resilient Design Standards and Guidelines* (BPWD Resilient Guidelines) to address flooding due to sea level rise and storm surge to protect public rights-of-way. The guidelines augment existing design practice by considering climate impacts and managing segmental shore-based flood protection projects over time; they also provide climate design adjustments for the design of flood barriers and a process for evaluating design, operations, maintenance, and cost considerations. Technical considerations include climate parameter design adjustments, boundary constraints, stormwater, structural, geotechnical, transportation and accessibility, utility, groundwater, and vegetative.

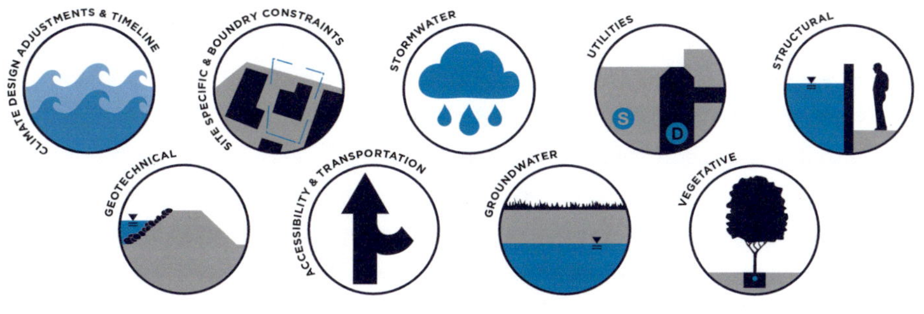

Figure 4. BPWD Resilient Guidelines Design Considerations. Courtesy Weston & Sampson.

Around the same time as the development of the new climate-resilient guidelines, the Boston Parks and Recreation Department was preparing designs for the Langone Park and Puopolo Playground Improvement Project. Through separate procurements, Weston & Sampson became the City's consultant for both projects—developing the climate resiliency standards and designing improvements to the waterfront park—and they capitalized on this unique opportunity to augment both projects.

Figure 5. Existing seawall conditions. Courtesy Weston & Sampson.

Figure 6. Park improvements rendering. Courtesy Weston & Sampson.

Resilient Improvements to the Site

Langone Park and Puopolo Playground is one of the oldest parks in the Boston Parks and Recreation System and the only recreational space with athletic fields in the North End and Downtown neighborhoods. The previous active recreational features included a multi-use field, Little League baseball field, bocce ball courts, basketball court, and playground. Located on Boston Harbor, the site is composed of filled tidelands. The Boston Harborwalk (Harborwalk) path extends along the park's waterfront and crosses through the park to Commercial Street. The approximately 770 linear feet of seawall along the Harborwalk includes various construction materials: granite block, steel sheet piling, and cast-in-place concrete. The existing steel sheet piling seawall is partially failing, and there are voids in the granite block portion of the seawall. Proposed improvements to the park included repairs to the seawall, reconstruction of the Little League field, relocation and expansion of the playground, expansion of the bocce courts, refurbishment of the basketball court, a synthetic turf multi-use field, a new memorial garden, and updated landscaping.

The useful life of an active recreation park improvement is approximately 30 years, and the design flood event is the 1% annual chance event (100-year flood). To design for climate resilience, the project team considered climate projections for 2050 and 2070 time horizons based on the useful life of the project. Site-specific climate data were not available for the 2050 time horizon, so data for the 2070 time horizon were used to inform design. Resiliency goals for the park improvement effort include:

- Elevating areas of the park to El. 19.5 (BCB) or greater to reduce risk of flooding and maintain recreational programming before, during, and after a storm event through useful life. The elevation planned increases include the multi-use field, Little League field, bocce courts, portions of the playground, memorial garden, and interior sections of the Harborwalk path. It should be noted that El. 19.5

(BCB) satisfies the 1% annual chance base flood elevation (BFE) for 2070, assuming 40 inches of sea level rise. The 2070 time horizon is greater than the useful design life of the project (2050), but downscaled climate parameters for the site were not available during design for the 2050 time horizon.

- Leaving portions of the site at lower grades to tie into adjacent properties and designating these areas to accommodate floodwater over time. The areas slated to remain at lower grades include the basketball courts and portions of the seaside sections of the Harborwalk path.
- Minimizing recovery time after a storm event when designated areas of the park may become inundated during future storm events.

The resulting design included raising the Harborwalk by constructing a new boardwalk cantilevered over the existing seawall and anchored by micropiles behind the existing wall, constructing a new retaining wall behind the existing seawall at the back of the boardwalk, raising grades 3 to 6 feet across the site behind the retaining wall, and implementing stormwater management system improvements. Weston & Sampson documented the resilient improvements and processes using the BPWD Resilient Guidelines, as shown in Table 1 (next page).

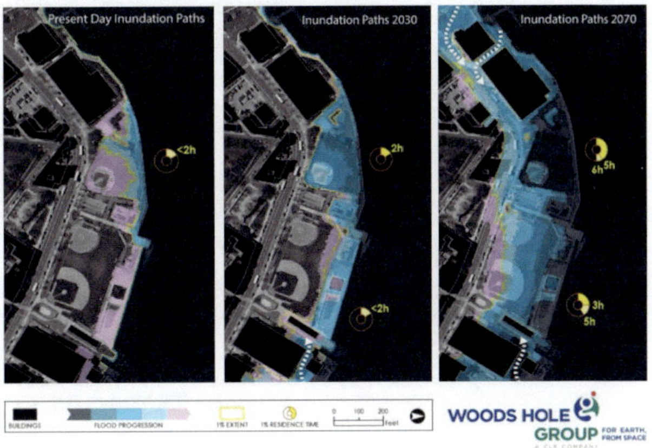

Figure 7. Flood Risk Maps. Courtesy Woods Hole Group.

Figures 8 and 9. Renderings of Langone Park and Puopolo Playground post construction. Courtesy Weston & Sampson.

Table 1. Design Considerations	

Climate Design Adjustments and Timeline

	• The site is within Boston Planning and Development Agency's (BPDA's) SLR-BFE zone – El. 19.5 to El. 19.7 BCB +/-. The Boston Harbor Flood Risk Model (BH-FRM) was available through the site. Woods Hole Group was subcontracted to provide the following data for existing conditions and design of improvements:
	  ◦ Probability of Flooding (map)
	  ◦ Flood Depth (map)
	  ◦ Duration of Flood (map)
	  ◦ Flood Pathways (map)
	  ◦ Probability Exceedance Curve Data (table)
	  ◦ Wave loads and current velocities
	• Extreme precipitation data were selected from the Boston Water and Sewer Commission (BWSC) "Comprehensive-Integrated Sustainable Wastewater and Storm Drainage System Facilities Plan Final Report," (June 2015) for the 2060 10% annual storm design.
	• Extreme heat events were not a central feature of the design, but vegetative considerations increased the tree canopy at the site during improvements.

Boundary Constraints 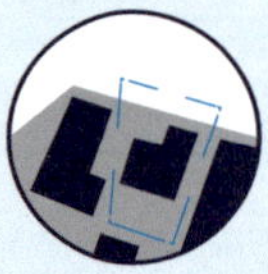	• Portions of the site are within the FEMA AE Zone and Chapter 91 Tidelands. The site is composed of Land Subject to Coastal Storm Flowage (LSCSF). There are no compensatory storage requirements for LSCSF per 310 CMR 10.04, and coastal flooding extents are not exacerbated by raised grades. • The surrounding site features (i.e., swimming pool, Coast Guard station, DCR skating rink) will remain at existing grades, and site improvements will tie into existing grades. Off-site flood pathways, as illustrated by the BH-FRM maps, pose a risk at the site (before park modifications). • The design maintains the public right-of-way and access to the waterfront via the Harborwalk, with walkability, connectivity, and social/neighborhood context a central part of the design.
Stormwater Considerations 	• Off-site flooding impacts are not anticipated from proposed grade changes. The back of the site along Commercial Street ranges from El. 19.5 ft. to El. 21 ft. (BCB), so raising grades near the waterfront does not create a "downstream" slope. Stormwater will not be trapped behind the barrier. • Stormwater is intended to collect into area drains via drainage swales located along the groin of grade changes at the edges of the site, and the synthetic turf field is designed for infiltration. • Stormwater will collect into a new 36-inch and 12-inch stormwater outfalls through the site. Refer to the Utility Considerations (below) for additional details.

Table continues →

<table>
<tr><th colspan="2">Table 1. Design Considerations</th></tr>
<tr><td>Structural Considerations
</td><td>
<ul>
<li>The existing seawall construction materials varied from granite block, steel sheet piling, and cast-in-place concrete. Portions of the steel sheet pile seawall have failed. The existing seawall is in fair-to-poor condition and cannot be modified without substantial reconstruction, which would have been expensive and triggered new, additional permitting. The geometry of the existing granite block wall is unknown; test pits performed in the area found shallow refusal on granite blocks and encountered utilities behind the wall. The proposed grade changes cannot add additional lateral forces on the existing seawall.</li>
<li>The Harborwalk is a timber boardwalk supported by micropiles to raise the boardwalk without raising grades over the existing seawall. The beams are cantilevered approximately 8 feet and are independent of the new floodwall on the inland side. The boardwalk is designed to withstand coastal forces for current and 2070 conditions, including tidal uplift, using the data provided by the Woods Hole Group and the BH-FRM.</li>
</ul>
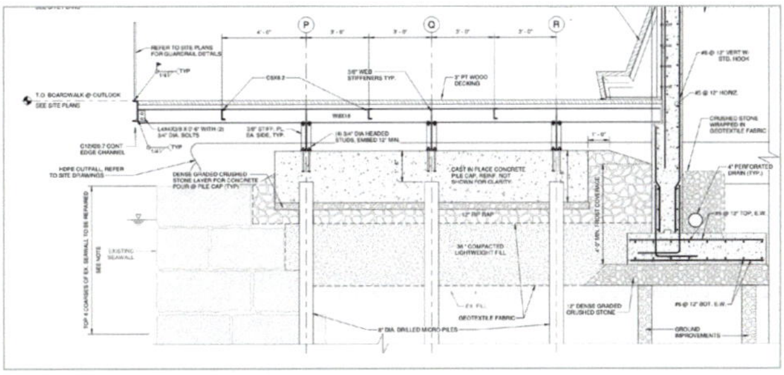

(Top) Sample structural section of new boardwalk; (bottom) construction of Langone Park and Puopolo Playground. Courtesy Weston & Sampson.
<ul>
<li>The boardwalk was originally going to be designed for small vehicular traffic loads (e.g., maintenance crews), but it was value engineered out of that design because the steel members were too large.</li>
<li>The new floodwall also serves as a retaining wall on the inland side of the raised boardwalk. The retaining wall will also function as a seat wall to provide seating along the new Harborwalk.</li>
</ul>
Refer to the Geotechnical Considerations (next) for additional details on design challenges.
</td></tr>
</table>

Table continues →

Table 1. Design Considerations	
Geotechnical Considerations 	• Geotechnical report identified subsurface conditions—urban fill (including debris possibly from the Great Molasses Flood) overlying organic silt over sands and clay, with glacial till approximately 62 feet below ground surface. • The site contains compressible soils prone to settlement and subsurface obstructions (abandoned foundations, utilities, etc.). Settlement is a primary concern due to proposed site grading and the proximity of existing seawall and DCR rink. • The site is designed so no additional lateral loads will be applied to existing structures (i.e., seawall) after grade increases and other improvements; it is designed for tolerable settlement. Lightweight aggregate fill wrapped in geotextile fabric was used behind the repaired seawall and playground. Alternatives considered geofoam blocks. Refer to the Groundwater Considerations (below) for additional details. • Ground improvement was selected to support fill and the new floodwall. Alternatives considered deep foundation support. • Riprap greater than 12 inches was recommended for erosion and scour protection below the raised timber Harborwalk based on coastal engineering analyses (current velocities for present and 2070 conditions based on data provided from Woods Hole Group from the BH-FRM).
Transportation and Accessibility Considerations 	• The existing seawall is being left in place, and the Harborwalk path is being raised via timber boardwalk. The new elevated Harborwalk along the waterfront will enable accessibility through the park. The Harborwalk must tie into existing grades in front of the DCR skating rink, and this area may be flooded during storm events without adjacent site improvements. • The new design maintains ADA accessibility, and sidewalks are graded up to El. 19.5 ft. (BCB); the design encourages walkability, livability, and waterfront connectivity for pedestrians and cyclists.
Utility Considerations 	• Electrical utilities were replaced, transformers are above El. 19.5 ft. (BCB), and water utilities were consolidated, and existing unused main lines were abandoned and removed. • Stormwater is gravity-fed to the ocean with a one-way tidal gate valve. A new 36-inch diameter outfall and 12-inch diameter outfall were designed in coordination with BWSC.
Groundwater Considerations 	• Groundwater was observed between 4.4 to 7.2 feet below ground surface during subsurface investigations; a monitoring well was constructed on site. The use of geofoam blocks was not recommended due to uplift concerns with current and future groundwater levels due to sea level rise. • Groundwater is tidally influenced at the site. Therefore, corrosion-resistant materials were selected for new structures.

Table continues →

Table 1. Design Considerations

Vegetative Considerations	• The design includes selection of synthetic turf multi-use field and salt-tolerant planting material for the site. • The design includes 58 new trees proposed for planting along the bocce ball courts, Harborwalk path, and playground. • The park design includes educational signage that explains flood risk at the site and highlights the strategies used for flood resilience.

Figure 10. Construction of Harborwalk and flood barrier (seat wall) behind Langone Park baseball field. Courtesy Weston & Sampson.

Figure 11. Completed Harborwalk and flood barrier (seat wall). Courtesy Weston & Sampson.

Figure 12. Completed renovated basketball courts (designed to allow flooding) and elevated athletic fields. Courtesy Weston & Sampson.

Figure 13. Sunset on the Harbor at Langone Park and Puopolo Playground. Courtesy Bret Clancy.

The park improvements finished construction in winter 2020–2021. A century later at the same site of the Great Molasses Flood, the park is ready for future flooding. Design of this park was the first project to test and implement the BPWD Resilient Guidelines and will be the first waterfront park in the City of Boston designed using 2070 climate projections. Langone Park and Puopolo Playground serves as a catalyst for resilient design and creates the pivotal first link in a future chain of flood protection along the Downtown Boston waterfront. When and if the faint aroma of molasses eventually fades, the site will continue to remind engineers and designers of its history and impact on our practice, and how it can progress in the future.

References

Boston Public Works Department. "Climate Resilient Design Standards & Guidelines for Protection of Public Rights-of-Way," October 17, 2018. https://www.boston.gov/sites/default/files/imce-uploads/2018-10/climate_resilient_design_standards_and_guidelines_for_protection_of_public_rights-of-way_no_appendices.pdf.

The Boston Research Advisory Group, City of Boston, and Boston Green Ribbon Commission. "Climate Change and Sea Level Rise Projections for Boston." Climate Ready Boston, June 1, 2016. https://www.boston.gov/sites/default/files/document-file-12-2016/brag_report_-_final.pdf.

Boston Water and Sewer Commission. "Comprehensive-Integrated Sustainable Wastewater and Storm Drainage System Facilities Plan Final Report," June 2015.

Cavanaugh, Ray. "How the Great Molasses Flood of 1919 Made the World a Little Bit Safer." *Time*, January 14, 2019. https://time.com/5500592/boston-great-molasses-flood-100/.

Mass Moments. "Great Molasses Flood." Mass Humanities, 2019. https://www.massmoments.org/moment-details/great-molasses-flood.html.

Puleo, Stephen. *Dark Tide: The Great Boston Molasses Flood of 1919*. Beacon Press, 2003.

Shanahan, Mark. "The Great Molasses Flood of 1919 Was Boston's Strangest Disaster." *Globe Magazine*, January 9, 2019, available online at https://www.bostonglobe.com/magazine/2019/01/09/the-great-molasses-flood-was-boston-strangest-disaster/VawySumFUf5vKCibM9PLtJ/story.html.

Woods Hole Group. "Boston Harbor Flood Risk Model (BH-FRM)." Boston: Massachusetts Department of Transportation (MassDOT), 2015.

CASE STUDY 4

Lake Delton Case Study

Raine L. Gardner

Introduction

What happens when the focal point of the number one tourist destination in the upper Midwest disappears? The sudden loss of an entire lake in the village of Lake Delton, Wisconsin, sent shockwaves through not only the community but also the state and overall region. Lake Delton was a man-made lake developed in 1927. It was the center of the community of Lake Delton, in what is known as the Wisconsin Dells area. The lake provided a wide range of recreational activities and tourist attractions.

In June 2008, record-breaking rains caused massive flooding, which resulted in a catastrophic event that drained the lake completely. Lake Delton overflowed, eroded an earthen dike, and flowed into the nearby Wisconsin River, taking out five homes, a county truck highway, forested land, and several underground utility lines. Repairs to the connecting roadway and utilities and replenishing the lake immediately became a priority for the community and the state. The peak of the tourist season had just begun; a rapid response was necessary to save the area economy and many vacationers' summer plans.

Overview of the Community

The Wisconsin Dells area is known for its waterparks and "fun in the sun" during the summer season. The community consists of the city of Wisconsin Dells on the north bank of the Wisconsin River and the village of Lake Delton across from it on the south bank; together, they have a total combined population of approximately 6,000. The village of Lake Delton encompasses the southern half of the community and was built around a 267-acre artificial Lake Delton. It is the focal point of the community, as many of the area attractions are located at the lake and several resorts are built on its shoreline. Over 4 million visitors come to the area each year, with approximately 100,000 people visiting each weekend in the summer months. The area and lake are critical to the state's economy and tourism. Lake Delton is home to the Tommy Bartlett water ski show and the famous "Duck" rides (World War II DUKW amphibious vehicles) given to tourists who want to see the area's beautiful landscape. The region has steep cliffs and several drainage pathways that flow down to Lake Delton and discharge into the lake, which the village continuously works to maintain.

The Wisconsin Dells area has unique topography and sandstone cliffs, which are only found in one other location in the world. Thousands of years ago, Glacial Lake Wisconsin formed after the nearby glacier that covered Wisconsin (but not the Dells area itself) retreated. Eventually the glacial lake burst its ice dam and the ensuing flood, which dropped the lake level by 100 feet, eroded the sandstone bluffs and produced the sandstone canyons and cliffs. In addition to the topography of the Wisconsin Dells, the area is known for its former logging operations of moving timber down the Wisconsin River.

Lake Delton is fed by Dell Creek, which enters the lake from the south and runs to the northeast. On the northeast side is a small dam that regulates the lake. The dam discharges its flow back into Dell Creek, which continues a short distance and then empties into the Wisconsin River. The river separates the village of Lake Delton from the city of Wisconsin Dells and is the corporate boundary between the two jurisdictions.

In 1927, the Lake Delton dam was completed and the lake was filled. The area became a popular tourist destination in the 1930s and 1940s when boats began to run on the Wisconsin River to showcase the rock formations, called "Dells" (named "dalles" by early French explorers). Then in the 1950s, the Tommy Bartlett water ski show developed for tourists to come and see. After that, tourism and economic development flourished.

Access to the dam is through Kaminski Park, one of the most popular in the village. The park houses several athletic fields and also a sanitary sewer lift station that pumps all of Lake Delton's sewage north under the Wisconsin River to the Wisconsin Dells side, where the communities share a wastewater treatment facility. All of this infrastructure is downstream of the Lake Delton dam. County truck highway (CTH) A also runs along the south side of the Wisconsin River, north of Kaminski Park and across the dam; it is considered a main transportation corridor into Lake Delton and the Dells area.

Weather Leading up to the Event

The spring of 2008 was typical until early June. Sustained rain began to fall on June 6 and continued through June 9, 2008. The accumulated rainfall totaled 14 inches, an amount never seen there before in that span of time. The heavy rainfall, paired with the other severe weather in the area, caused damage all across south-central Wisconsin. Other area rivers were at their highest flood stages, causing many people to evacuate; the area was in a state of emergency. Many roadways near the Dells were flooded or washed out, and access to and from the community was limited, causing major concerns for public safety. The lake had been rising and the community had been sandbagging to help protect their homes, businesses, and infrastructure from the flooding, but no one knew what would happen with Lake Delton until the morning of June 9.

Day of the Event

As the water level rose, the dam was not able to keep up by discharging the flow, and lake water began to flood lakeside homes and resorts in some areas. The village tried to protect the main lift station in Kaminski Park, which is critical for keeping its sewage system pumping to the wastewater treatment plant.

As the lake continued to rise, the water needed to find a new outlet because the dam was not allowing all of the water that was coming from the watershed to pass through. On the morning of June 9, 2008, the water finally found a new outlet on the north end of the lake, where five lakeside residential homes backed up against CTH A

[Left] View to the northwest of Lake Delton emptying through a 300-foot breach and cutting across County Truck Highway A. **[Right]** View from the south of empty Lake Delton, the area of the breach stripped of trees, and the natural channel of Dell Creek at the right, leading to the dam, which remained intact. In the background at the top is the Wisconsin River. *Photos:* Mead and Hunt.

[Left] A residence that was destroyed when floodwaters scoured the bank and undercut the foundations. **[Right]** County Truck Highway A, which runs northwest to southeast between Lake Delton and the Wisconsin River, was destroyed at the breach.

and the Wisconsin River. The water had migrated through the sandy soils to undermine the homes and roadway and vacant forested land that extended to the Wisconsin River. This caused a large washout of the land, infrastructure, and homes, which fell into the river in what became known as "the breach." The 300-foot-wide channel allowed Lake Delton to drain completely in just over two hours. The breach not only drained the lake but washed out CTH A, all underground mainline utilities for the community, five residential homes, and other privately owned land. Thousands of cubic yards of debris washed into the Wisconsin River, causing damage to this corridor's flow and sending debris downstream for other communities to deal with.

Right after the breach occurred, all efforts to protect the other infrastructure near the dam stopped and the focus turned toward the breached area, to repairing the lake, utilities, and roadway, and to restoring the destroyed homes. In addition, public safety quickly came to the forefront. Within hours, Governor Jim Doyle committed to replacing CTH A by December 2008 and to refilling Lake Delton by May 2009 so the area could get back to normal for the 2009 summer season, keeping in mind how important the area is to the state's economy.

Infrastructure Damage Assessment

After the breach, the village began its assessments to determine what needed to be repaired. Not in the breach area, Kaminski Park had been covered by a fair amount of overflowing water for hours as the dam held its position, and it was badly damaged. In addition, the entire eastern roadway along Lake Delton, called Canyon Road, was being reconstructed at the time and was exposed by the unfinished work. Those areas were severely washed out, requiring repairs above and beyond normal. Further damage assessments indicated the majority of the drainage ways down to the lake were also in great need of repair. Excessive scour from the rain caused shoreline to be lost, building and infrastructure damage, and unsafe conditions for the public.

Existing underground utilities were damaged, as they had been undermined by the breach and roadway flooding. Overall, the community was in a state of emergency and needed a plan for proper repairs as soon as possible.

Taking Action and Associated Repairs

After the governor's declaration, a design team was quickly assembled and began to work on various projects. The two main projects were to restore the lake and roadway in time for the 2009 tourist season. The team consisted of two main engineering firms, the Village, and public agencies such as the Wisconsin Department of Transportation and Wisconsin Department of Natural Resources. The secretaries of these two agencies organized a leadership team that was then backed by the Federal Highway Administration, the village of Lake Delton, Sauk County, and the design team. The leadership team of this nature allowed for real-time input and decision-making and expedited permitting.

Lake Delton took several steps to address the immediate conditions before they began proper repairs. The utilities were temporarily reconnected across the breach area with the use of a cable system to span the washed-out area. Then a channel was cut into the existing lake bottom to divert the Dell Creek water back to the dam and away from the breach site.

After those changes, repairs could begin. The team developed a plan that would be both sustainable for many years to come and resilient to similar events in the future. The team used a multiphase approach to expedite the reconstruction. The first step was to demolish the existing infrastructure no longer needed and to remove the remaining homes that required relocation (including the purchase of the real estate). Those were the most pressing needs. Once that was completed, the team focused on restoring the lake.

The next step was to redirect the creek to divert the water back toward the dam and away from the breach site to dry out the damaged area. Several designs and construction projects were deployed in this process. The projects included: building an embankment area at the breach site to allow this area to be re-established, dam upgrades to better regulate the lake in the future, re-establishment of the transportation corridor of CTH A, repairs to all the utilities around the lake and breach site, and refilling the lake, which included developing new fish habitats and restocking it with fish. Repairing the breach would include constructing a new utility corridor, a 100-foot-deep soil-bentonite slurry wall, and a new rim dike. Specialized equipment had to be brought in from around the country to construct these repairs. The existing dam was upgraded to include a new side-channel spillway and other modifications. All the natural drainage ways from the community that ultimately discharge into the lake required repairs to safely convey the stormwater until the lake was refilled.

Lastly, Canyon Road, one of the major roads connecting several of the eastern lake resorts, needed above-normal repairs to the roadway already under construction at the time of the breach. Several of the natural drainage ways run from Canyon Road down to the lake, and the run-off during the storm severely damaged not only

Aerial view of the breach area the following year. The road was repaired, the shoreline of the lake was reinforced, and the land between the road and the Wisconsin River was restored, all in a sustainable and resilient fashion. *Photo:* Mead and Hunt.

the roadway, but also adjacent resorts and the shoreline. These areas required major repairs and erosion armament to get them fully restored and prepared for future events.

The repairs incorporated sustainability and resiliency features to prevent this type of catastrophic damage from occurring again. At the time of the breach, no one thought this amount of rainfall and associated damage would occur again in their lifetimes, but unfortunately, as the climate has been changing, the Dells area has seen more elevated flooding events due to larger flash rain occurrences. These flash rain occurrences are becoming almost the norm for rain events in this area each spring and summer, causing more considerations and more intense designs to the area infrastructure.

The Breach Site Today

This event forever changed the area and is regularly remembered as the area develops. Resiliency is now included in the design of new infrastructure construction. After the repairs, the lake is now sustainable and resilient, and includes a system to monitor the water to prevent a breach from occurring again. The water quality has improved, it has a better fish habitat, and new recreational areas have been created in places that were damaged by the rainfall or destroyed by the breach. New parks are popular amenities for community residents and the many visitors.

Understanding for the Future

No one in the Wisconsin Dells–Lake Delton community wants to endure more of the events that occurred in June 2008. In recent years, there has been a continued increase in both the intensity and frequency of extreme rainfall events. This increase raises the overall risk of flooding and, therefore, the chance that a similar event could reoccur in the near future.

Conclusions

After such a catastrophic event, the community, the Village of Lake Delton, the engineering team, and public agencies acknowledged the concerns about our ever-changing climate. No one had anticipated that such an event could cause such destruction. "Sustainable" and "resilient" are terms now being used by the community in their planning and development activities. The memories of the Lake Delton breach are still in the backs of most community members' minds. The heightened rain events that have occurred since the Lake Delton breach have reminded the community that our climate has changed to the new normal, and these events continue to cause similar smaller-scale issues never seen before. Engineering professionals have begun to adapt their typical designs to account for such climate-related events.

CASE STUDY 5

Climate Change and Asset Management: Assessing and Protecting a Portfolio

Karin S. Holland

Introduction

For asset managers and property managers, climate change brings many new challenges, dramatically increasing the risks and associated costs to mitigate, recover, and adapt. Forward-thinking organizations are preparing their responses to everything from hurricanes, wildfires, and flooding to heatwaves. Being resilient means ensuring that an organization can not only bounce back but leap forward; business continuity is crucial.

Today, asset and property managers are making significant shifts in business practices. They're taking a more holistic approach to evaluating trade-offs between service, cost, and risk for built assets. Most industries have no choice but to look at asset management through a resilience lens, and their projects are reflecting that.

To illustrate, we asked asset management leaders across multiple industries about how they are incorporating resilience at their organizations, and about the issues surrounding their efforts. A cross-section of questions we posed is given below. Our discussions touched upon broad asset management topics and how to approach today's climate change–related challenges.

What are an organization's most valuable assets?

One thing that is common across all industries and sectors: when it comes to the most valuable assets, organizations all pointed to the health and safety of humans first, followed by the protection of infrastructure, including transportation infrastructure.

"It's always about the trained knowledge and strengthened awareness of people seeking actionable emergency approaches," said Akiya Simms, Sustainability Officer for the City of Atlanta Department of Aviation. Simms explained, "Although difficulties and complexities of emergency responses are often beyond human capacity and call for certain computer technologies, these challenges provide opportunities for people to improve interpersonal skills in any given teamwork capacity. Firms should err on the side of caution as the overreliance on technology could lead to the breakdown of organizational culture, trustful relationships, and eventually emergency response effectiveness. An applicable balance under these circumstances is only possible through improved communications, along with a strong safety culture, which informs people working to develop resilient operational assets."

It is essential to have well-informed and trained people to help organizations maintain crucial infrastructure. Anthony Kane, President and CEO at the Institute for Sustainable Infrastructure (ISI), noted that in addition to focusing on water infrastructure—including drinking water—transit systems are particularly important. "I've seen a lot of organizations focusing on transportation infrastructure," Kane said. "They're both fixing damage to existing transportation systems while also doing the best to invest in ensuring the resilience of roads and bridges."

Which assets are most vulnerable to climate change, and how do organizations prioritize which to protect?

Across all industries, knowing which assets are most vulnerable to climate change is crucial to identifying what to protect first, and how. Identifying and prioritizing what to safeguard is a critical step when building resilience to climate change. Frequent, comprehensive, enterprise-wide risk assessments are essential to make sure organizations are considering all risks and how fast the risks are changing. These assessments should involve everyone from finance and legal to operations and maintenance.

A director of sustainability at a major national railway company recognizes that flooding is the greatest risk for the organization. "Flooding has always been the biggest issue for our industry. Railroads always run along waterways, so they're always vulnerable to flooding," she said. "The first thing you need to do is perform a study to see where you are most vulnerable in terms of flooding and start devising ways to anticipate future vulnerabilities."

How does an organization protect assets from climate threats?

ISI's CEO Anthony Kane understands the strategies that organizations and municipalities use to protect their assets from climate and human-made threats. He stresses that it's integral for departments at organizations to work in unison, focused on the whole system. "At the project level, people are trying to increase the resilience of projects, but this doesn't always tie into a unified strategy, which is the first step to a project's success," Kane said.

"Part of developing this strategy is keeping a strict eye on the future. For example, every bridge project has to look 120 years into the future and consider hazards and wear and tear. Utility providers also need to take the long lens and focus on future-proofing assets for climate change and human-made climate issues. There are many productive conversations about decentralizing energy and renewable energy, and we have 100 U.S. cities committed to creating resilience plans. We need to keep having these conversations across industries."

Molly Parker, Dominion Energy's Director, Environmental Services, is passionate about alternative energy to increase resilience. Parker noted, "It's exciting to work for a company that's committed to being one of the most sustainable companies in the U.S." Nationwide, Dominion Energy owns the fourth-largest solar fleet among utility holding companies. The company also recently announced an expansion of its Align Renewable Natural Gas joint venture with Smithfield Foods and is working on the largest offshore wind and electric school bus projects in the country.

What are the most significant opportunities to increase resilience and protect assets?

While new technologies and studies are crucial, sometimes the biggest opportunities involve "small" things. Simms, an employee of Atlanta, the fifth most populous city in the United States, understands that if fundamental plans are not solid, the strategies which arise from such plans will also lack integrity.

Simms said, "As new technologies and studies become crucial to operational resilience, the most promising opportunities can often emerge from simplicity. Most of the time, the simplest things make a huge difference. Even throughout large project life cycles, we need diligent reference points for communicating basic safety management and performance procedures. For example, we may ask ourselves, 'Do we have sufficient maintenance controls in working order?' and, 'Are we continually detecting and optimizing our predefined risks in the smartest and most efficient way?' These types of concerns address why organizations need to actively expand communications to reflect policy and compliance as reasonably as possible." Simms concluded, "Sometimes a particular issue may seem minor to one department but may have a prioritized and holistic connection to another, including overarching goals."

ISI's Managing Director Melissa Peneycad explained that the human aspect of resilience goes beyond addressing the interests of internal employees and stakeholders.

"We need to understand how a project can support the resilience of the communities it serves. Too often, we limit our focus to ensuring the asset itself is more resilient," she explained. "We have to think about how the asset can contribute to community livability, empowerment, social cohesion, connectivity, and health which are all indicators of a resilient community. To do this, we must look beyond the project boundary. A successful infrastructure project provides an occasion to address broader community needs and opportunities."

What are the most significant barriers that prevent organizations from fully protecting their assets from climate change?

Creating broader opportunities and infrastructure to protect assets is a lot easier said than done. One doesn't have to be in the asset management field to understand that making and funding public policy changes are almost always uphill battles. The experts we spoke with agreed that resilience needs to be woven into an organization's DNA. The same experts also agreed on the biggest barriers that keep that from happening.

"One of the challenges is how to budget on preparedness and who should pay," ISI's Kane said. The problem when it comes to public spending on preparedness or recovery is limited funding. "How do you decide whether something should be more resilient versus not being there at all?" Kane pointed out. "Many cities have very limited funding already. It's difficult to advocate spending money on something that may or may not take place down the road."

Resilience can carry a perceived cost premium, but it is important to look at how a project or asset impacts the community as a whole. ISI's Peneycad cited changes to the Envision framework and how it was expanded to evaluate risk and resilience in order to further clarify how to undertake an effective assessment. Projects need to look wider at a community level, and the framework was fundamentally rewritten to encourage projects to do this.

How leaders are protecting their assets—and what other organizations can learn

Whether it's a municipality or a company, when it comes to protecting assets, some do not know where to start, even after a careful analysis. Those whose assets are most vulnerable to climate change are usually the most motivated in preparing for change, which means those less vulnerable can learn from their methods. Urban Land Institute's Vice President of Urban Resilience Katharine Burgess explained.

"In the U.S., many cities facing more obvious or immediate vulnerabilities have been proactive in terms of policy and establishing new funding sources for resilience through taxes, bonds, or other methods. Larger cities are also often ahead of the game in terms of policy," Burgess said, noting climate adaptation policies and projects underway in New York, Boston, and Washington.

Dominion Energy's Parker is proud to be working for a resilience leader. "We're definitely a leader," she noted. "We need to be leaders and do this right, while providing electricity reliably and affordably. In Virginia, we currently have a proposal to invest up to $3 billion over the next decade to modernize the electric grid and improve customer experience. Part of these efforts includes hardening the grid to be more resilient to weather events and increasing cyber and physical security of grid assets."

Climate change is here. Those charged with protecting assets, either in the public or private sector, have no choice but to be forward-thinking about our changing landscape. It's uncharted territory with a hefty price tag. The good news is that leaders today are recognizing that sustainability and resilience go hand-in-hand with asset protection—they know that the best way to cope with operational risk is to anticipate and avoid it.

Index

Page numbers followed by an f indicate a figure on that page; page numbers followed by a t indicate a table on that page.

C

D

E

M

N

S

Z

Made in the USA
Las Vegas, NV
10 November 2026

57851954R00217